SEMA SIMAI Springer Series

Volume 15

More information about this series at http://www.springer.com/series/10532

Daniele Antonio Di Pietro • Alexandre Ern •
Luca Formaggia
Editors

Numerical Methods for PDEs

State of the Art Techniques

Springer

Editors
Daniele Antonio Di Pietro
Institut Montpelliérain Alexander Grothendieck
CNRS
Université Montpellier
Montpellier, France

Alexandre Ern
Université Paris Est, CERMICS (ENPC)
Marne la Vallée, France

INRIA, Paris, France

Luca Formaggia
MOX - Dipartimento di Matematica
Politecnico di Milano
Milano, Italy

ISSN 2199-3041 ISSN 2199-305X (electronic)
SEMA SIMAI Springer Series
ISBN 978-3-030-06896-7 ISBN 978-3-319-94676-4 (eBook)
https://doi.org/10.1007/978-3-319-94676-4

This Springer imprint is published by the registered company Springer Nature Switzerland AG
The registered company address is: Gewerbestrasse 11, 6330 Cham, Switzerland

Preface

Numerical analysis applied to the approximate resolution of partial differential equations (PDEs) has become a key discipline in applied mathematics. One of the reasons for this success is that the wide availability of high-performance computational resources and the increase in the predictive capabilities of the models have significantly expanded the range of possibilities offered by numerical modeling.

With this in mind, in the fall of 2016 the editors of the present volume organized a thematic quarter at the Institut Henri Poincaré in Paris focusing on various aspects of numerical analysis. The quarter started with a 1-week introductory school comprising courses on the virtual element method, the hybridizable discontinuous Galerkin method, gradient schemes, mimetic spectral methods, low-rank and sparse tensor methods, reduced-basis methods, a posteriori error estimates, adaptive finite element methods, and interfacing models of different dimensions. During the quarter, additional advanced courses were offered on the hybrid high-order method, eigenvalue problems, ill-posed problems, direct, inverse, and reduced-order modeling, high-dimensional approximation of parametric PDEs, error control, and adaptivity.

This volume reflects many of the topics covered during the quarter and provides up-to-date reference material for graduate students, scientists, and engineers interested in advanced numerical techniques. Even if the material is of an introductory nature, it concerns rather state of the art methodologies, so the reader is expected to have a basic knowledge of the mathematical theory of PDEs and numerical methods. Additional material can be found at the address:

http://imag.edu.umontpellier.fr/event/ihp-nmpdes

We are thankful to the Scientific Committee of the Institut Henri Poincaré (and in particular to its Vice-President, Marc Herzlich) for their support of the quarter, to the administrative staff of the IHP for their help during the quarter, and to all participants for the lively and stimulating mathematical discussions.

Montpellier, France
Paris, France
Milano, Italy
January 2018

Daniele Antonio Di Pietro
Alexandre Ern
Luca Formaggia

Acknowledgements

We are grateful to the Scientific Committee of Institut Henri Poincaré (IHP) and its former director Cédric Villani for selecting our project, thus making possible the thematic quarter "Numerical methods for PDEs". We particularly wish to thank Marc Herzlich for pointing out this opportunity and for providing precious help during the initial stages of this project.

We gratefully acknowledge the financial and logistic support provided by Institut Henri Poincaré. We are especially thankful to the administrative staff of Centre Émile Borel (Sylvie Lhermitte, Delphine Lépissier, Nitdavanh Sriratanakoul) for their precious support throughout the thematic quarter. Our gratitude also goes to the administrative staff of Institut Montpelliérain Alexander Grothendieck (and particularly to Bernadette Lacan and François-David Collin) for their logistic support as well as for their help in keeping the webpage up to date. We also acknowledge the important financial support from (in alphabetic order) Agence Nationale de la Recherche through grant HHOMM (ANR-15-CE40-0005), Centre National de la Recherche Scientifique, Fondation des Sciences Mathématiques de Paris, Institut des Hautes Études Scientifiques, LabEx CARMIN through the common program IHP-IHES, Université Paris-Sud, and Université Pierre et Marie Curie.

We wish to thank Agence pour les mathématiques en interaction avec l'entreprise et la société, Société de Mathématiques Appliquées et Industrielles, and Società di Matematica Applicata e Industriale for their support in the organization of the workshop "Industry and Mathematics".

Finally, we wish to thank all of the participants in the thematic quarter, whose presence and involvement were crucial to its success.

Contents

About the Editors

Daniele Antonio Di Pietro is Full Professor of Numerical Analysis at the Institut Montpelliérain Alexander Grothendieck at the University of Montpellier. He received his Ph.D. in Computational Fluid Mechanics from the University of Bergamo. His research fields include the development and analysis of advanced numerical methods for partial differential equations, with applications to fluid- and solid mechanics and porous media. He is author of one book and over 50 research papers in peer-reviewed journals.

Alexandre Ern is a Senior Researcher at Ecole Nationale des Ponts et Chaussees (France) and at INRIA, and he holds a part-time Professor position in Numerical Analysis at Ecole Polytechnique (France). His contributions encompass mathematical modeling and numerical analysis in computational fluid and solid mechanics with applications in problems related to the environment and energy production. From 2006 to 2012, he has chaired the French National Research Program on Underground nuclear waste storage, and he has ongoing collaborations with several industrial partners. Alexandre Ern has authored three books and over 130 papers in peer-reviewed journals, and he has supervised about 20 PhD students working now in academia or industry.

Luca Formaggia is Professor of Numerical Analysis at the Modelling and Scientific Computing Laboratory (MOX) of the Department of Mathematics of Politecnico di Milano. His scientific work addresses the study of numerical methods for partial differential equations, scientific computing, computational fluid dynamics with applications to computational geosciences, biomedicine and industrial problems. He is the author of more than 90 articles and editor of several books. Currently he is the President of the Italian Society of Applied and Industrial Mathematics and was the Head of the MOX Laboratory from 2012 to 2016.

Chapter 1
An Introduction to Recent Developments in Numerical Methods for Partial Differential Equations

Daniele Antonio Di Pietro, Alexandre Ern, and Luca Formaggia

Abstract Numerical Analysis applied to the approximate resolution of Partial Differential Equations (PDEs) has become a key discipline in Applied Mathematics. One of the reasons for this success is that the wide availability of high-performance computational resources and the increase in the predictive capabilities of the models have significantly expanded the range of possibilities offered by numerical modeling.

Novel discretization methods, the solution of ill-posed and nonlinear problems, model reduction and adaptivity are main topics covered by the contributions of this volume. This introductory chapter provides a brief overview of the book and some related references.

Numerical Analysis applied to the approximate resolution of Partial Differential Equations (PDEs) has become a key discipline in Applied Mathematics. One of the reasons for this success is that the wide availability of high-performance computational resources and the increase in the predictive capabilities of the models have significantly expanded the range of possibilities offered by numerical modeling.

D. A. Di Pietro
Institut Montpelliérain Alexander Grothendieck, CNRS, Université Montpellier, Montpellier, France
e-mail: daniele.di-pietro@umontpellier.fr

A. Ern
Université Paris Est, CERMICS (ENPC), Marne la Vallée, France

INRIA, Paris, France
e-mail: alexandre.ern@enpc.fr

L. Formaggia (✉)
MOX, Dipartimento di Matematica, Politecnico di Milano, Milano, Italy
e-mail: luca.formaggia@polimi.it

D. A. Di Pietro et al. (eds.), *Numerical Methods for PDEs*,
SEMA SIMAI Springer Series 15, https://doi.org/10.1007/978-3-319-94676-4_1

This volume comprises nine chapters reflecting many of the topics covered by the Ph.D. level courses given during the Thematic Quarter *Numerical Methods for PDEs*, held at the Institut Henri Poincaré in the fall 2016.[1] These chapters can be loosely organized into three groups: (1) novel discretisation methods; (2) nonlinear and ill-posed problems; (3) model reduction and adaptivity.

Over the last few years, new paradigms have appeared to devise discretization methods supporting polytopal elements and arbitrary approximation orders. One key motivation is that the use of general element shapes provides an unprecedented flexibility in mesh generation, which is often the most time-consuming step in numerical modeling. Two examples of polytopal, arbitrary-order discretization methods are treated in this volume. On the one hand, the Hybridizable Discontinuous Galerkin (HDG) methods introduced in [9], where one central idea is the devising of local spaces to approximate the flux and the primal variable using the notion of M-decompositions from [11]. On the other hand, the Hybrid High-Order (HHO) methods introduced in [12, 13], where one central idea is the devising of the stabilization operator within a primal formulation. HDG and HHO methods have been recently bridged in [10]. Another important paradigm for the development of discretization methods is to reproduce exactly at the discrete level the fundamental properties of the model problem at hand, leading to so-called mimetic (or compatible, or structure-preserving) discretizations. This field, which is at the crossroads of differential geometry, algebraic topology and numerical analysis, has seen a lot of activity over the last decades; recent reviews with an historical perspective can be found in [2, 3, 8]. The contributions gathered in the first four chapters of this volume concern the theory of M-decomposition and its application to hybridisable discontinuous Galerkin and mixed methods; Mimetic Spectral Element method where the metric- and material-dependent Hodge operator is built as a mass matrix from tensor-product polynomials on Cartesian and deformed grids; an introduction to Hybrid High-Order methods able to deal with generally polytopal grids, with applications to the p-Laplace and diffusion-reaction equations.

The second group of three chapters concerns nonlinear and ill-posed problems. The first contribution concerns a numerical investigation of the Distributed Lagrange Multiplier method for fluid-structure interaction. This method, which has close links with the Immersed Boundary method [17] as well as with Fictitious Domain methods with a distributed Lagrange multiplier [14], has been recently developed and analyzed in [4]. The second contribution deals with the approximation of the spectrum of an elliptic operator and addresses the benefits of combining isogeometric analysis [15] with blending quadrature rules [1]. A Pythagorean theorem linking eigenvalue and eigenfunction errors, together with numerical results, are presented. The third contribution considers ill-posed problems as encountered, for instance, in the context of inverse and data assimilation problems. While state-of-the-art methods typically rely on the introduction of a regularization at the

[1] http://imag.edu.umontpellier.fr/event/ihp-nmpdes.

continuous level, one introduces here only a weakly-consistent regularization at the discrete level. Using very recent ideas on finite element stabilization [5, 6] leads to error estimates that are compatible with the (modest, yet provable) stability of the continuous problem at hand.

The third group of three chapters highlights recent advances in reduced-order modeling and adaptivity. The increased complexity of the physical models and the need to use PDE simulators in many-query scenarios (optimisation, inverse problems, real-time, etc.) has prompted the study of model reduction techniques such as the Reduced Basis (RB) method [18]. The present contribution, which focuses on elasticity problems in affinely parameterised geometries with (non-)compliant output error control [19], describes the RB approximation of such problems and presents various numerical examples. Finally, the numerical resolution of complex problems is often feasible only if the computation resources are used judiciously. This has prompted the study of adaptive resolution algorithms, often based on a posteriori estimates of the approximation error. Important advances have been accomplished over the last decade, as discussed among others in [7, 16] and in the recent textbook [20]. The present contribution develops a relatively less explored question, namely the adaptive approximation of a given univariate target function using mesh refinement by bisection. The last chapter gives an introduction on the possible treatment of defective boundary conditions, which typically appear in the coupling of PDE problems posed in domains of different geometrical dimensions.

References

1. Ainsworth, M., Wajid, H.A.: Optimally blended spectral-finite element scheme for wave propagation and nonstandard reduced integration. SIAM J. Numer. Anal. **48**(1), 346–371 (2010)
2. Arnold, D.N., Falk, R.S., Winther, R.: Finite element exterior calculus: from Hodge theory to numerical stability. Bull. Am. Math. Soc. (N.S.) **47**(2), 281–354 (2010)
3. Bochev, P., Hyman, J.M.: Principles of mimetic discretizations of differential operators. In: Arnold, D., Bochev, P., Lehoucq, R., Nicolaides, R.A., Shashkov, M. (eds.) Compatible Spatial Discretization. The IMA Volumes in Mathematics and Its Applications, vol. 142, pp. 89–120. Springer, New York (2005)
4. Boffi, D., Cavallini, N., Gastaldi, L.: The finite element immersed boundary method with distributed Lagrange multiplier. SIAM J. Numer. Anal. **53**(6), 2584–2604 (2015)
5. Burman, E.: Stabilized finite element methods for nonsymmetric, noncoercive, and ill-posed problems. Part I: elliptic equations. SIAM J. Sci. Comput. **35**(6), A2752–A2780 (2013)
6. Burman, E.: Stabilised finite element methods for ill-posed problems with conditional stability. In: Building Bridges: Connections and Challenges in Modern Approaches to Numerical Partial Differential Equations. Lecture Notes in Computational Science and Engineering, vol. 114, pp. 93–127. Springer, Cham (2016)
7. Carstensen, C., Feischl, M., Page, M., Praetorius, D.: Axioms of adaptivity. Comput. Math. Appl. **67**(6), 1195–1253 (2014)
8. Christiansen, S.H., Munthe-Kaas, H.Z., Owren, B.: Topics in structure-preserving discretization. Acta Numer. **20**, 1–119 (2011)

9. Cockburn, B., Gopalakrishnan, J., Lazarov, R.: Unified hybridization of discontinuous Galerkin, mixed, and continuous Galerkin methods for second order elliptic problems. SIAM J. Numer. Anal. **47**(2), 1319–1365 (2009)
10. Cockburn, B., Di Pietro, D.A., Ern, A.: Bridging the hybrid high-order and hybridizable discontinuous Galerkin methods. ESAIM: Math. Model Numer. Anal. **50**(3), 635–650 (2016)
11. Cockburn, B., Fu, G., Sayas, F.J.: Superconvergence by M-decompositions. Part I: general theory for HDG methods for diffusion. Math. Comput. **86**(306), 1609–1641 (2017)
12. Di Pietro, D.A., Ern, A.: A hybrid high-order locking-free method for linear elasticity on general meshes. Comput. Methods Appl. Mech. Eng. **283**, 1–21 (2015)
13. Di Pietro, D.A., Ern, A., Lemaire, S.: An arbitrary-order and compact-stencil discretization of diffusion on general meshes based on local reconstruction operators. Comput. Methods Appl. Mech. Eng. **14**(4), 461–472 (2014)
14. Girault, V., Glowinski, R.: Error analysis of a fictitious domain method applied to a Dirichlet problem. Jpn. J. Ind. Appl. Math. **12**(3), 487–514 (1995)
15. Hughes, T.J.R., Evans, J.A., Reali, A.: Finite element and NURBS approximations of eigenvalue, boundary-value, and initial-value problems. Comput. Methods Appl. Mech. Eng. **272**, 290–320 (2014)
16. Nochetto, R.H., Siebert, K.G., Veeser, A.: Theory of adaptive finite element methods: an introduction. In: Multiscale, Nonlinear and Adaptive Approximation, pp. 409–542. Springer, Berlin (2009)
17. Peskin, C.S.: The immersed boundary method. Acta Numer. **11**, 479–517 (2002)
18. Prud'homme, C., Rovas, D.V., Veroy, K., Patera, A.T.: A mathematical and computational framework for reliable real-time solution of parametrized partial differential equations. M2AN Math. Model. Numer. Anal. **36**(5), 747–771 (2002)
19. Rozza, G., Huynh, D.B.P., Patera, A.T.: Reduced basis approximation and a posteriori error estimation for affinely parametrized elliptic coercive partial differential equations: application to transport and continuum mechanics. Arch. Comput. Meth. Eng. **15**(3), 229–275 (2008)
20. Verfürth, R.: A posteriori error estimation techniques for finite element methods. In: Numerical Mathematics and Scientific Computation. Oxford University Press, Oxford (2013)

Chapter 2
An Introduction to the Theory of M-Decompositions

Bernardo Cockburn, Guosheng Fu, and Ke Shi

Abstract We provide a short introduction to the theory of M-decompositions in the framework of steady-state diffusion problems. This theory allows us to systematically devise hybridizable discontinuous Galerkin and mixed methods which can be proven to be superconvergent on unstructured meshes made of elements of a variety of shapes. The main feature of this approach is that it reduces such an effort to the definition, for each element K of the mesh, of the spaces for the flux, $\boldsymbol{V}(K)$, and the scalar variable, $W(K)$, which, roughly speaking, can be *decomposed* into suitably chosen orthogonal subspaces related to the space traces on ∂K of the scalar unknown, $M(\partial K)$. We begin by showing how a simple a priori error analysis motivates the notion of an M-decomposition. We then study the main properties of the M-decompositions and show how to actually construct them. Finally, we provide many examples in the two-dimensional setting. We end by briefly commenting on several extensions including to other equations like the wave equation, the equations of linear elasticity, and the equations of incompressible fluid flow.

2.1 Introduction

The theory of M-decompositions has been recently introduced as an effective tool to systematically find the local spaces defining hybridizable discontinuous Galerkin and mixed methods which can be proven to be superconvergent on unstructured

---

B. Cockburn (✉)
School of Mathematics, University of Minnesota, Minneapolis, MN, USA
e-mail: cockburn@math.umn.edu

G. Fu
Division of Applied Mathematics, Brown University, Providence, RI, USA
e-mail: guosheng_fu@brown.edu

K. Shi
Department of Mathematics & Statistics, Old Dominion University, Norfolk, VA, USA
e-mail: kshi@odu.edu

D. A. Di Pietro et al. (eds.), *Numerical Methods for PDEs*,
SEMA SIMAI Springer Series 15, https://doi.org/10.1007/978-3-319-94676-4_2

meshes made of elements of a variety of shapes. By "superconvergent" we mean that they can provide a new approximation, computed in an elementwise manner, which converges optimally and faster than the original approximation.

The general theory of M-decompositions was introduced in [14, 15, 27] in the framework of steady-state diffusion problems, as a refinement of the work done in [22]. Using some of these M-decompositions, new commutative diagrams for the deRham complex were presented in [16]. The extension to the Stokes system of incompressible fluid flow was done in [25], to the Navier-Stokes equations in [24], and to linear elasticity with symmetric approximate stresses in [13]. In this paper, we provide an introduction to the theory of M-decompositions.

We do this for HDG and mixed methods for the following steady-state diffusion problem:

$$\begin{aligned} c\boldsymbol{q} + \nabla u &= 0 && \text{in } \Omega, \\ \nabla \cdot \boldsymbol{q} &= f && \text{in } \Omega, \\ u &= g && \text{on } \partial\Omega, \end{aligned}$$

where $\Omega \subset \mathbb{R}^n$ $(n = 2, 3)$ is a bounded polyhedral domain, c is a uniformly bounded, uniformly positive definite symmetric matrix-valued function, $f \in L^2(\Omega)$ and $g \in H^{1/2}(\partial\Omega)$. The HDG methods have been thoroughly reviewed in [8]. Therein, the M-decompositions were briefly mentioned as a step in the development of the HDG methods. So, this paper can be considered to be a continuation of such review.

Our intention is to introduce the main ideas about M-decompositions as simply as possible; for a brief historical overview of the effort of devising superconvergent methods defined on unstructured meshes, see [27]. The material of this paper is based on three papers on the early development of M-decompositions. The first is the work done in [22], which provides general sufficient conditions for HDG and mixed methods to be superconvergent. The second is the work done in [27], which refines the previous work and introduces a general theory of M-decompositions for steady-state diffusion problems. The third is [14], which is devoted to the actual construction of M-decompositions in two-space dimensions.

The paper is organized as follows. In Sect. 2.2, we begin by placing the appearance of the idea of M-decompositions into historical perspective. In Sect. 2.3, we then introduce the notion of spaces admitting an M-decomposition and show how to use it to define hybridizable discontinuous Galerkin and mixed methods which can be proven to be superconvergent on unstructured meshes made of elements of a variety of shapes. In Sect. 2.4, we display our general construction of spaces admitting an M-decomposition, and in Sect. 2.5, we give concrete examples. We end in Sect. 2.6 by briefly describing past and ongoing extensions of this approach.

2.2 What Motivated the Appearance of the M-Decompositions?

Here, we briefly place the appearance of the M-decompositions into historical perspective. When the first wave of DG methods appeared around the end of last century, they were criticized because they could not be as efficiently implemented and could not provide as accurate approximations as the well-known hybridized version of the mixed methods. The HDG methods were then introduced in order to address the issue of efficient implementation. In addition, as these HDG methods were shown to be closely related to the mixed methods, a systematic effort started to devise HDG methods with the same superconvergence properties of the mixed methods. The theory of M-decompositions appeared as a tool to systematically do this.

2.2.1 DG Methods

To begin our discussion, let us define the DG methods for the model steady-state diffusion problem. Let $\mathcal{T}_h$ be a conforming mesh of Ω made of polygonal (n $=$ 2) or polyhedral (n $=$ 3) elements K. Let $\partial\Omega_h$ denote the set of boundaries ∂K of the elements $K \in \mathcal{T}_h$, $\mathcal{F}_h$ denote the set of faces F of the elements $K \in \mathcal{T}_h$, and $\mathcal{F}(K)$ denote the set of faces F of the element K. As usual, we write $(\eta\,,\,\zeta)_{\mathcal{T}_h} := \sum_{K\in\mathcal{T}_h}(\eta,\zeta)_K$, where $(\eta,\zeta)_D$ denotes the integral of $\eta\zeta$ over the domain $D \subset \mathbb{R}^n$. We also write $\langle\eta\,,\,\zeta\rangle_{\partial\mathcal{T}_h} := \sum_{K\in\mathcal{T}_h}\langle\eta\,,\,\zeta\rangle_{\partial K}$, where $\langle\eta\,,\,\zeta\rangle_D$ denotes the integral of $\eta\zeta$ over the 1-codimensional domain D. When vector-valued functions are involved, we use a similar notation.

The DG methods seek an approximation to $(u,\boldsymbol{q})$, $(u_h,\boldsymbol{q}_h)$, in the finite dimensional space $W_h \times \boldsymbol{V}_h$, where

$$\boldsymbol{V}_h := \{\boldsymbol{v} \in \boldsymbol{L}^2(\mathcal{T}_h) : \boldsymbol{v}|_K \in \boldsymbol{V}(K),\ \ K \in \mathcal{T}_h\},$$

$$W_h := \{w \in L^2(\mathcal{T}_h) : w|_K \in W(K),\ K \in \mathcal{T}_h\},$$

and determine it as the only solution of the following weak formulation:

$$\begin{aligned}(\mathrm{c}\,\boldsymbol{q}_h\,,\,\boldsymbol{v})_{\mathcal{T}_h} - (u_h\,,\,\nabla\cdot\boldsymbol{v})_{\mathcal{T}_h} + \langle\widehat{u}_h\,,\,\boldsymbol{v}\cdot\boldsymbol{n}\rangle_{\partial\mathcal{T}_h} &= 0,\\ -(\boldsymbol{q}_h\,,\,\nabla w)_{\mathcal{T}_h} + \langle\widehat{\boldsymbol{q}}_h\cdot\nabla\,,\,w\rangle_{\partial\mathcal{T}_h} &= (f\,,\,w)_{\mathcal{T}_h},\end{aligned}$$

for all $(w,\boldsymbol{v},\mu) \in W_h \times \boldsymbol{V}_h$, where the numerical traces $\widehat{u}_h$ and $\widehat{\boldsymbol{q}}_h\cdot\nabla$ are suitably defined functions of the unknown $(u_h,\boldsymbol{q}_h)$.

In the 2002 unified analysis of the DG methods [2], it was shown that, for elements of general shapes and $\boldsymbol{V}(K) \times W(K) := \boldsymbol{\mathcal{P}}_k(K) \times \mathcal{P}_k(K)$, the best orders of convergence for all the DG methods treated there in were k for the error in the flux $\|\boldsymbol{q} - \boldsymbol{q}_h\|_{L^2(\Omega)}$, which is suboptimal by 1, and $k + 1$ for the error in the scalar variable $\|u - u_h\|_{L^2(\Omega)}$, which is optimal. The same results can also be obtained with $\boldsymbol{V}(K) \times W(K) := \boldsymbol{\mathcal{P}}_{k-1}(K) \times \mathcal{P}_k(K)$.

These orders of converge are obtained, in particular, for the following choice of numerical traces:

$$\widehat{u}_h = \begin{cases} \{\!\{u_h\}\!\} - \boldsymbol{C}_{12} \cdot [\![u_h]\!] + C_{22} [\![\boldsymbol{q}_h]\!] & \text{in } \mathcal{F}_h \setminus \partial\Omega, \\ g & \text{in } \mathcal{F}_h \cap \partial\Omega, \end{cases}$$

$$\widehat{\boldsymbol{q}}_h = \begin{cases} \{\!\{\boldsymbol{q}_h\}\!\} + \boldsymbol{C}_{12} [\![\boldsymbol{q}_h]\!] + C_{11} [\![u_h]\!] & \text{in } \mathcal{F}_h \setminus \partial\Omega, \\ \boldsymbol{q}_h + C_{11}(u_h - g)\boldsymbol{n} & \text{in } \mathcal{F}_h \cap \partial\Omega. \end{cases},$$

and C_{11} positive, of order h^{-1}, $\boldsymbol{C}_{12}$ of order one, and $C_{22} = 0$, that is , for the LDG method [9]. When C_{11} and C_{22} are positive and of order one, and $\boldsymbol{C}_{12}$ is also of order one, it was shown in 2000 in [5] that the order of convergence of the flux increases to $k + 1/2$ and that of the scalar variable remains $k + 1$. In 2009 in [18], when the elements are restricted to be simplexes, it was shown in that, if $C_{11}, 1/C_{11}, C_{22}, 1/C_{22}, |\boldsymbol{C}_{12}|$ are positive and uniformly bounded, the order of the flux and that of the scalar variable are both $k + 1$ and that the error in the local averages superconverges with order $k + 2$, just as happens for the approximations of the well known RT_k and BDM_k mixed methods. This result was obtained by exploiting the relation between these DG methods and the corresponding HDG methods which we introduce next.

2.2.2 HDG Methods

The HDG methods were introduced in 2009 in [19] with the intention of obtaining DG methods for which static condensation was guaranteed. As argued in the 2016 review in [8], this resulted in a significant reduction of the number of globally-coupled degrees of freedom for the DG methods, highlighted the strong link between the HDG methods and the hybridized mixed methods, and led to new DG methods with better accuracy than all previously known DG methods.

The HDG methods seek an approximation to $(u, \boldsymbol{q}, u|_{\mathcal{F}_h})$, $(u_h, \boldsymbol{q}_h, \widehat{u}_h)$, in the finite dimensional space $W_h \times \boldsymbol{V}_h \times M_h$, where

$$\boldsymbol{V}_h := \{\boldsymbol{v} \in \boldsymbol{L}^2(\mathcal{T}_h) : \boldsymbol{v}|_K \in \boldsymbol{V}(K), \ \ K \in \mathcal{T}_h\},$$

$$W_h := \{w \in L^2(\mathcal{T}_h) : w|_K \in W(K), \ K \in \mathcal{T}_h\},$$

$$M_h := \{\mu \in L^2(\mathcal{F}_h) : \mu|_F \in M(F), \ \ F \in \mathcal{F}_h\},$$

and determine it as the only solution of the following weak formulation:

$$(c\,\boldsymbol{q}_h\,,\,\boldsymbol{v})_{\mathcal{T}_h} - (u_h\,,\,\nabla\cdot\boldsymbol{v})_{\mathcal{T}_h} + \langle\widehat{u}_h\,,\,\boldsymbol{v}\cdot\boldsymbol{n}\rangle_{\partial\mathcal{T}_h} = 0, \tag{2.1a}$$

$$-(\boldsymbol{q}_h\,,\,\nabla w)_{\mathcal{T}_h} + \langle\widehat{\boldsymbol{q}}_h\cdot\boldsymbol{n}\,,\,w\rangle_{\partial\mathcal{T}_h} = (f\,,\,w)_{\mathcal{T}_h}, \tag{2.1b}$$

$$\langle\widehat{\boldsymbol{q}}_h\cdot\boldsymbol{n},\,\mu\rangle_{\partial\mathcal{T}_h\setminus\partial\Omega} = 0, \tag{2.1c}$$

$$\langle\widehat{u}_h,\,\mu\rangle_{\partial\Omega} = \langle g,\,\mu\rangle_{\partial\Omega}, \tag{2.1d}$$

for all $(w,\,\boldsymbol{v},\,\mu)\in W_h\times\boldsymbol{V}_h\times M_h$, where

$$\widehat{\boldsymbol{q}}_h\cdot\boldsymbol{n} = \boldsymbol{q}_h\cdot\boldsymbol{n} + \alpha(u_h-\widehat{u}_h) \quad \text{on} \quad \partial\mathcal{T}_h. \tag{2.1e}$$

As pointed out in [19], by taking particular choices of the local spaces $\boldsymbol{V}(K)$, $W(K)$ and

$$M(\partial K) := \{\mu\in L^2(\partial K):\ \mu|_F\in M(F) \text{ for all } F\in\mathcal{F}(K)\},$$

and of the *linear local stabilization* function α, different HDG methods are obtained. If we can take α to be zero, we obtain nothing but the well-known hybridized version of the mixed methods. This establishes a strong link between the HDG methods, which use a non-zero stabilization α, and the mixed methods.

It can be shown, see [8, 19], that the very structure of the above weak formulation guarantees that the only globally-coupled degrees of freedom are those of the numerical trace $\widehat{u}_h$. This results in a very efficient implementation of the method which provides a significantly smaller stiffness matrix in comparison to that of all other DG methods.

It can also be shown that the HDG methods are strongly related to previously introduced DG methods. For example, if we take for $\boldsymbol{V}(K)\times W(K) := \boldsymbol{\mathcal{P}}_k(K)\times\mathcal{P}_k(K)$ and $M(F) := \mathcal{P}_k(K)$, and the stabilization function as $\alpha(\mu) := \tau\,\mu$, where τ is a constant on each face, it can be easily shown that the resulting HDG method is nothing but a classic DG methods with the following numerical traces:

$$\widehat{u}_h = \begin{cases} \frac{\tau^+}{\tau^++\tau^-}u_h^+ + \frac{\tau^-}{\tau^++\tau^-}u_h^- + \frac{1}{\tau^++\tau^-}[\![\boldsymbol{q}_h]\!] & \text{in } \mathcal{F}_h\setminus\partial\Omega,\\ g & \text{in } \mathcal{F}_h\cap\partial\Omega, \end{cases}$$

$$\widehat{\boldsymbol{q}}_h = \begin{cases} \frac{\tau^-}{\tau^++\tau^-}\boldsymbol{q}_h^+ + \frac{\tau^+}{\tau^++\tau^-}\boldsymbol{q}_h^- + \frac{\tau^+\tau^-}{\tau^++\tau^-}[\![u_h]\!] & \text{in } \mathcal{F}_h\setminus\partial\Omega,\\ \boldsymbol{q}_h+\tau(u_h-g)\boldsymbol{n} & \text{in } \mathcal{F}_h\cap\partial\Omega. \end{cases},$$

To illustrate the convergence properties of this method, let us consider the model problem

$$-\Delta u = f \quad \text{in } \Omega,$$
$$u = g \quad \text{on } \partial\Omega,$$

where Ω is a unit square, and the exact solution is $u(x, y) = \sin(2\pi x)\sin(2\pi y)$. In the table below, we display a history of convergence for the case $k = 1$ for three different types of meshes and $\tau = 1$. We display the $L^2(\Omega)$-norm of the error between the exact solution u and a local postprocessing u_h^*, see [32, 40, 41], defined on the element K as the polynomial of degree $k + 1$ such that

$$(\nabla u_h^*, \nabla w)_K = -(\mathrm{c}\,\boldsymbol{q}_h, \nabla w)_K \quad \forall\, w \in \mathcal{P}_{k+1}(K), \text{ and } (u_h^*, 1)_K = (u_h, 1)_K.$$

For this HDG method, $C_{11} = C_{22} = 1$, $\boldsymbol{C}_{12} = \mathbf{0}$. The results of the first column fully agree with the theoretical predictions in [18] which, for triangular meshes, ensures that the flux converges with order $k + 1$ and that the local averages superconverges with order $k + 2$; the local postprocessing thus must converge with order $k + 2$, as we see in the table. For polygonal meshes, we cannot rely on the theoretical predictions in [5] which only guarantee an order of convergence of the flux of $k + 1/2$ and that of the scalar variable is $k + 1$.

Thus, we see that the optimal order of convergence for u_h^* of $3 = k + 2$ holds only for triangular meshes and deteriorates as the number of sides of the element increases. This raises the question of how to achieve the superconvergence of the local averages independently of the shape of the elements.

	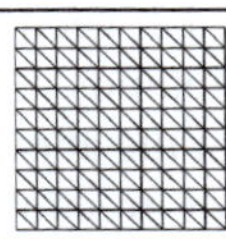		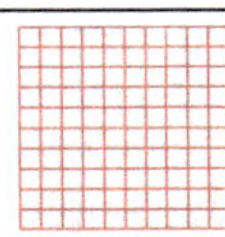		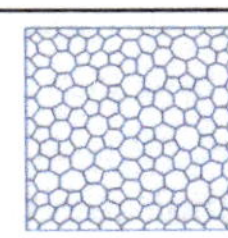	
h	$\Vert u - u_h^\star \Vert_{\mathcal{T}_h}$	Rate	$\Vert u - u_h^\star \Vert_{\mathcal{T}_h}$	Rate	$\Vert u - u_h^\star \Vert_{\mathcal{T}_h}$	Rate
			$\tau = 1$			
0.1	0.15E−2	–	0.83E−2	–	0.52E−2	–
0.05	0.18E−3	3.06	0.16E−2	2.36	0.10E−2	2.34
0.025	0.23E−4	3.03	0.28E−3	2.52	0.19E−3	2.43
0.0125	0.28E−5	**3.02**	0.44E−4	**2.68**	0.35E−4	**2.46**

2.2.3 Local Spaces or Stabilization Functions

The theory of M-decompositions allows us to answer to this question. Roughly speaking, this theory provides an explicit construction of the smallest number of basis functions one has to add to the local spaces of the approximate flux so that the resulting method becomes superconvergent. Once the new local spaces are found, the theory automatically constructs two mixed methods whose local spaces "sandwich" the new found spaces. Thus, we can also consider the theory of M-decompositions as a systematic way of constructing superconvergent mixed methods.

The emphasis of the approach based on M-decompositions is on the construction of the local spaces $\boldsymbol{V}(K) \times W(K)$ and the trace space $M(\partial K)$. It is *not* on the how to determine a stabilization function α which could render the resulting HDG method superconvergent. This second approach represents an complementary alternative to the theory of M-decompositions and is being currently developed. For more details, we refer the reader to before-the-last paragraph of the Introduction in [27].

Here, let us end by briefly mentioning the main contributions to this alternative. Lehrenfeld-Schöberl proposed a new, relatively simple stabilization function back in 2010 in [33, Remark 1.2.4]. The corresponding HDG method was then proven to be superconvergent by Oikawa in 2015 in [34]; see the extension to Stokes in [35]. In a parallel, independent effort, a new, sophisticated stabilization function α was identified in 2015 in [23] which is associated to the hybrid high-order (HHO) methods introduced in 2014 in [29] and in 2015 in [28] (for linear elasticity). See also [36] for an extension to the linear elasticity equations with strong symmetric approximate stresses, and [37] for the Navier-Stokes equations.

2.3 The M-Decompositions

In this section, we show that when the local spaces $\boldsymbol{V}(K) \times W(K)$ admit an $M(\partial K)$-decomposition for every element $K \in \mathcal{T}_h$, the associated HDG or mixed methods are superconvergent on unstructured meshes.

In what follows, to simplify the notation, when there is no possible confusion, we do not indicate the domain on which the functions of a given space are defined. For example, instead of $\boldsymbol{V}(K)$, we simply write $\boldsymbol{V}$.

2.3.1 Definition

To define the M-decomposition of the space

$$\boldsymbol{V} \times W \subset \{\boldsymbol{v} \in \boldsymbol{H}(\text{div}, K) : \boldsymbol{v} \cdot \boldsymbol{n}|_{\partial K} \in L^2(\partial K)\} \times H^1(K),$$

we need to consider the combined trace operator

$$\begin{aligned} \text{tr} : \boldsymbol{V} \times W &\longrightarrow L^2(\partial K) \\ (\boldsymbol{v}, w) &\longmapsto (\boldsymbol{v} \cdot \boldsymbol{n} + w)|_{\partial K} \end{aligned}$$

where $\boldsymbol{n} : \partial K \to \mathbb{R}^d$ is the unit outward pointing normal field on ∂K.

Definition 2.1 (The M-Decomposition [27]) We say that $\boldsymbol{V} \times W$ admits an M-decomposition when

(a) $\text{tr}(\boldsymbol{V} \times W) \subset M$,

and there exists a subspace $\widetilde{V} \times \widetilde{W}$ of $V \times W$ satisfying

(b) $\nabla W \times \nabla \cdot V \subset \widetilde{V} \times \widetilde{W}$,
(c) $\mathrm{tr} : \widetilde{V}^{\perp} \times \widetilde{W}^{\perp} \to M$ is an isomorphism.

Here $\widetilde{V}^{\perp}$ and $\widetilde{W}^{\perp}$ are the $L^2(K)$-orthogonal complements of $\widetilde{V}$ in V, and of $\widetilde{W}$ in W, respectively.

Although it can be proven that we must have $\widetilde{W} = \nabla \cdot V$, the space $\widetilde{V}$ is not unique. However, it is always possible to choose $\widetilde{V}$ as indicated in the following result which is expressed in terms of the following space of solenoidal, $H(\mathrm{div}, K)$-bubbles:

$$V_{\mathrm{sbb}} := \{v \in V : \nabla \cdot v = 0,\ v \cdot n|_{\partial K} = 0\}.$$

Proposition 2.1 (The Canonical M-Decomposition [27]) *If the space $V \times W$ admits an M-decomposition, then it admits an M-decomposition based on the subspaces*

$$\widetilde{V} = \nabla W \oplus V_{\mathrm{sbb}} \qquad \textit{(orthogonal sum)}, \qquad \widetilde{W} = \nabla \cdot V.$$

Of course, it is far from obvious that spaces $V \times W$ admitting M-decompositions can lead to superconvergent HDG and mixed methods. To see that, we need to carry out the error analysis of the methods with the help of a projection we define next.

2.3.2 The HDG-Projection

We define this auxiliary projection in terms of the $L^2(\partial K)$−projection into $M(\partial K)$, which we denote by P_M.

Definition 2.2 (The HDG-Projection [22]) Let (q, u) be smooth enough so that their boundary traces are in $L^2(\partial K)$. Let $V \times W$ admit an M-decomposition. Then, the pair $\Pi_h(q, u) = (\boldsymbol{\Pi}_V q, \Pi_W u) \in V \times W$ defined by the equations

(α) $(\Pi_W u, w)_K = (u, w)_K \qquad \forall w \in \widetilde{W}$,
(β) $(\boldsymbol{\Pi}_V q, v)_K = (q, v)_K \qquad \forall v \in \widetilde{V}$,
(γ) $\langle \boldsymbol{\Pi}_V q \cdot n + \alpha(\Pi_W u - P_M u), \mu\rangle_{\partial K} = \langle q \cdot n, \mu\rangle_{\partial K} \qquad \forall \mu \in M$,

is the HDG-projection associated to the M-decomposition and to the stabilization operator $\alpha : L^2(\partial K) \to L^2(\partial K)$.

Note that, when the stabilization function α is zero, we obtain nothing but the well-known projection used for the analysis of the mixed methods. The HDG-projection is thus an extension of such projection. Indeed, for any $w \in W$, we have

$$\begin{aligned}(\Pi_W \nabla \cdot \boldsymbol{q}, w)_K &= -(\boldsymbol{q}, \nabla w)_K + \langle \boldsymbol{q} \cdot \boldsymbol{n}, w\rangle_{\partial K} \\ &= -(\boldsymbol{\Pi}_V \boldsymbol{q}, \nabla w)_K + \langle \boldsymbol{\Pi}_V \boldsymbol{q} \cdot \boldsymbol{n} + \alpha(\Pi_W u - P_M u), w\rangle_{\partial K} \\ &= (\nabla \cdot \boldsymbol{\Pi}_V \boldsymbol{q}, w)_K + \langle \alpha(\Pi_W u - P_M u), w\rangle_{\partial K},\end{aligned}$$

and if we define $L_W(m)$ as the element of W such that

$$(L_W(m), w)_K = \langle m, w\rangle_{\partial K} \quad \forall\, w \in W,$$

we can write

$$\Pi_W \nabla \cdot \boldsymbol{q} = \nabla \cdot \boldsymbol{\Pi}_V \boldsymbol{q} + L_W(\alpha(\Pi_W u - P_M u)).$$

This extends to our framework the commutativity properties of the projections Π_W and $\boldsymbol{\Pi}_V$ for the mixed methods, that is, for the case in which we can take $\alpha = 0$.

Next, we provide a sufficient condition on the stabilization function α ensuring that the HDG-projection is actually well defined.

Proposition 2.2 (The HDG-Projection [22]) *Let $\boldsymbol{V} \times W$ admit an M-decomposition. Then the auxiliary HDG-projection Π_h is well defined if we take the linear stabilization operator $\alpha : L^2(\partial K) \to L^2(\partial K)$ such that*

$$w \in \widetilde{W}^{\perp}: \quad \langle \alpha(w), w\rangle_{\partial K} = 0 \quad \Longrightarrow \quad w = 0.$$

This result shows that we can take the stabilization function α equal to zero whenever $\widetilde{W}^{\perp} = \{0\}$. In this way, the stabilization function α can be linked to the gap between W and $\widetilde{W} = \nabla \cdot \boldsymbol{V}$. To measure such a gap, we introduce the following number, which is nonnegative because of the inclusion property (b).

Definition 2.3 (The S-Index) The S-index ("S" for stabilization) of the space $\boldsymbol{V} \times W$ is the number

$$I_S(\boldsymbol{V} \times W) := \dim W - \dim \nabla \cdot \boldsymbol{V}.$$

Note that by the inclusion condition (b), $I_S(\boldsymbol{V} \times W)$ is a natural number. It is zero if and only if $\widetilde{W}^{\perp} = \{0\}$ in which case we can take $\alpha = 0$.

Proof (of Proposition 2.2) Let us start by noting that the system defining the projection is square. The number of equations is $\dim \widetilde{\boldsymbol{V}} + \dim \widetilde{W} + \dim M$ and the number of unknowns is $\dim \boldsymbol{V} + \dim W$. Let us show that these numbers coincide. Since $\boldsymbol{V} \times W$ admits an M-decomposition, there are spaces $\widetilde{\boldsymbol{V}}$ and $\widetilde{W}$ satisfying property (c), and so

$$\dim M = \dim \widetilde{\boldsymbol{V}}^{\perp} + \dim \widetilde{W}^{\perp}.$$

This implies that $\dim \widetilde{\boldsymbol{V}} + \dim \widetilde{W} + \dim M = \dim \boldsymbol{V} + \dim W$, and so the system is square.

Now we only have to set $(\boldsymbol{q}, u) = (\boldsymbol{0}, 0)$ and prove that the only solution is the trivial one. In this case, we get that

$$\begin{aligned} (\Pi_W u, w)_K &= 0 && \forall w \in \widetilde{W}, \\ (\boldsymbol{\Pi}_V \boldsymbol{q}, \boldsymbol{v})_K &= 0 && \forall \boldsymbol{v} \in \widetilde{\boldsymbol{V}}, \\ \langle \boldsymbol{\Pi}_V \boldsymbol{q} \cdot \boldsymbol{n} + \alpha(\Pi_W u), \mu \rangle_{\partial K} &= 0 && \forall \mu \in M, \end{aligned}$$

which means that $\boldsymbol{\Pi}_V \boldsymbol{q} \in \widetilde{\boldsymbol{V}}^{\perp}$ and that $\Pi_W u \in \widetilde{W}^{\perp}$. Since, by property (a), $W|_{\partial K} \subset M$, we can take $\mu := \Pi_W u$ in the third equation defining the projection to get

$$\begin{aligned} \langle \alpha(\Pi_W u), \Pi_W u \rangle_{\partial K} &= -\langle \boldsymbol{\Pi}_V \boldsymbol{q} \cdot \boldsymbol{n}, \Pi_W u \rangle_{\partial K} \\ &= (\nabla \cdot \boldsymbol{\Pi}_V \boldsymbol{q}, \Pi_W u)_K + (\boldsymbol{\Pi}_V \boldsymbol{q}, \nabla \Pi_W u)_K \\ &= 0, \end{aligned}$$

by the inclusion properties (b), since $\nabla \cdot \boldsymbol{\Pi}_V \boldsymbol{q} \in \nabla \cdot \boldsymbol{V} \subset \widetilde{W}$ and since $\nabla \Pi_W u \in \nabla W \subset \widetilde{\boldsymbol{V}}$. Therefore, by the assumption on the stabilization function α, it follows that $\Pi_W u = 0$. Finally, by property (a), since $\boldsymbol{V} \cdot \boldsymbol{n}|_{\partial K} \subset M$, we can take $\mu := \boldsymbol{\Pi}_V \boldsymbol{q} \cdot \boldsymbol{n}$ in the third equation defining the projection to get

$$\langle \boldsymbol{\Pi}_V \boldsymbol{q} \cdot \boldsymbol{n}, \boldsymbol{\Pi}_V \boldsymbol{q} \cdot \boldsymbol{n} \rangle_{\partial K} = 0,$$

which implies, by property (c), that $\boldsymbol{\Pi}_V \boldsymbol{q} = \boldsymbol{0}$ since $\boldsymbol{\Pi}_V \boldsymbol{q} \in \widetilde{\boldsymbol{V}}^{\perp}$. This completes the proof. □

2.3.3 Estimates of the Projection of the Errors

Next, we find the equations of the projection of the errors:

$$\boldsymbol{e_q} := \boldsymbol{\Pi}_V \boldsymbol{q} - \boldsymbol{q}_h, \quad e_u := \Pi_W u - u_h, \quad \boldsymbol{e}_{\widehat{\boldsymbol{q}}} \cdot \boldsymbol{n} := P_M(\boldsymbol{q} \cdot \boldsymbol{n}) - \widehat{\boldsymbol{q}}_h \cdot \boldsymbol{n}, \quad e_{\widehat{u}} := P_M(u) - \widehat{u}_h.$$

We show that the definition of an M-decomposition and that of the HDG-projection are tailored to the numerical schemes under consideration.

Since the exact solution also satisfies the weak formulation defining the HDG method, we can write that

$$\begin{aligned} (\mathrm{c}\,(\boldsymbol{q} - \boldsymbol{q}_h), \boldsymbol{v})_{\mathcal{T}_h} - (u - u_h, \nabla \cdot \boldsymbol{v})_{\mathcal{T}_h} + \langle u - \widehat{u}_h, \boldsymbol{v} \cdot \boldsymbol{n} \rangle_{\partial \mathcal{T}_h} &= 0, \\ -(\boldsymbol{q} - \boldsymbol{q}_h, \nabla w)_{\mathcal{T}_h} + \langle \boldsymbol{q} \cdot \boldsymbol{n} - \widehat{\boldsymbol{q}}_h \cdot \boldsymbol{n}, w \rangle_{\partial \mathcal{T}_h} &= 0, \end{aligned}$$

$$\begin{aligned}\langle \boldsymbol{q}\cdot\boldsymbol{n}-\widehat{\boldsymbol{q}}_h\cdot\boldsymbol{n},\mu\rangle_{\partial\mathcal{T}_h\setminus\partial\Omega}&=0,\\ \langle u-\widehat{u}_h,\mu\rangle_{\partial\Omega}&=0,\end{aligned}$$

for all $(w, \boldsymbol{v}, \mu) \in W_h \times \boldsymbol{V}_h \times M_h$, where $\widehat{\boldsymbol{q}}_h \cdot \boldsymbol{n} = \boldsymbol{q}_h \cdot \boldsymbol{n} + \alpha(u_h - \widehat{u}_h)$ on $\partial\mathcal{T}_h$. But, we have that

$$\begin{aligned}
\langle u-\widehat{u}_h\,,\,\boldsymbol{v}\cdot\boldsymbol{n}\rangle_{\partial\mathcal{T}_h}&=\langle e_{\widehat{u}}\,,\,\boldsymbol{v}\cdot\boldsymbol{n}\rangle_{\partial\mathcal{T}_h} && \text{by property (a),}\\
\langle \boldsymbol{q}\cdot\boldsymbol{n}-\widehat{\boldsymbol{q}}_h\cdot\boldsymbol{n}\,,\,w\rangle_{\partial\mathcal{T}_h}&=\langle \boldsymbol{e}_{\widehat{\boldsymbol{q}}}\cdot\boldsymbol{n}\,,\,w\rangle_{\partial\mathcal{T}_h} && \text{by property (a),}\\
(u-u_h\,,\,\nabla\cdot\boldsymbol{v})_{\mathcal{T}_h}&=(e_u\,,\,\nabla\cdot\boldsymbol{v})_{\mathcal{T}_h} && \text{by properties }(\alpha)\text{ and (b),}\\
(\boldsymbol{q}-\boldsymbol{q}_h\,,\,\nabla w)_{\mathcal{T}_h}&=(\boldsymbol{e}_{\boldsymbol{q}}\,,\,\nabla w)_{\mathcal{T}_h} && \text{by properties }(\beta)\text{ and (b),}\\
\boldsymbol{e}_{\widehat{\boldsymbol{q}}}\cdot\boldsymbol{n}&=\boldsymbol{e}_{\boldsymbol{q}}\cdot\boldsymbol{n}+P_M\alpha(e_u-e_{\widehat{u}}) && \text{on }\partial\mathcal{T}_h,
\end{aligned}$$

by property (γ), and so, we get that

$$\begin{aligned}
-(e_u\,,\,\nabla\cdot\boldsymbol{v})_{\mathcal{T}_h}+\langle e_{\widehat{u}}\,,\,\boldsymbol{v}\cdot\boldsymbol{n}\rangle_{\partial\mathcal{T}_h}&=-(\mathrm{c}\,(\boldsymbol{q}-\boldsymbol{\Pi}_V\boldsymbol{q})\,,\,\boldsymbol{v})_{\mathcal{T}_h},\\
-(\boldsymbol{e}_{\boldsymbol{q}}\,,\,\nabla w)_{\mathcal{T}_h}+\langle \boldsymbol{e}_{\widehat{\boldsymbol{q}}}\cdot\boldsymbol{n}\,,\,w\rangle_{\partial\mathcal{T}_h}&=0,\\
\langle \boldsymbol{e}_{\widehat{\boldsymbol{q}}}\cdot\boldsymbol{n},\mu\rangle_{\partial\mathcal{T}_h\setminus\partial\Omega}&=0,\\
\langle e_{\widehat{u}},\mu\rangle_{\partial\Omega}&=0,
\end{aligned}$$

for all $(w, \boldsymbol{v}, \mu) \in W_h \times \boldsymbol{V}_h \times M_h$.

We immediately see that if the right-hand side of the first equation is zero, then the all the projection of the errors are zero. This means that all of them are controlled by the size of the approximation error $\boldsymbol{q} - \boldsymbol{\Pi}_V\boldsymbol{q}$. In particular, the standard energy argument, obtained by setting $(\boldsymbol{v}, w, \mu) := (\boldsymbol{e}_{\boldsymbol{q}}, e_u, e_{\widehat{u}})$ and adding the equations, and noting that $e_{\widehat{u}}|_{\partial\Omega} = 0$, gives that

$$(\mathrm{c}\,\boldsymbol{e}_{\boldsymbol{q}}\,,\,\boldsymbol{e}_{\boldsymbol{q}})_{\mathcal{T}_h}+\langle\alpha(e_u-e_{\widehat{u}})\,,\,e_u-e_{\widehat{u}}\rangle_{\partial\mathcal{T}_h}=-(\mathrm{c}\,(\boldsymbol{q}-\boldsymbol{\Pi}_V\boldsymbol{q})\,,\,\boldsymbol{e}_{\boldsymbol{q}})_{\mathcal{T}_h}.$$

In fact, it is possible to prove the following estimates.

Theorem 2.1 (A Priori Error Estimates) *Suppose that for every $K \in \mathcal{T}_h$, the space $\boldsymbol{V}(K) \times W(K)$ admits an $M(\partial K)$-decomposition and that the stabilization function α satisfies the following properties:*

(i) $w \in \widetilde{W}^{\perp}(K), \quad \langle\alpha(w), w\rangle_{\partial K} = 0 \quad\Longrightarrow\quad w = 0,$
(ii) $\langle\alpha(\mu), \mu\rangle_{\partial K} \geq 0$ *for all* $\mu \in M(\partial K)$,
(iii) $\langle\alpha(\lambda), \mu\rangle_{\partial K} = \langle\lambda, \alpha(\mu)\rangle_{\partial K}$, *for all* $\lambda, \mu \in M(\partial K)$.

Then, we have

$$\|\boldsymbol{e_q}\|_{\mathcal{T}_h} \leq C\, \|\boldsymbol{q} - \boldsymbol{\Pi}_V \boldsymbol{q}\|_{\mathcal{T}_h},$$
$$\|e_u\|_{\mathcal{T}_h} \leq C\, H\, \|\boldsymbol{q} - \boldsymbol{\Pi}_V \boldsymbol{q}\|_{\mathcal{T}_h},$$

where $H = 1$ for general polyhedral domains. For convex polyhedral domains, we have that $H = h$ provided

$$\mathcal{P}_0(K) \subset \nabla W(K)\ \forall\ K \in \mathcal{T}_h.$$

2.3.4 Local Postprocessing

Next, we define an elementwise postprocessing u_h^* defined to converge faster than the original approximation u_h; we follow [32, 40, 41]. We take the postprocessing u_h^* in the space

$$W_h^* := \{w \in L^2(\mathcal{T}_h) : w|_K \in W^*(K),\ K \in \mathcal{T}_h\},$$

and define it as follows. On each element $K \in \mathcal{T}_h$, the function u_h^* is the element of $W^*(K)$ such that

$$(\nabla u_h^*, \nabla w)_K = -\,(c\,\boldsymbol{q}_h, \nabla w)_K \quad \forall\ w \in \widetilde{W}^*(K)^\perp,$$
$$(u_h^*, w)_K = (u_h, w)_K \quad \forall\ w \in \widetilde{W}^*(K).$$

where $W^*(K) = \widetilde{W}^*(K) \oplus \widetilde{W}^*(K)^\perp$ and $\widetilde{W}^*(K)$ is any non-trivial subspace of $\widetilde{W}(K)$ containing constant functions. We have the following result which follows directly from the analysis carried out in [22].

Theorem 2.2 *Under the assumptions of the previous result, and if*

$$\mathcal{P}_0(K) \subset \nabla \cdot \boldsymbol{V}(K)\ \forall\ K \in \mathcal{T}_h,$$

then

$$\|u - u_h^*\|_{\mathcal{T}_h} \leq \|\Pi_W u - u_h\|_{\mathcal{T}_h} + C\, h\, (\|\boldsymbol{q} - \boldsymbol{\Pi}_V \boldsymbol{q}\|_{\mathcal{T}_h} + \inf_{\omega \in W_h^*} \|\nabla(u - \omega)\|_{\mathcal{T}_h}).$$

This result states that, once we find spaces $\boldsymbol{V} \times W$ spaces admitting M-decompositions, we still have to check the conditions

(J.1) $\mathcal{P}_0(K) \subset \nabla \cdot \boldsymbol{V}$,
(J.2) $\mathcal{P}_1(K) \subset W$,

in order to achieve the superconvergence of the elementwise averages and the optimal convergence of the elementwise postprocessing.

It remains to obtain the approximation properties of the HDG-projection. We do that next.

2.3.5 *Approximation Properties of the HDG-Projection*

Note that, in view of the second equation defining the auxiliary HDG-projection, one might think that its approximation properties depend on the choice of the subspace $\widetilde{V}$. This would be rather unpleasant given that, unlike the subspace $\widetilde{W}$, the subspace $\widetilde{V}$ of an M-decomposition is *not* uniquely defined. Fortunately, this is not so as we see in the next result which is a small variation of a similar result in [22]; for the sake of completeness, we include a proof in the Appendix. To state it, we need to introduce the quantities

$$a_{\widetilde{W}^\perp} := \begin{cases} \inf_{\mu \in \gamma \widetilde{W}^\perp \setminus \{0\}} \langle \alpha(\mu), \mu \rangle_{\partial K} / \|\mu\|^2_{\partial K} & \text{if } \widetilde{W}^\perp \neq \{0\}, \\ \infty & \text{if } \widetilde{W}^\perp = \{0\}, \end{cases}$$

and

$$\|\alpha\| := \sup_{\lambda, \mu \in M \setminus \{0\}} \langle \alpha(\lambda), \mu \rangle_{\partial K} / (\|\lambda\|_{\partial K} \|\mu\|_{\partial K}).$$

When $\widetilde{W}^\perp = \{0\}$, that is, when $\widetilde{W} = W$, we take $\alpha := 0$.

In what follows, P_S denotes the $L^2(\Omega)$−projection into the space S. We use this notation for $S := \boldsymbol{V}_h$, $S := W$ and $S := \widetilde{W}$.

Proposition 2.3 (Approximation Properties of the HDG-Projection) *Let $V \times W$ admit an M-decomposition, and let the stabilization function α satisfy the condition*

$$a_{\widetilde{W}^\perp} > 0.$$

Then, we have

$$\begin{aligned} \|\boldsymbol{q} - \boldsymbol{\Pi}_V \boldsymbol{q}\|_K \leq{}& \|(Id - P_V)\,\boldsymbol{q}\|_K + \mathsf{C}_1 h_K^{1/2} \|((Id - P_V)\boldsymbol{q}) \cdot \boldsymbol{n}\|_{\partial K} \\ &+ \mathsf{C}_2 h_K \|(Id - P_{\widetilde{W}})\nabla \cdot \boldsymbol{q}\|_K + \mathsf{C}_3 h_K^{1/2} \|(Id - P_W)u\|_{\partial K}, \\ \|u - \Pi_W u\|_K \leq{}& \|(Id - P_W)u\|_K + \mathsf{C}_4 h_K^{1/2} \|(Id - P_W)u\|_{\partial K} \\ &+ \mathsf{C}_5 h_K \|(Id - P_{\widetilde{W}})\nabla \cdot \boldsymbol{q}\|_K, \end{aligned}$$

where $\mathsf{C}_1 := C_{\widetilde{V}^\perp}$ *and*

$$\mathsf{C}_2 := \frac{C_{\widetilde{W}^\perp}}{a_{\widetilde{W}^\perp}} C_{\widetilde{V}^\perp} \|\alpha\|, \ \mathsf{C}_3 := \big(1 + \frac{\|\alpha\|}{a_{\widetilde{W}^\perp}}\big) C_{\widetilde{V}^\perp} \|\alpha\|, \ \mathsf{C}_4 := \frac{C_{\widetilde{W}^\perp}}{a_{\widetilde{W}^\perp}} \|\alpha\|, \ \mathsf{C}_5 := \frac{C^2_{\widetilde{W}^\perp}}{a_{\widetilde{W}^\perp}}.$$

Here

$$C_{\widetilde{V}^\perp} := \sup_{\boldsymbol{v}\in\widetilde{\boldsymbol{V}}^\perp\setminus\{\boldsymbol{0}\}} h_K^{-1/2} \|\boldsymbol{v}\|_K / \|\boldsymbol{v}\cdot\boldsymbol{n}\|_{\partial K}, \qquad C_{\widetilde{W}^\perp} := \sup_{w\in\widetilde{W}^\perp\setminus\{0\}} h_K^{-1/2} \|w\|_K / \|w\|_{\partial K},$$

Note that the fact that the coercivity constant $a_{\widetilde{W}^\perp}$ is positive implies the property of the stabilization function α used in Proposition 2.2: this is due to the third condition in the definition of M-decomposition. Note also that, if $W = \widetilde{W} = \nabla \cdot \ \boldsymbol{V}$, then $\mathsf{C}_i = 0$ for $i = 2, 3, 4, 5$ since in this case we are taking $\alpha = 0$ and $a_{\widetilde{W}^\perp} = \infty$.

2.4 A Construction of *M*-Decompositions

Here, we show how to use the notion of M-decompositions to actually construct spaces admitting M-decompositions. To do that, we begin by establishing a characterization of M-decompositions which is going to be the basis for the construction. We then apply it to show, given an element K, a space of traces $M(\partial K)$, and a the space $\boldsymbol{V}_g \times W_g$, how to systematically construct *three* spaces admitting an M-decomposition. One of them generates an HDG method whereas the other two generate mixed methods.

2.4.1 A Characterization of M-Decompositions

We begin by stating the main result of this section, namely, a characterization of the M-decompositions expressed solely in terms of the spaces $\boldsymbol{V} \times W$. Roughly speaking, it states that $\boldsymbol{V} \times W$ admits an M-decomposition if and only if the space M is the orthogonal sum of the traces of the kernels of $\nabla\cdot$ in $\boldsymbol{V}$ and of ∇ in W. It is expressed in terms of a special integer we define next.

Definition 2.4 (The *M*-Index) The M-index of the space $\boldsymbol{V} \times W$ is the number

$$\begin{aligned} I_M(\boldsymbol{V} \times W) := \ &\dim M - \dim\{\boldsymbol{v}\cdot\boldsymbol{n}|_{\partial K} \,:\, \boldsymbol{v} \in \boldsymbol{V}, \nabla\cdot\boldsymbol{v} = 0\} \\ &- \dim\{w|_{\partial K} \,:\, w \in W, \nabla w = 0\}. \end{aligned}$$

Theorem 2.3 (A Characterization of M-Decompositions) *For a given space of traces M, the space $\boldsymbol{V} \times W$ admits an M-decomposition if and only if*

(a) $tr(\boldsymbol{V} \times W) \subset M$,
(b) $\nabla W \times \nabla \cdot \boldsymbol{V} \subset \boldsymbol{V} \times W$,
(c) $I_M(\boldsymbol{V} \times W) = 0$.

In this case, we have the so-called the kernels' trace decomposition *identity*

$$M = \{\boldsymbol{v} \cdot \boldsymbol{n}|_{\partial K} : \boldsymbol{v} \in \boldsymbol{V}, \nabla \cdot \boldsymbol{v} = 0\} \oplus \{w|_{\partial K} : w \in W, \nabla w = 0\},$$

where the sum is orthogonal.

Note that the subspaces $\widetilde{\boldsymbol{V}}$ and $\widetilde{W}$ appearing in the definition of an M-decomposition, which were strongly associated to the very form of the HDG methods under consideration, are not present anymore in this characterization. This suggests that the M-decomposition can be considered to be associated to the operators $(\nabla\cdot, \nabla)$ rather than to a specific numerical method.

Note also that the above result states that, if the space $\boldsymbol{V} \times W$ satisfies the inclusion conditions (a) and (b), we have that

$$M = C_M \oplus \{\boldsymbol{v} \cdot \boldsymbol{n}|_{\partial K} : \boldsymbol{v} \in \boldsymbol{V}, \nabla \cdot \boldsymbol{v} = 0\} \oplus \{w|_{\partial K} : w \in W, \nabla w = 0\},$$

for some subspace C_M of M. This means that the dimension of C_M is nothing but $I_M(\boldsymbol{V} \times W)$ and that $\boldsymbol{V} \times W$ admits an M-decomposition if and only if $C_M = \{0\}$, that is, if and only if $I_M(\boldsymbol{V} \times W) = 0$.

2.4.2 The General Construction

Here, we show how to use the above result to construct spaces admitting M-decompositions. We proceed as follows. First, given the element K and the space of traces $M(\partial K)$, we pick our favorite space $\boldsymbol{V}_g \times W_g$ satisfying the inclusion properties (a) and (b) of Theorem 2.3. Then, we construct three of spaces admitting an M-decomposition as follows.

Step 1. We find a space $\delta \boldsymbol{V}_{\text{fillM}}$ such that

(a) $\delta \boldsymbol{V}_{\text{fillM}} \cdot \boldsymbol{n}|_{\partial K} = C_M$,
(b) $\nabla \cdot \delta \boldsymbol{V}_{\text{fillM}} = \{0\}$,
(c) $\dim \delta \boldsymbol{V}_{\text{fillM}} = I_M(\boldsymbol{V}_g \times W_g)$.

Then, we can verify that $(\boldsymbol{V}_g \oplus \delta \boldsymbol{V}_{\text{fillM}}) \times W_g$ admits an M-decomposition.

Step 2. The space $(V_g \oplus \delta V_{\text{fillM}}) \times \nabla \cdot V_g$ immediately admits an M-decomposition provided

$$\{w|_{\partial K} : w \in W_g, \nabla w = 0\} = \{w|_{\partial K} : w \in \nabla \cdot V_g, \nabla w = 0\}.$$

In this case, we can take the stabilization function α equal to zero and so the corresponding method is a mixed method.

Step 3. Finally, if $W_g = C_W \oplus \nabla \cdot V_g$, we find a space δV_{fillW} such that

(a) $\delta V_{\text{fillW}} \cdot n|_{\partial K} \subset M$,
(b) $\nabla \cdot \delta V_{\text{fillW}} = C_W$,
(c) $\dim \delta V_{\text{fillW}} = I_S(V_g \times W_g)$.

Then we immediately have that $(V_g \oplus \delta V_{\text{fillM}} \oplus \delta V_{\text{fillW}}) \times W_g$ admits an M-decomposition. Moreover, we can take the stabilization function α equal to zero and so the corresponding method is a mixed method.

We summarize our construction of spaces admitting M-decompositions in Tables 2.1, 2.2 and 2.3.

Table 2.1 Construction of spaces $V \times W$ admitting an M-decomposition, where the space of traces $M(\partial K)$ includes the constants

V	W	$\nabla \cdot V$
$V_g \oplus \delta V_{\text{fillM}} \oplus \delta V_{\text{fillW}}$	W_g (if $\supset \mathcal{P}_0(K)$)	W_g
$V_g \oplus \delta V_{\text{fillM}}$	W_g (if $\supset \mathcal{P}_0(K)$)	$\subset W_g$
$V_g \oplus \delta V_{\text{fillM}}$	$\nabla \cdot V_g$ (if $\supset \mathcal{P}_0(K)$)	$\nabla \cdot V_g$

The given space $V_g \times W_g$ satisfies the inclusion properties (a) and (b)

Table 2.2 The properties of the spaces δV

δV	$\nabla \cdot \delta V$	$\delta V \cdot n|_{\partial K}$	$\dim \delta V$
δV_{fillM}	$\{0\}$	C_M	$I_M(V_g \times W_g)$
δV_{fillW}	C_W	$\subset M$	$I_S(V_g \times W_g)$

The computation of the space C_W is fairly simple and, usually, independent of the shape of the element. In contrast, the computation of the space C_M is the most difficult part of the construction

Table 2.3 The spaces $\widetilde{V} \times \widetilde{W}$ defining the canonical decomposition of each space $V \times W$ in terms of the space $V_g \times W_g$

$\widetilde{V}$	$\widetilde{W}$
$\nabla W_g \oplus V_{g,\text{sbb}}$	W_g
$\nabla W_g \oplus V_{g,\text{sbb}}$	$\nabla \cdot V_g$
$\nabla(\nabla \cdot V_g) \oplus V_{g,\text{sbb}}$	$\nabla \cdot V_g$

Here $V_{g,\text{sbb}} := \{v \in V_g : \nabla \cdot v = 0,\ v \cdot n|_{\partial K} = 0\}$

2.5 Examples

Here, we give examples of this construction. We only present the spaces that can be concisely described and so we restrict ourselves to the two-dimensional case. First, we show the computation by hand of the whole construction in a very simple case. We then consider triangular, rectangular and quadrilateral elements and show the old and new spaces that result from our construction. Finally, we describe and briefly discuss the case of a general polygonal element.

2.5.1 *An Illustration of the Construction*

Let us illustrate the general construction just sketched in a very simple case, namely, when K is the unit square and

$$\begin{aligned} M(\partial K) &:= \{\mu \in L^2(\partial K) : \mu|_F \in \mathcal{P}_0(F) \text{ for all faces } F \text{ of } K\}, \\ \boldsymbol{V}_g \times W_g &:= \boldsymbol{\mathcal{P}}_0(K) \times \mathcal{P}_0(K). \end{aligned}$$

Here, $\mathcal{P}_0$ denotes the space constant functions, and $\boldsymbol{\mathcal{P}}_0$ the space of vectors whose components lie on $\mathcal{P}_0$. Since, it is clear that the inclusion properties (a) and (b) are satisfied, we can now proceed.

Step 1. Since

$$\{\boldsymbol{v}\cdot\boldsymbol{n}|_{\partial K} : \boldsymbol{v} \in \boldsymbol{V}_g, \nabla\cdot\boldsymbol{v} = 0\} = \text{span}\{\begin{smallmatrix} & 0 & \\ -1 & \square & 1 \\ & 0 & \end{smallmatrix}, \begin{smallmatrix} & 1 & \\ 0 & \square & 0 \\ & -1 & \end{smallmatrix}\},$$

$$\{w|_{\partial K} : w \in W_g, \nabla w = 0\} = \text{span}\{\begin{smallmatrix} & 1 & \\ 1 & \square & 1 \\ & 1 & \end{smallmatrix}\},$$

$$M(\partial K) = \text{span}\{\begin{smallmatrix} & 0 & \\ 0 & \square & 1 \\ & 0 & \end{smallmatrix}, \begin{smallmatrix} & 0 & \\ 1 & \square & 0 \\ & 0 & \end{smallmatrix}, \begin{smallmatrix} & 1 & \\ 0 & \square & 0 \\ & 0 & \end{smallmatrix}, \begin{smallmatrix} & 0 & \\ 1 & \square & 0 \\ & 0 & \end{smallmatrix}\}$$

we have that $I_M(\boldsymbol{V}_g \times W_g) = 4 - 2 - 1 = 1$ and we can take $C_M = \text{span}\{\begin{smallmatrix} & -1 & \\ 0 & \square & 1 \\ & 0 & \end{smallmatrix}\}$. So, we can take

$$\boldsymbol{V}_{\text{fillM}} := \text{span}\{(x, -y)\}.$$

This means that

$$(\boldsymbol{V}_g \oplus \delta\boldsymbol{V}_{\text{fillM}}) \times W_g = \text{span}\{(1,0), (0,1), (x,-y)\} \times \text{span}\{1\},$$

admits an $\mathcal{P}_0(\partial K)$-decomposition.

Step 2. The space constructed in this step, namely,

$$(\boldsymbol{V}_g \oplus \delta \boldsymbol{V}_{\text{fillM}}) \times \nabla \cdot \boldsymbol{V}_g = \text{span}\{(1,0), (0,1), (x,-y)\} \times \{0\},$$

does **not** admit an $\mathcal{P}_0(\partial K)$-decomposition because

$$\{w|_{\partial K} : w \in W_g, \nabla w = 0\} = \text{span}\{1\} \neq \{0\} = \{w|_{\partial K} : w \in \nabla \cdot \boldsymbol{V}_g, \nabla w = 0\}.$$

Step 3. Finally, we note that $\nabla \cdot \boldsymbol{V}_g = \{0\}$ and so $I_M(\boldsymbol{V}_g \times W_g) = 1 - 0 = 1$ and $C_W = W_g$. We can then take

$$\boldsymbol{V}_{\text{fillW}} := \text{span}\{(x, y)\}.$$

This means that the space

$$(\boldsymbol{V}_g \oplus \delta \boldsymbol{V}_{\text{fillM}} \oplus \delta \boldsymbol{V}_{\text{fillW}}) \times W_g = \text{span}\{(1,0), (0,1), (x,-y), (x,y)\} \times \text{span}\{1\},$$

also admits an $\mathcal{P}_0(\partial K)$-decomposition. This completes the construction.

2.5.2 Triangular and Quadrilateral Elements

Let us now consider triangular and quadrilateral elements, $M := \mathcal{P}_k(\partial K)$ and two cases of the spaces $\boldsymbol{V}_g \times W_g$. The first is only associated with rectangles, $\boldsymbol{V}_g \times W_g := \boldsymbol{\mathcal{Q}}_k \times \mathcal{Q}_k$; $\mathcal{Q}_k$ denotes the space of tensor product polynomials of degree at most k, and $\boldsymbol{\mathcal{Q}}_k$ denotes the space of vectors whose components lie on $\mathcal{Q}_k$. The second is $\boldsymbol{V}_g \times W_g := \boldsymbol{\mathcal{P}}_k \times \mathcal{P}_k$; $\mathcal{P}_k$ denotes the space polynomials of degree at most k, and $\boldsymbol{\mathcal{P}}_k$ denotes the space of vectors whose components lie on $\mathcal{P}_k$. The results are displayed in Table 2.4 taken from [14].

In Table 2.4, we use the notation $\mathbf{curl}\, p := (-p_y, p_x)$. We also need to define the linear function λ_i and the rational function ξ_i associated to the definition of the spaces for quadrilaterals. Let $\{\mathbf{v}_i\}_{i=1}^4$ be the set of vertices of the quadrilateral K which we take to be counter-clockwise ordered. Let $\{\mathbf{e}_i\}_{i=1}^4$ be the set of edges of K where the edge $\mathbf{e}_i$ connects the vertices $\mathbf{v}_i$ and $\mathbf{v}_{i+1}$, where we set $\mathbf{v}_5 = \mathbf{v}_1$. Then, for $1 \leq i \leq 4$, we define λ_i to be the linear function that vanishes on edge $\mathbf{e}_i$ and reaches maximum value 1 in the closure of K, and ξ_i to be a rational function such that $\xi_i|_{\mathbf{e}_i} \in \mathcal{P}_1(\mathbf{e}_i)$ and $\xi_i(\mathbf{v}_j) = \delta_{ij}$, where δ_{ij} is the Kronecker delta. A particular choice of ξ_i is given as follows:

$$\xi_i := \eta_{i-1} \frac{\lambda_{i-2}}{\lambda_{i-2}(\mathbf{v}_i)} + \eta_i \frac{\lambda_{i+1}}{\lambda_{i+1}(\mathbf{v}_i)}, \qquad \text{where} \qquad \eta_i := \Pi_{\substack{j=1 \\ j \neq i}}^4 \frac{\lambda_j}{\lambda_j + \lambda_i}.$$

Table 2.4 Spaces $V \times W$ admitting an $M(\partial K)$-decomposition, where $M = \mathcal{P}_k(\partial K)$

V	W	Method
K is a square and $\boldsymbol{V}_g \times W_g = \mathcal{Q}_k \times \mathcal{Q}_k$		
$\mathcal{Q}_k \oplus \mathbf{curl}\ \mathrm{span}\{x^{k+1}y, x\,y^{k+1}\} \oplus\ \mathrm{span}\{\boldsymbol{x}\,x^k y^k\}$	$\mathcal{Q}_k$	$\mathbf{TNT}_{[k]}$ [22]
$\mathcal{Q}_k \oplus \mathbf{curl}\ \mathrm{span}\{x^{k+1}y, x\,y^{k+1}\}$	$\mathcal{Q}_k$	$\mathbf{HDG}^{Q}_{[k]}$[22]
$\mathcal{Q}_k \oplus \mathbf{curl}\ \mathrm{span}\{x^{k+1}y, x\,y^{k+1}\}$	$\mathcal{Q}_k \setminus \{x^k\,y^k\}$	$\mathbf{BDM}_{[k]}$
K is a triangle and $\boldsymbol{V}_g \times W_g = \boldsymbol{\mathcal{P}}_k \times \mathcal{P}_k$		
$\boldsymbol{\mathcal{P}}_k \oplus \boldsymbol{x}\,\widetilde{\mathcal{P}}_k$	$\mathcal{P}_k$	$\mathbf{RT}_k$ [38]
$\boldsymbol{\mathcal{P}}_k$	$\mathcal{P}_k$	$\mathbf{HDG}_k$[22]
$\boldsymbol{\mathcal{P}}_k$	$\mathcal{P}_{k-1}$	$\mathbf{BDM}_k$ [4]
K is a square and $\boldsymbol{V}_g \times W_g = \boldsymbol{\mathcal{P}}_k \times \mathcal{P}_k$		
$\boldsymbol{\mathcal{P}}_k \oplus \mathbf{curl}\ \mathrm{span}\{x^{k+1}y, x\,y^{k+1}\} \oplus\ \boldsymbol{x}\,\widetilde{\mathcal{P}}_k$	$\mathcal{P}_k$	(new)
$\boldsymbol{\mathcal{P}}_k \oplus \mathbf{curl}\ \mathrm{span}\{x^{k+1}y, x\,y^{k+1}\}$	$\mathcal{P}_k$	(new)
$\boldsymbol{\mathcal{P}}_k \oplus \mathbf{curl}\ \mathrm{span}\{x^{k+1}y, x\,y^{k+1}\}$	$\mathcal{P}_{k-1}$	$\mathbf{BDM}_{[k]}$ [4]
K is a quadrilateral and $\boldsymbol{V}_g \times W_g = \boldsymbol{\mathcal{P}}_k \times \mathcal{P}_k$		
$\boldsymbol{\mathcal{P}}_k \oplus_{i=1}^{ne} \mathbf{curl}\ \mathrm{span}\{\xi_4\,\lambda_3^k, \xi_4\,\lambda_4^k\} \oplus \boldsymbol{x}\,\widetilde{\mathcal{P}}_k$	$\mathcal{P}_k$	(new)
$\boldsymbol{\mathcal{P}}_k \oplus_{i=1}^{ne} \mathbf{curl}\ \mathrm{span}\{\xi_4\,\lambda_3^k, \xi_4\,\lambda_4^k\}$	$\mathcal{P}_k$	(new)
$\boldsymbol{\mathcal{P}}_k \oplus_{i=1}^{ne} \mathbf{curl}\ \mathrm{span}\{\xi_4\,\lambda_3^k, \xi_4\,\lambda_4^k\}$	$\mathcal{P}_{k-1}$	(new)

The rational function η_i is constructed in such a way that its trace on ∂K is zero except on the edge $\mathbf{e}_i$, where it is equal to one.

2.5.3 General Polygonal Elements

For general polygonal elements, we have the following result.

Theorem 2.4 ([14]) *Let K be a polygonal of ne edges such that no consecutive edges lie on the same line. Then, for $M := \mathcal{P}_k(\partial K)$ and $\boldsymbol{V}_g \times W_g = \boldsymbol{\mathcal{P}}_k(K) \times \mathcal{P}_k(K)$, we have that*

$$I_M(\boldsymbol{V}_g \times W_g) = (ne-3)(\theta+1) - \frac{1}{2}\theta(\theta-1), \qquad \text{and} \qquad I_S(\boldsymbol{V}_g \times W_g) = k+1,$$

where $\theta := \min\{k, ne-3\}$. Moreover, we have

$$\delta \boldsymbol{V}_{\text{fillM}} := \oplus_{i=1}^{ne} \mathbf{curl}\,\Psi_i,$$
$$\delta \boldsymbol{V}_{\text{fillW}} := \boldsymbol{x}\,\widetilde{\mathcal{P}}_k.$$

Here

$$\Psi_i = \begin{cases} \{0\} & if i = 1, 2, \\ \mathrm{span}\{\xi_{i+1}\lambda_{i+1}^b;\ \max\{k+3-i, 0\} \le b \le k\} & if\ 3 \le i \le ne-1, \\ \mathrm{span}\{\xi_{i+1}\lambda_{i+1}^b;\ \max\{k+4-i, 1\} \le b \le k\} & if\ i = ne. \end{cases}$$

The functions $\{\xi_i\}_{i=1}^{ne} \subset H^1(K)$ *are lifting functions that satisfy*

(L.1) $\xi_i|_{\mathbf{e}_j} \in \mathcal{P}_1(\mathbf{e}_j),\ j = 1, \ldots, ne,$

(L.2) $\xi_i(\mathbf{v}_j) = \delta_{i,j},\ j = 1, \ldots, ne,$

where $\delta_{i,j}$ *is the Kronecker delta.*

Thus results gives us an explicit, ready-to-implement description of the three spaces of our construction.

It is interesting to see how the dimension of these spaces changes when we fix the polynomial degree k and let the number of edges of the element K, ne, vary. Indeed, although the space $\delta\boldsymbol{V}_{\mathrm{fillW}}$ remains unchanged, this is not true for $\delta\boldsymbol{V}_{\mathrm{fillM}}$. In fact, when $k \le ne-3$, for each additional edge in the element, the above result states that we have to add $k+1$ new basis functions to $\delta\boldsymbol{V}_{\mathrm{fillM}}$. In particular, if $k=1$, the dimension of $\delta\boldsymbol{V}_{\mathrm{fillM}}$ is $2\,(ne-3)$.

Next, we test the convergence properties of one of them. In the table below, we retake our earlier example and instead of using $\boldsymbol{V}(K) \times W(K) = \boldsymbol{\mathcal{P}}_k(K) \times \mathcal{P}_k(K)$ and $M(\partial K) = \mathcal{P}(\partial K)$ as local spaces for elements of all shapes, we consider the local spaces

$$\boldsymbol{V}(K) \times W(K) = (\boldsymbol{\mathcal{P}}_k(K) \oplus \delta\boldsymbol{V}_{\mathrm{fillM}}) \times \mathcal{P}_k(K),$$

which, by the previous result, admit an $M(\partial K) = \mathcal{P}(\partial K)$–decomposition. We now obtain the optimal convergence order of $3 = k+2$. This is in full agreement with our theoretical error estimates of Theorems 2.2 and 2.1, given that the approximation errors of the HDG-projection of Proposition 2.3 are both of order $k+1$ for smooth solutions.

	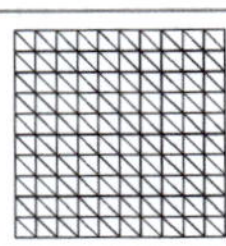		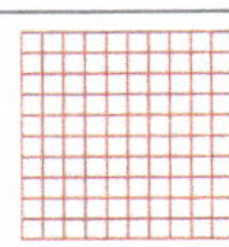		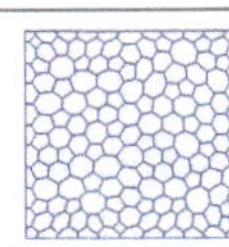	
h	$\|u-u_h^\star\|_{\mathcal{T}_h}$	Rate	$\|u-u_h^\star\|_{\mathcal{T}_h}$	Rate	$\|u-u_h^\star\|_{\mathcal{T}_h}$	Rate
	$\tau = 1$					
0.1	0.15E−2	–	0.26E−2	–	0.17E−2	–
0.05	0.18E−3	3.06	0.31E−3	3.06	0.21E−3	3.02
0.025	0.23E−4	3.03	0.38E−4	3.03	0.27E−4	2.95
0.0125	0.28E−5	**3.02**	0.47E−5	**3.02**	0.35E−5	**2.96**

2.6 Extensions

We end by describing extensions of the work presented here.

Curved Elements Note that our general theory of M-decompositions for diffusion problems can be *easily* extended to curved elements by following the work done in [21].

Hanging Nodes Although in Theorem 2.4, we restricted ourselves to the case of elements with no consecutive edges in the same line, two-dimensional elements with hanging nodes can be treated by applying the general theory by simply considering that an edge with a hanging node is in fact two different edges. The three-dimensional case can be similarly treated. The case of a triangle with a hanging node is considered in [14, Section 4.2].

Local Postprocessing of the Flux By using our construction, we can locally compute *two* $H(\text{div})$-conforming approximate fluxes, see [27, Section 6.3], for the HDG approximation. This elementwise postprocessing extends the postprocessing obtained back in 2003 by Bastian and Rivière [3] (see the variations proposed, for simplicial meshes, in 2005 [17], in 2007 [31] and in 2010 in [20]). As was argued therein, see also [1, Section 2.2], $H(\text{div})$-conforming fluxes seem to be preferable to the original DG-like approximation, even if both approximations are of the same accuracy, when used on other convection-diffusion problems in which the fluxes drive the convection.

2D Versus 3D The three-dimensional case is significantly more involved than the two-dimensional case, essentially because of the computation of the space

$$\{\boldsymbol{v} \in \boldsymbol{V}_g : \nabla \cdot \boldsymbol{v} = 0, \boldsymbol{v} \cdot \boldsymbol{n}|_{\partial K} = 0\},$$

which is very simple in 2D but very complicated in 3D. This reflects the fact that, although M-decompositions were explicitly obtained for arbitrary polygonal elements [14], in the three dimensional case, the explicit construction of M-decompositions has been done for tetrahedra, prisms, pyramids and hexahedra [15]. The *automatic* construction of M-decompositions for three-dimensional polyhedral elements of arbitrary shape constitutes the subject of ongoing research.

New Discrete H^1-Inequalities In [24], new H^1-discrete inequalities were introduced which extend to *all* spaces admitting M-decompositions similar inequalities obtained in [30, Proposition 3.2], for the well known Raviart-Thomas spaces for simplexes, and, for smaller spaces, in [7, Theorem 3.2] for the Staggered DG method.

Other Equations As pointed out in [27], this work can be extended to devise superconvergent HDG and mixed methods for the heat equation, by following [6], to the wave equation by following [12], see [26] for a Stormer-Numerov time-marching method and [39] for symplectic methods, to the velocity gradient-velocity-pressure formulation of the Stokes problem by following [10], see [25], and for methods for the the equations of linear elasticity with weakly symmetric stress approximations by following [11]. The extension to methods for the equations of linear elasticity with strongly symmetric stresses was carried out in [13]—the actual construction of the local spaces in 3D is still an open problem though. The extension to the incompressible Navier-Stokes equations was done in [24].

The theory of M-decompositions Maxwell equations constitute subject of ongoing research.

Acknowledgements The authors would like to express their gratitude to Daniele Antonio Di Pietro, Alexander Ern and Luca Formaggia for their invitation to write this paper. The first author would thank them for the invitation to give a couple of lectures on HDG methods as part of the Introductory School (to the IHP quarter on Numerical Methods for PDEs) they organized in September 5–9, 2016, at the Institut d'Études Scientifiques de Cargèse, in Corse, France. Part of the material of those lectures is further developed here.

The author "Bernardo Cockburn" was partially supported by National Science Foundation grant DMS 1522657.

Appendix: Proof of the Characterization of M-Decompositions

In this Appendix, we provide a proof Theorem 2.3, as it sheds light on the nature of M-decompositions. We closely follow the proof given in [27], and use the existence of the so-called *canonical* decomposition of Proposition 2.1.

Step 1. We take $\widetilde{\boldsymbol{V}} \times \widetilde{W}$ given by the canonical M-decomposition and begin by showing that

$$\dim \widetilde{\boldsymbol{V}}^{\perp} \cdot \boldsymbol{n}|_{\partial K} = \dim \widetilde{\boldsymbol{V}}^{\perp} \quad \text{and} \quad \dim \widetilde{W}^{\perp}|_{\partial K} = \dim \widetilde{W}^{\perp}.$$

Let us prove the first equality. If $\widetilde{\boldsymbol{v}}^{\perp} \in \widetilde{\boldsymbol{V}}^{\perp}$ is such that $\widetilde{\boldsymbol{v}}^{\perp} \cdot \boldsymbol{n}|_{\partial K} = 0$, for any $w \in W$, we have that

$$0 = \langle w, \widetilde{\boldsymbol{v}}^{\perp} \cdot \boldsymbol{n} \rangle_{\partial K} = (\nabla w, \widetilde{\boldsymbol{v}}^{\perp})_K + (\widetilde{w}^{\perp}, \nabla \cdot \widetilde{\boldsymbol{v}}^{\perp})_K = (\widetilde{w}^{\perp}, \nabla \cdot \widetilde{\boldsymbol{v}}^{\perp})_K$$

since $\nabla w \subset \widetilde{\boldsymbol{V}}$. Since $W \supset \nabla \cdot \boldsymbol{V}$, we can take $w := \nabla \cdot \widetilde{\boldsymbol{v}}^{\perp}$ and conclude that $\nabla \cdot \widetilde{\boldsymbol{v}}^{\perp} = 0$, which means that $\widetilde{\boldsymbol{v}}^{\perp} \in \boldsymbol{V}_{\text{sbb}}$, which means that $\widetilde{\boldsymbol{v}}^{\perp} = \boldsymbol{0}$. Thus, the first equity holds.

Now, let us prove the second equality. If $\widetilde{w}^{\perp} \in \widetilde{W}^{\perp}$ and is zero on ∂K, then, for any $\boldsymbol{v} \in \boldsymbol{V}$, we have

$$0 = \langle \widetilde{w}^{\perp}, \boldsymbol{v} \cdot \boldsymbol{n} \rangle_{\partial K} = (\nabla \widetilde{w}^{\perp}, \boldsymbol{v})_K + (\widetilde{w}^{\perp}, \nabla \cdot \boldsymbol{v})_K = (\nabla \widetilde{w}^{\perp}, \boldsymbol{v})_K$$

since $\widetilde{W} = \nabla \cdot \boldsymbol{V}$. Since $\boldsymbol{V} \supset \nabla W$, we can now take $\boldsymbol{v} := \nabla \widetilde{w}^{\perp}$ and conclude that $\widetilde{w}^{\perp}$ is a constant on K. As a consequence $\widetilde{w}^{\perp} = 0$, and the second equality follows.

Step 2. Next, we show that

$$\dim \operatorname{tr}(\widetilde{\boldsymbol{V}}^{\perp} \times \widetilde{W}^{\perp}) = \dim \widetilde{\boldsymbol{V}}^{\perp} \cdot \boldsymbol{n}|_{\partial K} + \dim \widetilde{W}^{\perp}|_{\partial K}.$$

To do that, we only need to show that $\widetilde{\boldsymbol{V}}^{\perp} \cdot \boldsymbol{n}|_{\partial K} \cap \widetilde{W}^{\perp}|_{\partial K} = \{0\}$. So, if $(\widetilde{\boldsymbol{v}}^{\perp}, \widetilde{w}^{\perp}) \in \widetilde{\boldsymbol{V}}^{\perp} \times \widetilde{W}^{\perp}$ we get that

$$\langle \widetilde{w}^{\perp}, \widetilde{\boldsymbol{v}}^{\perp} \cdot \boldsymbol{n} \rangle_{\partial K} = (\nabla \widetilde{w}^{\perp}, \widetilde{\boldsymbol{v}}^{\perp})_K + (\widetilde{w}^{\perp}, \nabla \cdot \widetilde{\boldsymbol{v}}^{\perp})_K = 0,$$

because $\nabla \widetilde{w}^{\perp} \in \nabla W \subset \widetilde{\boldsymbol{V}}$ and because $\nabla \cdot \widetilde{\boldsymbol{v}}^{\perp} \in \nabla \cdot \widetilde{\boldsymbol{V}} \subset \widetilde{W}$.

Step 3. By the inclusion property (a), the number

$$\begin{aligned} I :=&\ \dim M - \dim \widetilde{\boldsymbol{V}}^{\perp} - \dim \widetilde{W}^{\perp} \\ =&\ \dim M - \dim \widetilde{\boldsymbol{V}}^{\perp} \cdot \boldsymbol{n}|_{\partial K} - \dim \widetilde{W}^{\perp}|_{\partial K}. \end{aligned}$$

is always nonnegative and is equal to zero if and only if property (c) holds. Next, we show that $I = I_M(\boldsymbol{V} \times W)$; this is the *key* computation of the proof. Indeed, we have

$$\begin{aligned} I :=&\dim M - \dim \widetilde{\boldsymbol{V}}^{\perp} - \dim \widetilde{W}^{\perp} \\ =&\dim M - (\dim \boldsymbol{V} - \dim \widetilde{\boldsymbol{V}}) - (\dim W - \dim \widetilde{W}) \\ =&\dim M - (\dim \boldsymbol{V} - \dim \nabla W - \dim \boldsymbol{V}_{\text{sbb}}) - (\dim W - \dim \nabla \cdot \boldsymbol{V}) \\ =&\dim M - (\dim \boldsymbol{V} - \dim \nabla \cdot \boldsymbol{V} - \dim \boldsymbol{V}_{\text{sbb}}) - (\dim W - \dim \nabla W) \\ =&\dim M - (\dim\{\boldsymbol{v} \in \boldsymbol{V} : \nabla \cdot \boldsymbol{v} = 0\} - \dim\{\boldsymbol{v} \in \boldsymbol{V} : \nabla \cdot \boldsymbol{v} = 0, \boldsymbol{v} \cdot \boldsymbol{n}|_{\partial K} = 0\}) \\ &- \dim\{w \in W : \nabla w = 0\} \\ =&\dim M - \dim\{\boldsymbol{v} \cdot \boldsymbol{n}|_{\partial K} : \boldsymbol{v} \in \boldsymbol{V}, \nabla \cdot \boldsymbol{v} = 0\} - \dim\{w|_{\partial K} : w \in W, \nabla w = 0\} \\ =&I_M(\boldsymbol{V} \times W). \end{aligned}$$

Step 4. Now, by the inclusion property (a), we have that

$$\{\boldsymbol{v} \cdot \boldsymbol{n}|_{\partial K} : \boldsymbol{v} \in \boldsymbol{V}, \nabla \cdot \boldsymbol{v} = 0\} \oplus \{w|_{\partial K} : w \in W, \nabla w = 0\} \subset M,$$

where the sum is $L^2(\partial K)$-orthogonal since

$$\langle \boldsymbol{v} \cdot \boldsymbol{n}, w \rangle_{\partial K} = (\nabla \cdot \boldsymbol{v}, w)_K + (\boldsymbol{v}, \nabla w)_K = 0$$

if $\nabla \cdot \boldsymbol{v} = 0$ and $\nabla w = 0$. Finally, since the M-index $I_M(\boldsymbol{V} \times W)$ is zero by property (c), the equality holds. This completes the proof of the characterization Theorem 2.3.

References

1. Arbogast, T., Xiao, H.: Two-level mortar domain decomposition mortar preconditioners for heterogeneous elliptic problems. Comput. Methods Appl. Mech. Eng. **292**, 221–242 (2015)
2. Arnold, D.N., Brezzi, F., Cockburn, B., Marini, L.D.: Unified analysis of discontinuous Galerkin methods for elliptic problems. SIAM J. Numer. Anal. **39**, 1749–1779 (2002)
3. Bastian, P., Rivière, B.: Superconvergence and $H(\text{div})$ projection for discontinuous Galerkin methods. Int. J. Numer. Methods Fluids **42**, 1043–1057 (2003)
4. Brezzi, F., Douglas, J. Jr., Marini, L.D.: Two families of mixed finite elements for second order elliptic problems. Numer. Math. **47**, 217–235 (1985)
5. Castillo, P., Cockburn, B., Perugia, I., Schötzau, D.: An a priori error analysis of the local discontinuous Galerkin method for elliptic problems. SIAM J. Numer. Anal. **38**, 1676–1706 (2000)
6. Chabaud, B., Cockburn, B.: Uniform-in-time superconvergence of HDG methods for the heat equation. Math. Comput. **81**, 107–129 (2012)
7. Chung, E.T., Engquist, B.: Optimal discontinuous Galerkin methods for the acoustic wave equation in higher dimensions. SIAM J. Numer. Anal. **47**(5), 3820–3848 (2009)
8. Cockburn, B.: Static condensation, hybridization, and the devising of the HDG methods. In: Barrenechea, G.R., Brezzi, F., Cagniani, A., Georgoulis, E.H. (eds.) Building Bridges: Connections and Challenges in Modern Approaches to Numerical Partial Differential Equations. Lecture Notes in Computational Science and Engineering, vol. 114, pp. 129–177. Springer, Berlin (2016). LMS Durham Symposia funded by the London Mathematical Society. Durham, July 8–16, 2014
9. Cockburn, B., Shu, C.-W.: The local discontinuous Galerkin method for time-dependent convection-diffusion systems. SIAM J. Numer. Anal. **35**, 2440–2463 (1998)
10. Cockburn, B., Shi, K.: Conditions for superconvergence of HDG methods for stokes flow. Math. Comput. **82**, 651–671 (2013)
11. Cockburn, B., Shi, K.: Superconvergent HDG methods for linear elasticity with weakly symmetric stresses. IMA J. Numer. Anal. **33**, 747–770 (2013)
12. Cockburn, B., Quenneville-Bélair, V.: Uniform-in-time superconvergence of HDG methods for the acoustic wave equation. Math. Comput. **83**, 65–85 (2014)
13. Cockburn, B., Fu, G.: Devising superconvergent HDG methods with symmetric approximate stresses for linear elasticity by M-decomposition. IMA J. Numer. Anal. **38**(2), 566–604 (2018)
14. Cockburn, B., Fu, G., Superconvergence by M-decompositions. Part II: construction of two-dimensional finite elements. ESAIM Math. Model. Numer. Anal. **51**(1), 165–186 (2017)
15. Cockburn, B., Fu, G.: Superconvergence by M-decompositions. Part III: construction of three-dimensional finite elements. ESAIM Math. Model. Numer. Anal. **51**(1), 365–398 (2017)
16. Cockburn, B., Fu, G.: A systematic construction of finite element commuting exact sequences. SIAM J. Numer. Anal. **55**(4), 1650–1688 (2017)
17. Cockburn, B., Kanschat, G., Schötzau, D.: A locally conservative LDG method for the incompressible Navier-Stokes equations. Math. Comput. **74**, 1067–1095 (2005)
18. Cockburn, B., Guzmán, J., Wang, H.: Superconvergent discontinuous Galerkin methods for second-order elliptic problems. Math. Comput. **78**, 1–24 (2009)
19. Cockburn, B., Gopalakrishnan, J., Lazarov, R.: Unified hybridization of discontinuous Galerkin, mixed and continuous Galerkin methods for second order elliptic problems. SIAM J. Numer. Anal. **47**, 1319–1365 (2009)
20. Cockburn, B., Gopalakrishnan, J., Sayas, F.-J.: A projection-based error analysis of HDG methods. Math. Comput. **79**, 1351–1367 (2010)
21. Cockburn, B., Qiu, W., Shi, K.: Conditions for superconvergence of HDG methods on curvilinear elements for second-order elliptic problems. SIAM J. Numer. Anal. **50**, 1417–1432 (2012)
22. Cockburn, B., Qiu, W., Shi, K.: Conditions for superconvergence of HDG methods for second-order elliptic problems. Math. Comput. **81**, 1327–1353 (2012)

23. Cockburn, B., Di-Pietro, D.A., Ern, A.: Bridging the hybrid high-order and hybridizable discontinuous Galerkin methods. ESAIM Math. Model. Numer. Anal. **50**, 635–650 (2016)
24. Cockburn, B., Fu, G., Qiu, W.: Discrete H^1-inequalities for spaces admitting M-decompositions (2017, submitted)
25. Cockburn, B., Fu, G., Qiu, W.: A note on the devising of superconvergent HDG methods for stokes flow by M-decompositions. IMA J. Numer. Anal. **37**(2), 730–749 (2017)
26. Cockburn, B., Fu, X., Hungria, A., Ji, L., Sánchez, M.A., Sayas, F.-J.: Stormer-Numerov HDG methods for the acoustic waves. J. Sci. Comput. **75**(2), 597–624 (2018)
27. Cockburn, B., Fu, G., Sayas, F.J.: Superconvergence by M-decompositions. Part I: general theory for HDG methods for diffusion. Math. Comput. **86**(306), 1609–1641 (2017)
28. Di-Pietro, D.A., Ern, A.: A hybrid high-order locking-free method for linear elasticity on general meshes. Comput. Methods Appl. Mech. Eng. **283**, 1–21 (2015)
29. Di-Pietro, D.A., Ern, A., Lemaire, S.: An arbitrary-order and compact-stencil discretization of diffusion on general meshes based on local reconstruction operators. Comput. Meth. Appl. Math. **14**(4), 461–472 (2014)
30. Egger, H., Schöberl, J.: A hybrid mixed discontinuous Galerkin finite-element method for convection-diffusion problems. IMA J. Numer. Anal. **30**(4), 1206–1234 (2010)
31. Ern, A., Nicaise, S., Vohralík, M.: An accurate **H**(div) flux reconstruction for discontinuous Galerkin approximations of elliptic problems. C. R. Math. Acad. Sci. Paris **345**, 709–712 (2007)
32. Gastaldi, L., Nochetto, R.H.: Sharp maximum norm error estimates for general mixed finite element approximations to second order elliptic equations. RAIRO Modél. Math. Anal. Numér. **23**, 103–128 (1989)
33. Lehrenfeld, C.: Hybrid discontinuous Galerkin methods for solving incompressible flow problems. Ph.D. thesis, Diplomigenieur Rheinisch-Westfalishen Technischen Hochchule Aachen (2010)
34. Oikawa, I.: A hybridized discontinuous Galerkin method with reduced stabilization. J. Sci. Comput. **65**, 327–340 (2015)
35. Oikawa, I.: Analysis of a reduced-order HDG method for the stokes equations. J. Sci. Comput. **67**(2), 475–492 (2016)
36. Qiu, W., Shen, J., Shi, K.: An HDG method for linear elasticity with strong symmetric stresses. Math. Comput. **87**, 69–93 (2018)
37. Qui, W., Shi, K.: A superconvergent HDG method for the incompressible Navier-Stokes equations on general polyhedral meshes. IMA J. Numer. Anal. **36**, 1943–1967 (2016)
38. Raviart, P.A., Thomas, J.M.: A mixed finite element method for second order elliptic problems. In: Galligani, I., Magenes, E. (eds.) Mathematical Aspects of Finite Element Method. Lecture Notes in Mathematics, vol. 606, pp. 292–315. Springer, New York (1977)
39. Sánchez, M.A., Ciuca, C., Nguyen, N.C., Peraire, J., Cockburn, B.: Symplectic Hamiltonian HDG methods for wave propagation phenomena. J. Comput. Phys. **350**, 951–973 (2018)
40. Stenberg, R.: A family of mixed finite elements for the elasticity problem. Numer. Math. **53**, 513–538 (1988)
41. Stenberg, R.: Postprocessing schemes for some mixed finite elements. RAIRO Modél. Math. Anal. Numér. **25**, 151–167 (1991)

Chapter 3
Mimetic Spectral Element Method for Anisotropic Diffusion

Marc Gerritsma, Artur Palha, Varun Jain, and Yi Zhang

Abstract This chapter addresses the topological structure of steady, anisotropic, inhomogeneous diffusion problems. Differential operators are represented by sparse incidence matrices, while weighted mass matrices play the role of metric-dependent Hodge matrices. The resulting mixed formulation is point-wise divergence-free if the right hand side function $\boldsymbol{f} = 0$. The method is inf-sup stable; no stabilization is required and the method displays optimal convergence on orthogonal and deformed grids.

3.1 Introduction

Anisotropic and inhomogeneous diffusion appears in many applications such as heat transfer [15], flow through porous media [87], turbulent fluid flow [116], image processing [98] or plasma physics [112]. In 2D, steady, anisotropic diffusion is governed by the following elliptic partial differential equation

$$-\nabla \cdot (\mathbb{K}\nabla p) = f \ . \tag{3.1}$$

Here, p is the flow potential, f the source term, with $p = \bar{p}$ along Γ_p and $(\mathbb{K}\nabla p, \boldsymbol{n}) = \bar{u}_n$ along Γ_u. Here, for all $\boldsymbol{x}$, $\mathbb{K}(\boldsymbol{x})$ is a symmetric, positive definite tensor.

M. Gerritsma (✉) · V. Jain · Y. Zhang
Faculty of Aerospace Engineering, Delft University of Technology, Delft, The Netherlands
e-mail: m.i.gerritsma@tudelft.nl; v.jain@tudelft.nl; y.zhang-14@tudelft.nl

A. Palha
Department of Mechanical Engineering, Eindhoven University of Technology, Eindhoven, The Netherlands
e-mail: a.palha@tue.nl

D. A. Di Pietro et al. (eds.), *Numerical Methods for PDEs*,
SEMA SIMAI Springer Series 15, https://doi.org/10.1007/978-3-319-94676-4_3

In the presence of *strong anisotropy*, i.e. large ratio between the smallest and largest eigenvalues of the diffusion tensor, the construction of robust and efficient discretizations becomes particularly challenging. Under these conditions, the convergence rates of the discretization error can be considerably reduced; this effect is commonly referred in the literature as *locking effect*, see for example [4, 5, 12, 84]. For sufficiently refined discretizations, the deterioration of the convergence rates eventually disappears. Unfortunately, this may occur only when the grid cell size is prohibitively small.

Another important aspect is mesh flexibility. In many applications of diffusion equations, particularly in porous media flow, typical grids are highly irregular. In many of these situations the results obtained are strongly dependent on the grid type, see [11] for a discussion of the use and properties of different grids in reservoir modelling.

3.1.1 Overview of Standard Discretizations

In order to overcome these limitations and improve the efficiency and robustness of the discretization of the anisotropic diffusion equations, several approaches have been proposed.

The discretization of the anisotropic diffusion equations in complex media in many situations is still a trade-off between, e.g. [89]:

- Accuracy in the representation of the medium (complex grids).
- Accuracy in the discretization of the equations.

The need for such a choice is rooted in the use of numerical schemes based on *two-point flux approximations* (TPFA), see for example, [3, 89, 120]. These methods produce good approximations on orthogonal grids when the diffusion tensor $\mathbb{K}$ is diagonal, but are known to introduce significant discretization errors in the presence of a non-diagonal diffusion tensor. This introduces severe limitations into the possible grid choices. Under these conditions, the geometric flexibility introduced by *perpendicular bisector* (PEBI) grids, [11, 67, 90], is considerably limited, for example.

It has been known that the discretization error is related to the misalignment between the grid and the principal directions of the diffusion tensor $\mathbb{K}$. In fact, Aavatsmark showed in [3] that for TPFA this misalignment leads to the discretization of the wrong diffusion tensor.

These ideas initially led to the construction of grids aligned with the principal axis of the diffusion tensor, so called $\mathbb{K}$-*orthogonal grids*, see for example [65, 67]. This approach significantly improves the performance of the numerical method but substantially limits the geometric flexibility.

More recently, *multipoint flux-approximation* (MPFA) schemes have been introduced specifically to address these limitations, see e.g. the initial works by Aavatsmark [4, 5] or a more recent presentation [2], and by Edwards and Rogers [57]. This method is based on a cell-centred finite volume formulation and introduces a dual grid in order to generate shared sub-cells and sub-interfaces. This in turn produces a discretization of the flux between two cells that involves a linear combination of several adjacent cells. This method is robust and locally conservative but does not guarantee a resulting symmetric discrete diffusion operator. More recently, this work has been connected to the mixed finite element method, [56].

Alternative approaches based on the finite element formulation have also been proposed by several authors. We briefly mention the work on the control-volume finite element discretization by Forsyth [60] and Durlofsky [54], on nodal Galerkin finite elements by Young [122], and on mixed finite elements by Durlofsky [53].

3.1.2 Overview of Mimetic Discretizations

Over the years, the development of numerical schemes that preserve some of the structures of the differential models they approximate has been identified as an important ingredient of numerical analysis. One of the contributions of the formalism of mimetic methods is to identify differential geometry as the proper language in which to encode these structures/symmetries. Another novel aspect of mimetic discretizations is the identification and separation of physical field laws into two sets: (1) topological relations (metric-free), and (2) constitutive relations (metric dependent). Topological relations are intimately related to conservation laws and can (and should) be exactly represented on the computational grid. Constitutive relations include all material properties and therefore are approximate relations. For this reason, all numerical discretization error should be included in these equations. A general introduction and overview of spatial and temporal mimetic/geometric methods can be found in [38, 42, 66, 100].

The relation between differential geometry and algebraic topology in physical theories was first established by Tonti [117]. Around the same time Dodziuk [52] set up a finite difference framework for harmonic functions based on Hodge theory. Both Tonti and Dodziuk introduce differential forms and cochain spaces as the building blocks for their theory. The relation between differential forms and cochains is established by the Whitney map (k-cochains $\rightarrow$ k-forms) and the de Rham map (k-forms $\rightarrow$ k-cochains). The interpolation of cochains to differential forms on a triangular grid was already established by Whitney, [119]. These generalized interpolatory forms are now known as *Whitney forms*.

Hyman and Scovel [74] set up the discrete framework in terms of cochains, which are the natural building blocks of finite volume methods. Later, Bochev and Hyman [18] extended this work and derived discrete operators such as the discrete wedge product, the discrete codifferential, and the discrete inner products.

Robidoux, Hyman, Steinberg and Shashkov, [75–78, 107, 108, 111, 113, 114] used symmetry considerations to construct discretizations on rough grids, within the finite difference/volume setting . In a more recent paper by Robidoux and Steinberg [110] a finite difference discrete vector calculus is presented. In that work, the differential operators grad, curl and div are exactly represented at the discrete level and the numerical approximations are all contained in the constitutive relations, which are already polluted by modeling and experimental error. For mimetic finite differences, see also the work of Brezzi et al. [31, 36] and Beirão da Veiga et al. [45].

The application of mimetic ideas to unstructured triangular staggered grids has been extensively studied by Perot, [99, 101–103, 123], specially in [100] where the rationale of preserving symmetries in numerical algorithms is well described. The most *geometric approach* is presented in the work by Desbrun et al. [49, 58, 86, 97] and the thesis by Hirani [72].

The *Japanese papers* by Bossavit, [25–29], serve as an excellent introduction and motivation for the use of differential forms in the description of physics and the use in numerical modeling. The field of application is electromagnetism, but these papers are sufficiently general to extend to other physical theories.

In a series of papers by Arnold, Falk and Winther, [8–10], a finite element exterior calculus framework is developed. Higher order methods are described by Rapetti [104, 105] and Hiptmair [71]. Possible extensions to spectral methods were described by Robidoux, [109]. A different approach for constructing arbitrary order mimetic finite elements has been proposed by the authors [30, 64, 92, 94], with applications to advection problems [95], Stokes' flow [81], MHD equilibrium [96], Navier-Stokes [93], and within a Least-Squares finite element formulation [16, 62, 63, 91].

Extensions of these ideas to polyhedral meshes have been proposed by Ern, Bonelle and co-authors in [22–24, 40], by Di Pietro and co-authors in [50, 51], by Brezzi and co-authors in [37], and by Beirão da Veiga and co-authors in [44, 46–48]. These approaches provide more geometrical flexibility while maintaining fundamental structure preserving properties.

Mimetic isogeometric discretizations have been introduced by Buffa et al. [39], Evans and Hughes [59], and Hiemstra et al. [70].

Another approach to develop a discretization of the physical field laws is based on a discrete variational principle for the discrete Lagrangian action. This approach has been used in the past to construct variational integrators for Lagrangian systems, e.g. [79, 85]. Kraus and Maj [80] have used the method of formal Lagrangians to derive generalized Lagrangians for non-Lagrangian systems of equations. This allows to apply variational techniques to construct structure preserving discretizations on a much wider range of systems. Recently, Bauer and Gay-Balmaz presented variational integrators for elastic and pseudo-incompressible flows [14].

Due to the inherent challenges in discretizing the diffusion equations with anisotropic diffusion tensor $\mathbb{K}$, several authors have explored different mimetic discretizations of these equations. Focussing on generalized diffusion equations we highlight [13, 69, 75–78, 102, 107, 108, 111, 113, 114] for a finite-difference/finite-volume setting, [24, 33–35] for polyhedral discretizations, and [19, 20, 94, 106, 121] for a finite element/mixed finite element setting. For applications to Darcy flow equations and reservoir modelling see for example [1, 6, 7, 55, 73, 83, 89].

3.1.3 Outline of Chapter

In Sect. 3.2 the topological structure of anisotropic diffusion problems is discussed. In Sect. 3.3 spectral basis functions are introduced which are compatible with the topological structure introduced in Sect. 3.2. In Sect. 3.4 transformation to curvilinear elements is discussed. Results of the proposed method are presented in Sect. 3.5.

3.2 Anisotropic Diffusion/Darcy Problem

Let $\Omega \subset \mathbb{R}^d$ be a contractible domain with Lipschitz continuous boundary $\partial\Omega = \Gamma_p \cup \Gamma_u$, $\Gamma_p \cap \Gamma_u = \varnothing$. The steady anisotropic diffusion problem is given by

$$-\nabla \cdot (\mathbb{K}\nabla p) = f \;, \tag{3.2}$$

with $p = \bar{p}$ along Γ_p and $(-\mathbb{K}\nabla p, \boldsymbol{n}) = \bar{u}_n$ along Γ_u. Here, for all $\boldsymbol{x}$, $\mathbb{K}(\boldsymbol{x})$ is a symmetric, positive definite tensor, i.e. there exist constants $\alpha, C > 0$ such that

$$\alpha \boldsymbol{\xi}^T \boldsymbol{\xi} \leq \boldsymbol{\xi}^T \mathbb{K}(\boldsymbol{x}) \boldsymbol{\xi} \leq C \boldsymbol{\xi}^T \boldsymbol{\xi} \;.$$

If $\Gamma_p \neq \varnothing$, then (3.2) has a unique solution. If $\Gamma_p = \varnothing$ then (3.2) only possesses solutions if

$$\int_{\partial\Omega} \bar{u}_n \, \mathrm{d}S = \int_{\Omega} f \, \mathrm{d}\Omega \;,$$

in which case the solution, p, is determined up to a constant.

An equivalent first order system is obtained by introducing $\boldsymbol{u} = -\mathbb{K}\nabla p$ in which case (3.2) can be written as

$$\begin{cases} \boldsymbol{u} + \mathbb{K}\nabla p = 0 \text{ in } \Omega \\ \nabla \cdot \boldsymbol{u} = f \quad\quad \text{in } \Omega \end{cases} \text{with} \quad \begin{cases} (\boldsymbol{u}, \boldsymbol{n}) = \bar{u}_n \text{ along } \Gamma_u \\ p = \bar{p} \quad\quad\quad \text{along } \Gamma_p \end{cases} . \tag{3.3}$$

An alternative first-order formulation is given by

$$\begin{cases} \boldsymbol{v} - \nabla p = 0 \text{ in } \Omega \\ \boldsymbol{u} + \mathbb{K}\boldsymbol{v} = 0 \text{ in } \Omega \\ \nabla \cdot \boldsymbol{u} = f \quad\;\; \text{in } \Omega \end{cases} \text{with} \quad \begin{cases} (\boldsymbol{u}, \boldsymbol{n}) = \bar{u}_n \text{ along } \Gamma_u \\ p = \bar{p} \quad\quad\quad \text{along } \Gamma_p \end{cases} . \tag{3.4}$$

Formulation (3.3) is generally referred to as the *Darcy problem*, while the relation $\boldsymbol{u} = -\mathbb{K}\nabla p$ is called *Darcy's law*, [87]. The Darcy problem plays an important role in reservoir engineering. In this case $\boldsymbol{u}$ is the flow velocity in a porous medium and p denotes the pressure.

While the formulations (3.2)–(3.4) are equivalent, (3.2) only has 1 unknown, p, (3.3) has $(d + 1)$ unknowns, p and the d components of $\boldsymbol{u}$, and (3.4) has $(2d + 1)$ unknowns. Formulation (3.4) is of special interest, because it decomposes the anisotropic diffusion problem into two topological conservation laws and one constitutive law.[1] By making a suitable choice *where* and *how* to represent the unknowns on a grid, the topological relations, $\boldsymbol{v} - \nabla p = 0$ and $\nabla \cdot \boldsymbol{u} = f$ reduce to extremely simple algebraic relations which depend only on the topology of the mesh and are independent of the mesh size, independent of the shape of the mesh, and independent of the order of the numerical scheme. We will refer to such discretizations as *exact discrete representations*.

3.2.1 Gradient Relation

Consider two points $A, B \in \Omega$ and a curve $\mathcal{C}$ which connects these two points, then

$$\boldsymbol{v} - \nabla p = 0 \quad \Longrightarrow \quad \bar{\boldsymbol{v}}_{\mathcal{C}} := \int_{\mathcal{C}} \boldsymbol{v} \cdot \mathrm{d}l = \int_A^B \boldsymbol{v} \cdot \mathrm{d}l = \int_A^B \nabla p \cdot \mathrm{d}l = p(B) - p(A) \,,$$

where $\mathrm{d}l$ is a small increment along the curve $\mathcal{C}$.

Suppose that we take another curve $\tilde{\mathcal{C}}$ which connects the two points A and B then we also have

$$\bar{\boldsymbol{v}}_{\tilde{\mathcal{C}}} := \int_{\tilde{\mathcal{C}}} \boldsymbol{v} \cdot \mathrm{d}l = p(B) - p(A) \,, \tag{3.6}$$

[1] An even more extended system is, see for instance [16]

$$\begin{cases} \boldsymbol{v} - \nabla p = 0 & \text{in } \Omega \\ \boldsymbol{u} + \mathbb{K}\boldsymbol{v} = 0 & \text{in } \Omega \\ \nabla \cdot \boldsymbol{u} - \psi = 0 & \text{in } \Omega \\ \psi = f & \text{in } \Omega \end{cases} \quad \text{with} \quad \begin{cases} (\boldsymbol{u}, \boldsymbol{n}) = \bar{u}_n & \text{along } \Gamma_u \\ p = \bar{p} & \text{along } \Gamma_p \end{cases} . \tag{3.5}$$

This seems an unnecessarily complicated system. If we eliminate ψ from (3.5) we obtain (3.4). The usefulness of this system lies in the fact that by introducing ψ, the conservation $\nabla \cdot \boldsymbol{u} = f$ becomes independent of the data of the PDE, in this case the right hand side function. A similar situation occurs when $\mathbb{K} = \mathbb{I}$, the identity tensor, then the equation $\boldsymbol{u} + \mathbb{K}\boldsymbol{v} = 0$ in (3.4) seems redundant, but we have good reason to keep this seemingly redundant equation as we will show in this paper.

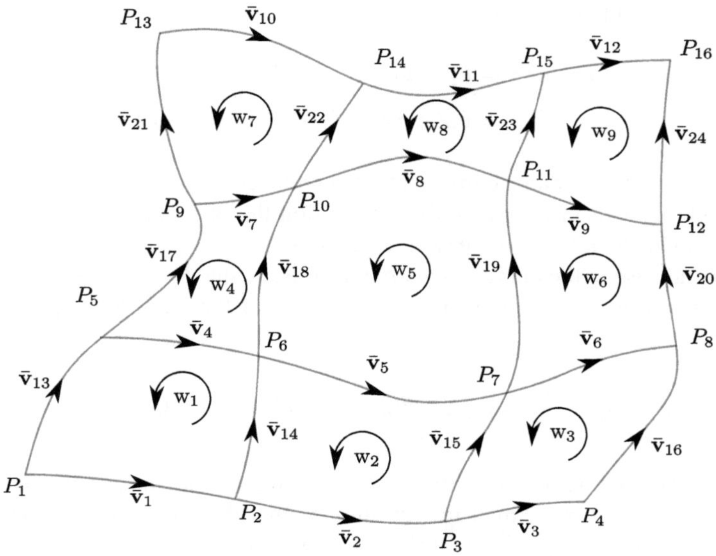

Fig. 3.1 Relation between pressure in points, integrated velocity along line segments and vorticity in surfaces

The integral along $\mathcal{C}$ is equal to the integral along $\tilde{\mathcal{C}}$. We will refer to $\bar{\boldsymbol{v}}$ as an *integral value*, since it denotes an integral and not a point-wise evaluation of $\boldsymbol{v}$. The advantages of integral values are:

1. The velocity-gradient relation is exact. It is not obtained by truncated Taylor-series expansions or does not depend on the choice of basis functions/interpolations.
2. Does not depend on mesh parameters. The mesh size h does not appear in (3.6). Whether the curve which connects two points is straight or curved is irrelevant in this relation, therefore this relation is directly applicable on curved domains.
3. Integral quantities are additive.

Consider the points and lines segments as shown in Fig. 3.1. In this figure the arrow along the curves indicates the direction in which $\boldsymbol{v}$ is integrated.[2] Application of (3.6) shows, for instance, that

$$\bar{\boldsymbol{v}}_{14} = P_6 - P_2 \ .$$

[2]The points in the grid shown in Fig. 3.1 are also 'oriented', in the sense that when we 'move into a point following the integration direction' we assign a positive value and when we 'leave a point' we assign a negative value. That is why we have *plus* $P(B)$ and *minus* $P(A)$ in (3.6). This is just a convention. Without loss of generality we could change this sign convention.

The additivity property implies that

$$\begin{aligned} P_7 - P_2 &= \bar{\boldsymbol{v}}_2 + \bar{\boldsymbol{v}}_{15} = P_3 - P_2 + P_7 - P_3 \\ &= \bar{\boldsymbol{v}}_{14} + \bar{\boldsymbol{v}}_5 = P_6 - P_2 + P_7 - P_6 \,, \end{aligned}$$

and even more paths can be constructed that connect P_2 and P_7. The independence of the path depends critically on the assumption that the space is contractible, i.e. there are no holes in the domain (Poincaré's Lemma).

A special case is the curve from a point to itself, say $P_2 \to P_2$ in Fig. 3.1. This integral is zero and if the integral is independent of the path this implies that, for instance,

$$0 = \bar{\boldsymbol{v}}_2 + \bar{\boldsymbol{v}}_{15} - \bar{\boldsymbol{v}}_5 - \bar{\boldsymbol{v}}_{14} = \oint \boldsymbol{v} \cdot \mathrm{d}l = \iint \nabla \times \boldsymbol{v} \cdot \mathrm{d}\boldsymbol{S} = \mathsf{w}_2 \,, \tag{3.7}$$

where we once again use the additivity property. We see that the circulation vanishes if $\boldsymbol{v}$ is a potential flow, which in turn implies that the circulation of the velocity field over the boundary of any surface vanishes. Or, using Stokes' theorem, the integrated vorticity w vanishes. Here the vorticity w is represented as the integral over a surface.

We can collect all the integrated velocity fields and pressures in Fig. 3.1 in the following form

$$\begin{pmatrix} \bar{\boldsymbol{v}}_1 \\ \bar{\boldsymbol{v}}_2 \\ \bar{\boldsymbol{v}}_3 \\ \bar{\boldsymbol{v}}_4 \\ \bar{\boldsymbol{v}}_5 \\ \bar{\boldsymbol{v}}_6 \\ \bar{\boldsymbol{v}}_7 \\ \bar{\boldsymbol{v}}_8 \\ \bar{\boldsymbol{v}}_9 \\ \bar{\boldsymbol{v}}_{10} \\ \bar{\boldsymbol{v}}_{11} \\ \bar{\boldsymbol{v}}_{12} \\ \bar{\boldsymbol{v}}_{13} \\ \bar{\boldsymbol{v}}_{14} \\ \bar{\boldsymbol{v}}_{15} \\ \bar{\boldsymbol{v}}_{16} \\ \bar{\boldsymbol{v}}_{17} \\ \bar{\boldsymbol{v}}_{18} \\ \bar{\boldsymbol{v}}_{19} \\ \bar{\boldsymbol{v}}_{20} \\ \bar{\boldsymbol{v}}_{21} \\ \bar{\boldsymbol{v}}_{22} \\ \bar{\boldsymbol{v}}_{23} \\ \bar{\boldsymbol{v}}_{24} \end{pmatrix} = \begin{pmatrix} -1 & 1 & 0 & 0 & 0 & 0 & 0 & 0 & 0 & 0 & 0 & 0 & 0 & 0 & 0 & 0 \\ 0 & -1 & 1 & 0 & 0 & 0 & 0 & 0 & 0 & 0 & 0 & 0 & 0 & 0 & 0 & 0 \\ 0 & 0 & -1 & 1 & 0 & 0 & 0 & 0 & 0 & 0 & 0 & 0 & 0 & 0 & 0 & 0 \\ 0 & 0 & 0 & 0 & -1 & 1 & 0 & 0 & 0 & 0 & 0 & 0 & 0 & 0 & 0 & 0 \\ 0 & 0 & 0 & 0 & 0 & -1 & 1 & 0 & 0 & 0 & 0 & 0 & 0 & 0 & 0 & 0 \\ 0 & 0 & 0 & 0 & 0 & 0 & -1 & 1 & 0 & 0 & 0 & 0 & 0 & 0 & 0 & 0 \\ 0 & 0 & 0 & 0 & 0 & 0 & 0 & 0 & -1 & 1 & 0 & 0 & 0 & 0 & 0 & 0 \\ 0 & 0 & 0 & 0 & 0 & 0 & 0 & 0 & 0 & -1 & 1 & 0 & 0 & 0 & 0 & 0 \\ 0 & 0 & 0 & 0 & 0 & 0 & 0 & 0 & 0 & 0 & -1 & 1 & 0 & 0 & 0 & 0 \\ 0 & 0 & 0 & 0 & 0 & 0 & 0 & 0 & 0 & 0 & 0 & 0 & -1 & 1 & 0 & 0 \\ 0 & 0 & 0 & 0 & 0 & 0 & 0 & 0 & 0 & 0 & 0 & 0 & 0 & -1 & 1 & 0 \\ 0 & 0 & 0 & 0 & 0 & 0 & 0 & 0 & 0 & 0 & 0 & 0 & 0 & 0 & -1 & 1 \\ -1 & 0 & 0 & 0 & 1 & 0 & 0 & 0 & 0 & 0 & 0 & 0 & 0 & 0 & 0 & 0 \\ 0 & -1 & 0 & 0 & 0 & 1 & 0 & 0 & 0 & 0 & 0 & 0 & 0 & 0 & 0 & 0 \\ 0 & 0 & -1 & 0 & 0 & 0 & 1 & 0 & 0 & 0 & 0 & 0 & 0 & 0 & 0 & 0 \\ 0 & 0 & 0 & -1 & 0 & 0 & 0 & 1 & 0 & 0 & 0 & 0 & 0 & 0 & 0 & 0 \\ 0 & 0 & 0 & 0 & -1 & 0 & 0 & 0 & 1 & 0 & 0 & 0 & 0 & 0 & 0 & 0 \\ 0 & 0 & 0 & 0 & 0 & -1 & 0 & 0 & 0 & 1 & 0 & 0 & 0 & 0 & 0 & 0 \\ 0 & 0 & 0 & 0 & 0 & 0 & -1 & 0 & 0 & 0 & 1 & 0 & 0 & 0 & 0 & 0 \\ 0 & 0 & 0 & 0 & 0 & 0 & 0 & -1 & 0 & 0 & 0 & 1 & 0 & 0 & 0 & 0 \\ 0 & 0 & 0 & 0 & 0 & 0 & 0 & 0 & -1 & 0 & 0 & 0 & 1 & 0 & 0 & 0 \\ 0 & 0 & 0 & 0 & 0 & 0 & 0 & 0 & 0 & -1 & 0 & 0 & 0 & 1 & 0 & 0 \\ 0 & 0 & 0 & 0 & 0 & 0 & 0 & 0 & 0 & 0 & -1 & 0 & 0 & 0 & 1 & 0 \\ 0 & 0 & 0 & 0 & 0 & 0 & 0 & 0 & 0 & 0 & 0 & -1 & 0 & 0 & 0 & 1 \end{pmatrix} \begin{pmatrix} P_1 \\ P_2 \\ P_3 \\ P_4 \\ P_5 \\ P_6 \\ P_7 \\ P_8 \\ P_9 \\ P_{10} \\ P_{11} \\ P_{12} \\ P_{13} \\ P_{14} \\ P_{15} \\ P_{16} \end{pmatrix} .$$

If we store all $\bar{v}_i$ in a vector v and all P_j in a vector P and denote the matrix by $\mathbb{E}^{1,0}$, we have

$$\mathsf{v} = \mathbb{E}^{1,0}\mathsf{P} \; .$$

If we now also collect all the integrated vorticities, w_i, we can relate them to the integrated velocities in the following way

$$\begin{pmatrix} \mathsf{w}_1 \\ \mathsf{w}_2 \\ \mathsf{w}_3 \\ \mathsf{w}_4 \\ \mathsf{w}_5 \\ \mathsf{w}_6 \\ \mathsf{w}_7 \\ \mathsf{w}_8 \\ \mathsf{w}_9 \end{pmatrix} = \begin{pmatrix} 1 & 0 & 0 & -1 & 0 & 0 & 0 & 0 & 0 & 0 & 0 & 0 & -1 & 1 & 0 & 0 & 0 & 0 & 0 & 0 & 0 & 0 & 0 & 0 \\ 0 & 1 & 0 & 0 & -1 & 0 & 0 & 0 & 0 & 0 & 0 & 0 & 0 & -1 & 1 & 0 & 0 & 0 & 0 & 0 & 0 & 0 & 0 & 0 \\ 0 & 0 & 1 & 0 & 0 & -1 & 0 & 0 & 0 & 0 & 0 & 0 & 0 & 0 & -1 & 1 & 0 & 0 & 0 & 0 & 0 & 0 & 0 & 0 \\ 0 & 0 & 0 & 1 & 0 & 0 & -1 & 0 & 0 & 0 & 0 & 0 & 0 & 0 & 0 & 0 & -1 & 1 & 0 & 0 & 0 & 0 & 0 & 0 \\ 0 & 0 & 0 & 0 & 1 & 0 & 0 & -1 & 0 & 0 & 0 & 0 & 0 & 0 & 0 & 0 & 0 & -1 & 1 & 0 & 0 & 0 & 0 & 0 \\ 0 & 0 & 0 & 0 & 0 & 1 & 0 & 0 & -1 & 0 & 0 & 0 & 0 & 0 & 0 & 0 & 0 & 0 & -1 & 1 & 0 & 0 & 0 & 0 \\ 0 & 0 & 0 & 0 & 0 & 0 & 1 & 0 & 0 & -1 & 0 & 0 & 0 & 0 & 0 & 0 & 0 & 0 & 0 & 0 & -1 & 1 & 0 & 0 \\ 0 & 0 & 0 & 0 & 0 & 0 & 0 & 1 & 0 & 0 & -1 & 0 & 0 & 0 & 0 & 0 & 0 & 0 & 0 & 0 & 0 & -1 & 1 & 0 \\ 0 & 0 & 0 & 0 & 0 & 0 & 0 & 0 & 1 & 0 & 0 & -1 & 0 & 0 & 0 & 0 & 0 & 0 & 0 & 0 & 0 & 0 & -1 & 1 \end{pmatrix} \begin{pmatrix} \bar{v}_1 \\ \bar{v}_2 \\ \bar{v}_3 \\ \bar{v}_4 \\ \bar{v}_5 \\ \bar{v}_6 \\ \bar{v}_7 \\ \bar{v}_8 \\ \bar{v}_9 \\ \bar{v}_{10} \\ \bar{v}_{11} \\ \bar{v}_{12} \\ \bar{v}_{13} \\ \bar{v}_{14} \\ \bar{v}_{15} \\ \bar{v}_{16} \\ \bar{v}_{17} \\ \bar{v}_{18} \\ \bar{v}_{19} \\ \bar{v}_{20} \\ \bar{v}_{21} \\ \bar{v}_{22} \\ \bar{v}_{23} \\ \bar{v}_{24} \end{pmatrix} .$$

If we store all vorticity integrals, w_i in the vector w, then we can write this as

$$\mathsf{w} = \mathbb{E}^{2,1}\mathsf{v} \; . \tag{3.8}$$

The matrices $\mathbb{E}^{1,0}$ and $\mathbb{E}^{2,1}$ are called *incidence matrices*. We have $\mathbb{E}^{2,1} \cdot \mathbb{E}^{1,0} \equiv 0$. This identity holds for this particular case, but is generally true; it holds when we would have used triangles or polyhedra instead of quadrilaterals and it holds in any space dimension d. If $\mathbb{E}^{1,0}$ represents the gradient operation and $\mathbb{E}^{2,1}$ represents the curl operation, then $\mathbb{E}^{2,1} \cdot \mathbb{E}^{1,0} \equiv 0$ is the discrete analogue of the vector identity $\nabla \times \nabla \equiv 0$, [22, 23, 25, 26, 49, 88, 110].

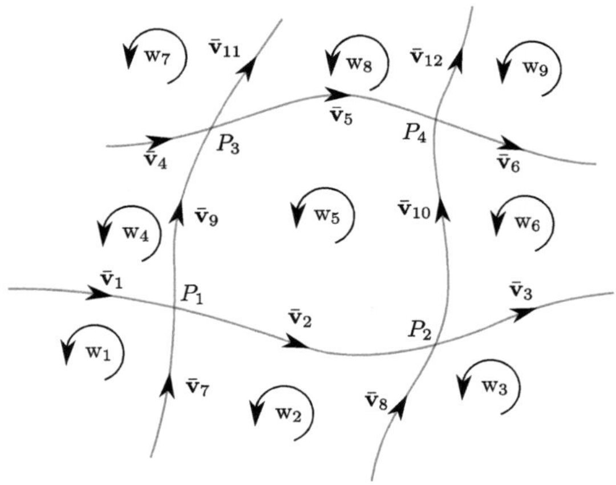

Fig. 3.2 Relation between pressure in points and integrated velocity along line segments in case $\Gamma_p = \partial\Omega$

If boundary conditions for p are prescribed along $\partial\Omega$, then these degrees of freedom can be removed from the grid in Fig. 3.1.

If p is known along the boundary then the integral of $\boldsymbol{v}$ is also known along the boundary, so the degrees of freedom for v can also be removed. Relabeling the remaining unknowns gives the geometric degrees of freedom as shown in Fig. 3.2.

$$\begin{pmatrix} \bar{\boldsymbol{v}}_1 \\ \bar{\boldsymbol{v}}_2 \\ \bar{\boldsymbol{v}}_3 \\ \bar{\boldsymbol{v}}_4 \\ \bar{\boldsymbol{v}}_5 \\ \bar{\boldsymbol{v}}_6 \\ \bar{\boldsymbol{v}}_7 \\ \bar{\boldsymbol{v}}_8 \\ \bar{\boldsymbol{v}}_9 \\ \bar{\boldsymbol{v}}_{10} \\ \bar{\boldsymbol{v}}_{11} \\ \bar{\boldsymbol{v}}_{12} \end{pmatrix} = \begin{pmatrix} 1 & 0 & 0 & 0 \\ -1 & 1 & 0 & 0 \\ 0 & -1 & 0 & 0 \\ 0 & 0 & 1 & 0 \\ 0 & 0 & -1 & 1 \\ 0 & 0 & 0 & -1 \\ 1 & 0 & 0 & 0 \\ 0 & 1 & 0 & 0 \\ -1 & 0 & 1 & 0 \\ 0 & -1 & 0 & 1 \\ 0 & 0 & -1 & 0 \\ 0 & 0 & 0 & -1 \end{pmatrix} \begin{pmatrix} P_1 \\ P_2 \\ P_3 \\ P_4 \end{pmatrix} . \tag{3.9}$$

3.2.2 Divergence Relation

Consider a bounded, contractible volume $\mathcal{V} \subset \Omega$ then we have

$$\nabla \cdot \boldsymbol{u} = f \quad \Longrightarrow \quad \int_{\partial \mathcal{V}} \boldsymbol{u} \cdot \boldsymbol{n} \, \mathrm{d}S = \int_{\mathcal{V}} f \, \mathrm{d}\mathcal{V} \, .$$

If the boundary $\partial \mathcal{V}$ can be partitioned into n sub-boundaries, $\partial \mathcal{V} = \bigcup_i \Gamma_i$ and $\bigcap_i \Gamma_i = 0$, we have

$$\pm \sum_{i=1}^{n} \bar{\mathbf{u}}_i = \pm \sum_{i=1}^{n} \int_{\Gamma_i} \boldsymbol{u} \cdot \boldsymbol{n} \, \mathrm{d}S = \int_{\mathcal{V}} f \, \mathrm{d}\mathcal{V} =: f_{\mathcal{V}} \, ,$$

where we have the convention that the fluxes, $\bar{\mathbf{u}}_i$, are *positive* when the flow leaves the volume and *negative* when the flow enters the volume. For a 2D case the integral flux degrees of freedom, $\bar{\mathbf{u}}_i$ are depicted in Fig. 3.3. The arrow in this figure indicates the positive default direction of the fluxes. The integrated values of source function f are shown in the 2D volumes in Fig. 3.3 as f_i. The topological relation between the fluxes and the integrated source values f_i, for the situation shown in Fig. 3.3, is

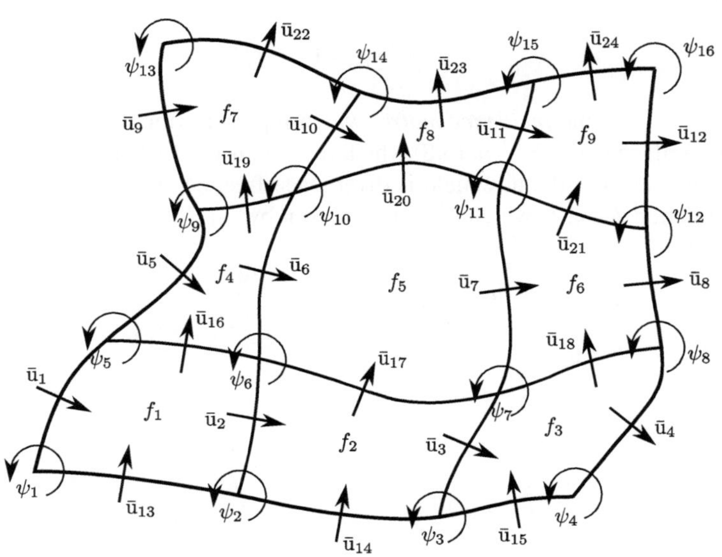

Fig. 3.3 Stream function, fluxes and the divergence degrees of freedom

given by

$$
\begin{pmatrix}
-1 & 1 & 0 & 0 & 0 & 0 & 0 & 0 & 0 & 0 & 0 & 0 & -1 & 0 & 0 & 1 & 0 & 0 & 0 & 0 & 0 & 0 & 0 & 0 \\
0 & -1 & 1 & 0 & 0 & 0 & 0 & 0 & 0 & 0 & 0 & 0 & 0 & -1 & 0 & 0 & 1 & 0 & 0 & 0 & 0 & 0 & 0 & 0 \\
0 & 0 & -1 & 1 & 0 & 0 & 0 & 0 & 0 & 0 & 0 & 0 & 0 & 0 & -1 & 0 & 0 & 1 & 0 & 0 & 0 & 0 & 0 & 0 \\
0 & 0 & 0 & 0 & -1 & 1 & 0 & 0 & 0 & 0 & 0 & 0 & 0 & 0 & 0 & -1 & 0 & 0 & 1 & 0 & 0 & 0 & 0 & 0 \\
0 & 0 & 0 & 0 & 0 & -1 & 1 & 0 & 0 & 0 & 0 & 0 & 0 & 0 & 0 & 0 & -1 & 0 & 0 & 1 & 0 & 0 & 0 & 0 \\
0 & 0 & 0 & 0 & 0 & 0 & -1 & 1 & 0 & 0 & 0 & 0 & 0 & 0 & 0 & 0 & 0 & -1 & 0 & 0 & 1 & 0 & 0 & 0 \\
0 & 0 & 0 & 0 & 0 & 0 & 0 & 0 & -1 & 1 & 0 & 0 & 0 & 0 & 0 & 0 & 0 & 0 & -1 & 0 & 0 & 1 & 0 & 0 \\
0 & 0 & 0 & 0 & 0 & 0 & 0 & 0 & 0 & -1 & 1 & 0 & 0 & 0 & 0 & 0 & 0 & 0 & 0 & -1 & 0 & 0 & 1 & 0 \\
0 & 0 & 0 & 0 & 0 & 0 & 0 & 0 & 0 & 0 & -1 & 1 & 0 & 0 & 0 & 0 & 0 & 0 & 0 & 0 & -1 & 0 & 0 & 1
\end{pmatrix}
\begin{pmatrix}
\bar{u}_1 \\ \bar{u}_2 \\ \bar{u}_3 \\ \bar{u}_4 \\ \bar{u}_5 \\ \bar{u}_6 \\ \bar{u}_7 \\ \bar{u}_8 \\ \bar{u}_9 \\ \bar{u}_{10} \\ \bar{u}_{11} \\ \bar{u}_{12} \\ \bar{u}_{13} \\ \bar{u}_{14} \\ \bar{u}_{15} \\ \bar{u}_{16} \\ \bar{u}_{17} \\ \bar{u}_{18} \\ \bar{u}_{19} \\ \bar{u}_{20} \\ \bar{u}_{21} \\ \bar{u}_{22} \\ \bar{u}_{23} \\ \bar{u}_{24}
\end{pmatrix}
=
\begin{pmatrix}
f_1 \\ f_2 \\ f_3 \\ f_4 \\ f_5 \\ f_6 \\ f_7 \\ f_8 \\ f_9
\end{pmatrix} .
$$

Collecting all fluxes and source terms in vectors u and f, respectively, we can write this equation as

$$\tilde{\mathbb{E}}^{2,1}\mathsf{u} = \mathsf{f} . \tag{3.10}$$

The matrix $\tilde{\mathbb{E}}^{2,1}$ is the *incidence matrix* which represents the divergence operator, not to be confused with $\mathbb{E}^{2,1}$ in (3.8) which represents the curl operator.

If, in the 2D case, the flow field is divergence-free, i.e. $f = 0$, we know that a stream function ψ exists which is connected to $\mathbf{u}$ by

$$u_x = \frac{\partial \psi}{\partial y} , \quad u_y = -\frac{\partial \psi}{\partial x} .$$

If we represent the stream function in the nodes of the grid shown in Fig. 3.3, then we have the exact topological equation

$$
\begin{pmatrix} \bar{u}_1 \\ \bar{u}_2 \\ \bar{u}_3 \\ \bar{u}_4 \\ \bar{u}_5 \\ \bar{u}_6 \\ \bar{u}_7 \\ \bar{u}_8 \\ \bar{u}_9 \\ \bar{u}_{10} \\ \bar{u}_{11} \\ \bar{u}_{12} \\ \bar{u}_{13} \\ \bar{u}_{14} \\ \bar{u}_{15} \\ \bar{u}_{16} \\ \bar{u}_{17} \\ \bar{u}_{18} \\ \bar{u}_{19} \\ \bar{u}_{20} \\ \bar{u}_{21} \\ \bar{u}_{22} \\ \bar{u}_{23} \\ \bar{u}_{24} \end{pmatrix}
=
\begin{pmatrix}
-1 & 0 & 0 & 0 & 1 & 0 & 0 & 0 & 0 & 0 & 0 & 0 & 0 & 0 & 0 & 0 \\
0 & -1 & 0 & 0 & 0 & 1 & 0 & 0 & 0 & 0 & 0 & 0 & 0 & 0 & 0 & 0 \\
0 & 0 & -1 & 0 & 0 & 0 & 1 & 0 & 0 & 0 & 0 & 0 & 0 & 0 & 0 & 0 \\
0 & 0 & 0 & -1 & 0 & 0 & 0 & 1 & 0 & 0 & 0 & 0 & 0 & 0 & 0 & 0 \\
0 & 0 & 0 & 0 & -1 & 0 & 0 & 0 & 1 & 0 & 0 & 0 & 0 & 0 & 0 & 0 \\
0 & 0 & 0 & 0 & 0 & -1 & 0 & 0 & 0 & 1 & 0 & 0 & 0 & 0 & 0 & 0 \\
0 & 0 & 0 & 0 & 0 & 0 & -1 & 0 & 0 & 0 & 1 & 0 & 0 & 0 & 0 & 0 \\
0 & 0 & 0 & 0 & 0 & 0 & 0 & -1 & 0 & 0 & 0 & 1 & 0 & 0 & 0 & 0 \\
0 & 0 & 0 & 0 & 0 & 0 & 0 & 0 & -1 & 0 & 0 & 0 & 1 & 0 & 0 & 0 \\
0 & 0 & 0 & 0 & 0 & 0 & 0 & 0 & 0 & -1 & 0 & 0 & 0 & 1 & 0 & 0 \\
0 & 0 & 0 & 0 & 0 & 0 & 0 & 0 & 0 & 0 & -1 & 0 & 0 & 0 & 1 & 0 \\
0 & 0 & 0 & 0 & 0 & 0 & 0 & 0 & 0 & 0 & 0 & -1 & 0 & 0 & 0 & 1 \\
1 & -1 & 0 & 0 & 0 & 0 & 0 & 0 & 0 & 0 & 0 & 0 & 0 & 0 & 0 & 0 \\
0 & 1 & -1 & 0 & 0 & 0 & 0 & 0 & 0 & 0 & 0 & 0 & 0 & 0 & 0 & 0 \\
0 & 0 & 1 & -1 & 0 & 0 & 0 & 0 & 0 & 0 & 0 & 0 & 0 & 0 & 0 & 0 \\
0 & 0 & 0 & 0 & 1 & -1 & 0 & 0 & 0 & 0 & 0 & 0 & 0 & 0 & 0 & 0 \\
0 & 0 & 0 & 0 & 0 & 1 & -1 & 0 & 0 & 0 & 0 & 0 & 0 & 0 & 0 & 0 \\
0 & 0 & 0 & 0 & 0 & 0 & 1 & -1 & 0 & 0 & 0 & 0 & 0 & 0 & 0 & 0 \\
0 & 0 & 0 & 0 & 0 & 0 & 0 & 0 & 1 & -1 & 0 & 0 & 0 & 0 & 0 & 0 \\
0 & 0 & 0 & 0 & 0 & 0 & 0 & 0 & 0 & 1 & -1 & 0 & 0 & 0 & 0 & 0 \\
0 & 0 & 0 & 0 & 0 & 0 & 0 & 0 & 0 & 0 & 1 & -1 & 0 & 0 & 0 & 0 \\
0 & 0 & 0 & 0 & 0 & 0 & 0 & 0 & 0 & 0 & 0 & 0 & 1 & -1 & 0 & 0 \\
0 & 0 & 0 & 0 & 0 & 0 & 0 & 0 & 0 & 0 & 0 & 0 & 0 & 1 & -1 & 0 \\
0 & 0 & 0 & 0 & 0 & 0 & 0 & 0 & 0 & 0 & 0 & 0 & 0 & 0 & 1 & -1
\end{pmatrix}
\begin{pmatrix} \psi_1 \\ \psi_2 \\ \psi_3 \\ \psi_4 \\ \psi_5 \\ \psi_6 \\ \psi_7 \\ \psi_8 \\ \psi_9 \\ \psi_{10} \\ \psi_{11} \\ \psi_{12} \\ \psi_{13} \\ \psi_{14} \\ \psi_{15} \\ \psi_{16} \end{pmatrix} .
$$

We can write this in terms of incidence matrices as[3]

$$
\mathsf{u} = \tilde{\mathbb{E}}^{1,0} \boldsymbol{\psi} . \tag{3.11}
$$

[3]Note that if we performed the same steps in 3D, then the divergence relation (3.10) would be

$$
\tilde{\mathbb{E}}^{3,2} \mathsf{u} = \mathsf{f} ,
$$

and the 2D stream function becomes the 3D stream vector field and we would have

$$
\mathsf{u} = \tilde{\mathbb{E}}^{2,1} \boldsymbol{\psi} .
$$

So clearly the incidence matrices $\tilde{\mathbb{E}}$ depend on the dimension of the space d in which the problem is posed. Note that this is not the case for the incidence matrices $\mathbb{E}$. Alternatively, we could refer to the dimension-dependent incidence matrices as

$$
\mathbb{E}^{d,d-1} = \begin{cases} \tilde{\mathbb{E}}^{2,1} & \text{if } d = 2 \\ \tilde{\mathbb{E}}^{3,2} & \text{if } d = 3 \end{cases} \quad \text{and} \quad \mathbb{E}^{d-1,d-2} = \begin{cases} \tilde{\mathbb{E}}^{1,0} & \text{if } d = 2 \\ \tilde{\mathbb{E}}^{2,1} & \text{if } d = 3 \end{cases} ,
$$

in which case it is immediately clear that these matrices depend on the d. From now on we will use the incidence matrices with the d, because then the results are valid for any space dimension d.

If the flux $\mathbf{u}$ is prescribed along the Γ_u the associated edges (2D) or surfaces (3D) can be eliminated from the system $\mathbb{E}^{d,d-1}\mathsf{u} = \mathsf{f}$ and transferred to the right hand side.

For the discretization of (3.4) the first and last equation in that system can be represented on the mesh by

$$\begin{cases} \mathsf{v} - \mathbb{E}^{1,0}\mathsf{p} = 0 \\ \tilde{\mathbb{E}}^{d,d-1}\mathsf{u} = \mathsf{f} \end{cases} .$$

Prescription of boundary conditions p along Γ_u and $\mathbf{u}$ along Γ_u can be done strongly. The degrees of freedom can be eliminated and transferred to the right hand side. The equation between p and $\boldsymbol{v}$ is exact on any grid and the discrete divergence relation between $\mathbf{u}$ and f is exact on any grid. Note the $(\boldsymbol{v}, p)$-grid is not necessarily the $(\mathbf{u}, f)$-grid, so in principle we can use different grids for both equations.

Unfortunately, neither of the two problems, $\boldsymbol{v} = \nabla p$ and $\nabla \cdot \mathbf{u} = f$ has a unique solution on their respective grids. It is the final equation in (3.4), $\mathbf{u} = -\mathbb{K}\boldsymbol{v}$, that couples the solution on the two grids and renders a unique solution. It is also in this equation that the numerical approximation is made; the more accurate we approximate this algebraic equation, the more accurate the solution to the first order system (3.4) will be.

For many numerical methods[4] well-posedness requires that the number of discrete degrees of freedom $\bar{\boldsymbol{v}}_i$ is equal to the discrete number of degrees of freedom $\bar{\mathbf{u}}_j$, or more geometrically, that the number of k-dimensional geometric objects on one grid is equal to the number of $(d-k)$-dimensional geometric objects on the other grid. Here $k = 0$ refers to points in the grid, $k = 1$ to edges in the grid, $k = 2$ the faces in the grid, and $k = 3$ the volumes in the grid.

The requirement $\#k = \#(d-k)$ cannot be accomplished on a single grid, so this requires two different grids which are constructed in such a way that $\#k = \#(d-k)$ holds, [22, 23, 49, 82, 88, 110].

A dual grid complex is shown in Fig. 3.4. The integral quantities $(\boldsymbol{v}, p)$ can be represented on the gray grid. If p is prescribed along the entire boundary, then those degrees of freedom are eliminated (including the gray edges along the boundary for which the integral value $\boldsymbol{v}$ is then known also), see for instance Fig. 3.2. In that case flux $\mathbf{u}$ along the boundary cannot be prescribed. In Fig. 3.4, the number of points in the gray grid, 9, equals the number of surfaces in the black grid, the number of edges in the grey grid is equal to the number of edges on the black grid, 24, and the number of surfaces on the gray grid equals the number of points in the black grid, 16, therefore, we have $\#k = \#(d-k)$ for $d = 2$.

[4] A notable exception is the class of least-squares formulations which aims to *minimize* the expression $\mathbf{u} + \mathbb{K}\boldsymbol{v}$ [17].

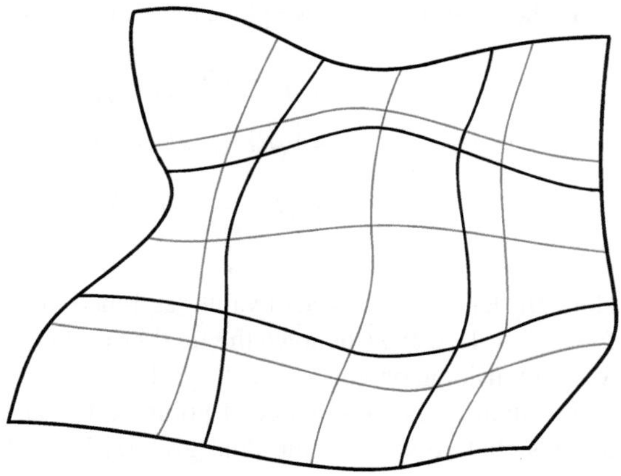

Fig. 3.4 The primal grid (thin gray) where $(\boldsymbol{v}, p)$ are represented and the dual grid (thick black) where $(\mathbf{u}, f)$ are represented. Note that $\Gamma_p = \partial\Omega$ and consequently $\Gamma_u = \varnothing$

Alternatively, we could have represented $(\mathbf{u}, f)$ on the gray grid with $\mathbf{u}$ and the stream function ψ prescribed and $(\boldsymbol{v}, p)$ on the black grid. In this case $\Gamma_u = \partial\Omega$ and $\Gamma_p = \varnothing$.

3.2.3 Dual Grids

If dual grids, such as described above, are employed then we have two properties:

1. There exists a square, invertible matrix $\mathbb{H}_{\mathbb{K}}^{d-1,1}$ such that $\mathsf{u} = \mathbb{H}_{\mathbb{K}}^{d-1,1}\mathsf{v}$.
2. The incidence matrices on the primal and dual grid satisfy[5]

$$\mathbb{E}^{d-k,d-k-1} = \left(\mathbb{E}^{k,k-1}\right)^T .$$

If we use dual grids and these properties hold, we can write (3.4) as

$$\begin{cases} \mathsf{v} - \mathbb{E}^{1,0}\mathsf{p} = 0 \\ \mathsf{u} - \mathbb{H}_{\mathbb{K}}^{d-1,1}\mathsf{v} = 0 \\ \mathbb{E}^{1,0^T}\mathsf{u} = \mathsf{f} \end{cases} , \tag{3.12}$$

where the vectors p, v, u and f contain the integral quantities in the mesh as discussed in the previous sections.

[5]This relation is true if the orientations on primal and dual grid agree. This is not always the case and then the relation reads $\mathbb{E}^{d-k,d-k-1} = -\mathbb{E}^{k,k-1^T}$. A well known example is the duality between grad and div.

In the diagram below, we place the various integral values in appropriate ‘spaces’

$$
\begin{array}{ccccc}
\mathsf{p} \in H_0 & \xrightarrow{\mathbb{E}^{1,0}} & \mathsf{v} \in H_1 & \xrightarrow{\mathbb{E}^{2,1}} & \xi \in H_2 \\
\mathbb{H}^{d,0} \Big\downarrow \Big\uparrow \mathbb{H}^{0,d} & & \mathbb{H}_{\mathbb{K}}^{d-1,1} \Big\downarrow \Big\uparrow \mathbb{H}_{\mathbb{K}^{-1}}^{1,d-1} & & \\
\mathsf{f} \in \tilde{H}_d & \xleftarrow[\mathbb{E}^{1,0^T}]{} & \mathsf{u} \in \tilde{H}_{d-1} & \xleftarrow[\mathbb{E}^{2,1^T}]{} & \psi \in \tilde{H}_{d-2}
\end{array}
$$

Here H_k denotes the space of values assigned to k-dimensional objects in the H-grid for $k = 0, 1, 2$. If $\tilde{H}$ denotes the dual grid, then $\tilde{H}_l$ is the space of values assigned to l-dimensional objects in the $\tilde{H}$-grid.

For dual grids the number of points in the H-grid is equal to the number of d-dimensional volumes in the dual grid $\tilde{H}$. Let $\mathbb{H}^{d,0}$ and $\mathbb{H}^{0,d}$ be square, invertible matrices which map between H_0 and $\tilde{H}_d$ as shown in the diagram above.

If we eliminate v and u from (3.12) we have

$$
\mathbb{E}^{1,0^T} \mathbb{H}_{\mathbb{K}}^{d-1,1} \mathbb{E}^{1,0} \mathsf{p} = \mathsf{f} \,. \tag{3.13}
$$

This discretization corresponds to (3.2). We will refer to this formulation as the *direct formulation*.

If $\mathsf{p} \in \tilde{H}_d$ we can set up the diffusion problem as

$$
\begin{cases}
-\mathbb{H}_{\mathbb{K}^{-1}}^{1,d-1} \mathsf{u} + \mathbb{E}^{d,d-1^T} \mathbb{H}^{0,d} \mathsf{p} = 0 \\
\mathbb{H}^{0,d} \mathbb{E}^{d,d-1} \mathsf{u} \qquad\qquad\quad = \mathsf{f}
\end{cases} . \tag{3.14}
$$

This formulation, where we solve for p and $\mathbf{u}$ simultaneously, resembles (3.3), and will be called the *mixed formulation*, [32].

3.3 Mimetic Spectral Element Method

The incidence matrices introduced in the previous section are generic and only depend on the grid topology. The matrices $\mathbb{H}$ which switch between the primal and the dual grid representation explicitly depend on the numerical method that is used. In this section we will introduce spectral element functions which interpolate the integral values in a grid. With these functions we can construct the $\mathbb{H}$-matrices, which turn out to be (weighted) finite element mass matrices. The derivation in this section will be on an orthogonal grid. The extension to curvilinear grids will be discussed in the next section.

3.3.1 One Dimensional Spectral Basis Functions

Consider the interval $[-1, 1] \subset \mathbb{R}$ and the Legendre polynomials, $L_N(\xi)$, of degree N, $\xi \in [-1, 1]$. The $(N+1)$ roots, ξ_i, of the polynomial $(1-\xi^2)L'_N(\xi)$ satisfy $-1 \leq \xi_i \leq 1$. Here $L'_N(\xi)$ is the derivative of the Legendre polynomial. The roots ξ_i, $i = 0, \ldots, N$, are called the *Gauss-Lobatto-Legendre (GLL) points*, [41]. Let $h_i(\xi)$ be the Lagrange polynomial through the GLL points such that

$$h_i(\xi_j) = \begin{cases} 1 & \text{if } i = j \\ 0 & \text{if } i \neq j \end{cases} \qquad i, j = 0, \ldots N \, . \tag{3.15}$$

The explicit form of the Lagrange polynomials in terms of the Legendre polynomials is given by

$$h_i(\xi) = \frac{(1-\xi^2)L'_N(\xi)}{N(N+1)L_N(\xi_i)(\xi_i - \xi)} \, . \tag{3.16}$$

Let $f(\xi)$ be a function defined for $\xi \in [-1, 1]$ by

$$f(\xi) = \sum_{i=0}^{N} a_i h_i(\xi) \, . \tag{3.17}$$

Using property (3.15) we see that $f(\xi_j) = a_j$, so the expansion coefficients in (3.17) coincide with the value of f in the GLL nodes. We will refer to this expansion as a *nodal expansion*, because the expansion coefficients, a_i in (3.17) are the value of $f(\xi)$ in the *nodes* ξ_i. The basis functions $h_i(\xi)$ are polynomials of degree N.

From the nodal basis functions, define the polynomials $e_i(\xi)$ by

$$e_i(\xi) = -\sum_{k=0}^{i-1} \frac{dh_k(\xi)}{d\xi} \, . \tag{3.18}$$

The functions $e_i(\xi)$ are polynomials of degree $(N-1)$. These polynomials satisfy, [61, 82, 94]

$$\int_{\xi_{j-1}}^{\xi_j} e_i(\xi) = \begin{cases} 1 & \text{if } i = j \\ 0 & \text{if } i \neq j \end{cases} \qquad i, j = 1, \ldots N \, . \tag{3.19}$$

Let a function $g(\xi)$ be expanded in these functions

$$g(\xi) = \sum_{i=1}^{N} b_i e_i(\xi) \, , \tag{3.20}$$

then using (3.19)

$$\int_{\xi_{j-1}}^{\xi_j} g(\xi) = b_j \, .$$

So the expansion coefficients b_i in (3.20) coincide with the integral of g over the edge $[\xi_{i-1}, \xi_i]$. We will call these basis functions *edge functions* and refer to the expansion (3.20) as an *edge expansion*, see for instance [16, 82, 94] for examples of nodal and edge expansions.

Let $f(\xi)$ be expanded in terms Lagrange polynomials as in (3.17), then the derivative[6] of f is given by, [61, 82, 94]

$$f'(\xi) = \sum_{i=0}^{N} a_i h_i'(\xi) = \sum_{i=1}^{N} (a_i - a_{i-1}) e_i(\xi) \, . \tag{3.21}$$

If we collect all the expansion coefficients in a column vector and all the basis functions in a row vector we have

$$f(\xi) = [h_0 \; h_1 \; \dots \; h_N] \begin{bmatrix} a_0 \\ \vdots \\ a_N \end{bmatrix} , \tag{3.22}$$

then the derivative is given by[7] (3.21)

$$f'(\xi) = [e_1 \; \dots \; e_N] \begin{pmatrix} -1 & 1 & 0 & \dots & & 0 \\ & \ddots & \ddots & & & 0 \\ & & -1 & 1 & & 0 \\ & & & \ddots & \ddots & \\ 0 & & \dots & 0 & -1 & 1 \end{pmatrix} \begin{bmatrix} a_0 \\ \vdots \\ a_N \end{bmatrix} = [e_1 \; \dots \; e_N] \mathbb{E}^{1,0} \begin{bmatrix} a_0 \\ \vdots \\ a_N \end{bmatrix} . \tag{3.23}$$

[6]Note that the set of polynomials $\{h_i'\}$, $i = 0, \dots, N$ is linearly dependent and therefore does not form a basis, while the set $\{e_i\}$, $i = 1, \dots, N$ is linearly independent and therefore forms a basis for the derivatives of the nodal expansion (3.17).

[7]The matrix $\mathbb{E}^{1,0}$ is the *incidence matrix* as was discussed in Sects. 3.2.1 and 3.2.2. It takes the nodal expansion coefficients and maps them to the edge expansion coefficients. The incidence matrix is the topological part of the derivative. It is independent of the order of the method (the polynomial degree N) and the size or the shape of the mesh. The incidence matrix only depends on the topology and orientation of the grid, see [18, 81, 82].

So taking the derivative essentially consists of two step: Apply the matrix $\mathbb{E}^{1,0}$ to the expansion coefficients and expand in a new basis.

3.3.2 Two Dimensional Expansions

3.3.2.1 Expanding p (Direct Formulation)

In finite element methods the direct finite element formulation for the anisotropic diffusion problem is given by: For $(\mathbb{K}\nabla p, \boldsymbol{n}) = 0$ along Γ_u and $f \in H^{-1}(\Omega)$, find $p \in H^1_{0,\Gamma_p}(\Omega)$ such that

$$(\nabla \tilde{p}, \mathbb{K}\nabla p) = (\tilde{p}, f)\,, \quad \forall \tilde{p} \in H^1_{0,\Gamma_p}(\Omega)\,. \tag{3.24}$$

where $H^1_{0,\Gamma_p} = \{p \in H^1(\Omega) | p = 0 \text{ on } \Gamma_p\}$.

Consider $[-1, 1]^2 \subset \mathbb{R}^2$ and let $p(\xi, \eta)$ be expanded as

$$p(\xi, \eta) = \sum_{i=0}^{N}\sum_{j=0}^{N} p_{i,j} h_i(\xi) h_j(\eta)\,. \tag{3.25}$$

From (3.15) it follows that $p_{i,j} = p(\xi_i, \eta_j)$. If we take the gradient of p using (3.21) we have

$$\nabla p = \begin{pmatrix} \sum_{i=1}^{N}\sum_{j=0}^{N}(p_{i,j} - p_{i-1,j}) e_i(\xi) h_j(\eta) \\ \sum_{i=0}^{N}\sum_{j=1}^{N}(p_{i,j} - p_{i,j-1}) h_i(\xi) e_j(\eta) \end{pmatrix} \tag{3.26}$$

$$= \begin{pmatrix} e_1(\xi)h_0(\eta) \dots e_N(\xi)h_N(\eta) & 0 & \dots & 0 \\ 0 & \dots & 0 & h_0(\xi)e_1(\eta) \dots h_N(\xi)e_N(\eta) \end{pmatrix} \mathbb{E}^{1,0} \begin{bmatrix} p_{0,0} \\ \vdots \\ p_{N,N} \end{bmatrix}$$

$$= \begin{pmatrix} e_1(\xi)h_0(\eta) \dots e_N(\xi)h_N(\eta) & 0 & \dots & 0 \\ 0 & \dots & 0 & h_0(\xi)e_1(\eta) \dots h_N(\xi)e_N(\eta) \end{pmatrix} \mathbb{E}^{1,0}\mathsf{p}\,. \tag{3.27}$$

If we insert this in (3.24), we have

$$\left(\mathbb{E}^{1,0}\right)^T \mathbb{M}^{(1)}_{\mathbb{K}} \mathbb{E}^{1,0}\mathsf{p} = \mathsf{f}\,, \tag{3.28}$$

where

$$\mathbb{M}_{\mathbb{K}}^{(1)} = \iint_{\Omega} \begin{pmatrix} e_1(\xi)h_0(\eta) & 0 \\ \vdots & \vdots \\ e_N(\xi)h_N(\eta) & 0 \\ 0 & h_0(\xi)e_1(\eta) \\ \vdots & \vdots \\ 0 & h_N(\xi)e_N(\eta) \end{pmatrix}$$

$$\times \mathbb{K} \begin{pmatrix} e_1(\xi)h_0(\eta) \dots e_N(\xi)h_N(\eta) & 0 & \dots & 0 \\ 0 & \dots & 0 & h_0(\xi)e_1(\eta) \dots h_N(\xi)e_N(\eta) \end{pmatrix} \mathrm{d}\Omega \,, \tag{3.29}$$

and p is the vector which contains the expansion coefficients of $p(\xi, \eta)$ in (3.25).

The vector f in (3.28) is given by

$$\mathsf{f} = \iint_{\Omega} \begin{pmatrix} h_0(\xi)h_0(\eta) \\ \vdots \\ h_N(\xi)h_N(\eta) \end{pmatrix} f(\xi, \eta) \, \mathrm{d}\Omega \,.$$

If we compare (3.28) with (3.13), we see that the $\mathbb{H}_{\mathbb{K}}^{d-1,1}$-matrix from (3.13) is represented in the finite element formulation by the weighted mass matrix $\mathbb{M}_{\mathbb{K}}^{(1)}$ given by (3.29), see also [18, 115].

3.3.2.2 Expanding u and *p* (Mixed Formulation)

The mixed formulation for the anisotropic steady diffusion problem is given by: For $p = 0$ along Γ_p and for $f \in L^2(\Omega)$, find $u \in H_{0,\Gamma_n}(div; \Omega)$ such that

$$\begin{cases} -(\tilde{\mathbf{u}}, \mathbb{K}^{-1}\mathbf{u}) + (\nabla \cdot \tilde{\mathbf{u}}, p) = & 0 \quad \forall \tilde{\mathbf{u}} \in H_{0,\Gamma_u}(div; \Omega) \\ (\tilde{p}, \nabla \cdot \mathbf{u}) & = (\tilde{p}, f) \quad \forall \tilde{p} \in L^2(\Omega) \end{cases} , \tag{3.30}$$

where, $H_{0,\Gamma_u}(div; \Omega) = \{u \in H(div; \Omega) | u \cdot n = 0 \text{ along } \Gamma_u\}$.

In contrast to the pressure expansion in Sect. 3.3.2.1 in the direct formulation, (3.25), in the mixed formulation the pressure is expanded in terms of edge functions

$$p(\xi,\eta) = \sum_{i=1}^{N}\sum_{j=1}^{N} p_{i,j} e_i(\xi) e_j(\eta) \,. \tag{3.31}$$

The velocity $\mathbf{u}$ is expanded as

$$\mathbf{u} = \begin{pmatrix} u \\ v \end{pmatrix} = \begin{pmatrix} \sum_{i=0}^{N}\sum_{j=1}^{N} u_{i,j} h_i(\xi) e_j(\eta) \\ \sum_{i=1}^{N}\sum_{j=0}^{N} v_{i,j} e_i(\xi) h_j(\eta) \end{pmatrix} \tag{3.32}$$

$$= \begin{pmatrix} h_0(\xi)e_1(\eta) \dots h_N(\xi)e_N(\eta) & 0 & \dots & 0 \\ 0 & \dots & 0 & e_1(\xi)h_0(\eta) \dots e_N(\xi)h_N(\eta) \end{pmatrix} \begin{pmatrix} u_{0,1} \\ \vdots \\ u_{N,N} \\ v_{1,0} \\ \vdots \\ v_{N,N} \end{pmatrix} .$$

Application of the divergence operator to (3.32) and using (3.21) we obtain

$$\nabla \cdot \mathbf{u} = \sum_{i=1}^{N}\sum_{j=1}^{N} (u_{i,j} - u_{i-1,j} + v_{i,j} - v_{i,j-1}) e_i(\xi) e_j(\eta) \tag{3.33}$$

$$= \big(e_1(\xi)e_1(\eta) \dots e_N(\xi)e_N(\eta) \big) \mathbb{E}^{d,d-1} \begin{pmatrix} u_{0,1} \\ \vdots \\ u_{N,N} \\ v_{1,0} \\ \vdots \\ v_{N,N} \end{pmatrix}$$

$$= \big(e_1(\xi)e_1(\eta) \dots e_N(\xi)e_N(\eta) \big) \mathbb{E}^{d,d-1} \mathsf{u} \,.$$

Note that $\mathbb{E}^{d,d-1}$ is the incidence matrix which also appeared in (3.10) and footnote 3.

If we insert the expansion (3.32) in $(\tilde{\mathbf{u}}, \mathbb{K}^{-1}\mathbf{u})$ we obtain

$$(\tilde{\mathbf{u}}, \mathbb{K}^{-1}\mathbf{u}) = \tilde{\mathsf{u}}^T \mathbb{M}^{(d-1)}_{\mathbb{K}^{-1}} \mathsf{u} \,, \tag{3.34}$$

with

$$\mathbb{M}^{(d-1)}_{\mathbb{K}^{-1}} = \tag{3.35}$$

$$\iint_{\Omega} \begin{pmatrix} h_0(\xi)e_1(\eta) & 0 \\ \vdots & \vdots \\ h_N(\xi)e_N(\eta) & 0 \\ 0 & e_1(\xi)h_0(\eta) \\ \vdots & \vdots \\ 0 & e_N(\xi)h_N(\eta) \end{pmatrix}$$

$$\times\mathbb{K}^{-1} \begin{pmatrix} h_0(\xi)e_1(\eta) \ldots h_N(\xi)e_N(\eta) & 0 & \ldots & 0 \\ 0 \quad \ldots \quad 0 & e_1(\xi)h_0(\eta) & \ldots & e_N(\xi)h_N(\eta) \end{pmatrix} \mathrm{d}\Omega \,. \tag{3.36}$$

Note that pressure is expanded in the same basis as the divergence of the velocity field, (3.31) and (3.33), therefore we can write

$$(\tilde{p}, \nabla \cdot \mathbf{u}) = \tilde{\mathsf{p}}^T \mathbb{M}^{(d)} \mathbb{E}^{d,d-1} \mathsf{u} \,, \tag{3.37}$$

with

$$\mathbb{M}^{(d)} = \iint_{\Omega} \begin{pmatrix} e_1(\xi)e_1(\eta) \\ \vdots \\ e_N(\xi)e_N(\eta) \end{pmatrix} \left(e_1(\xi)e_1(\eta) \ldots e_N(\xi)e_N(\eta) \right) \mathrm{d}\Omega \,.$$

With (3.34) and (3.37) we can write (3.30) as

$$\begin{cases} -\mathbb{M}^{(d-1)}_{\mathbb{K}^{-1}} \mathsf{u} + \mathbb{E}^{d,d-1^T} \mathbb{M}^{(d)} \mathsf{p} = 0 \\ \mathbb{M}^{(d)} \mathbb{E}^{d,d-1} \mathsf{u} \qquad\qquad = \mathbb{M}^{(d)} \mathsf{f} \end{cases} , \tag{3.38}$$

with

$$\mathsf{f} = \iint_{\Omega} \begin{pmatrix} e_1(\xi)e_1(\eta) \\ \vdots \\ e_N(\xi)e_N(\eta) \end{pmatrix} f(\xi, \eta) \, \mathrm{d}\Omega \,.$$

Comparison of (3.38) with (3.14) shows that the topological incidence matrices also appear in the finite element formulation and that the (weighted) mass matrices $\mathbb{M}^{(d-1)}_{\mathbb{K}^{-1}}$ and $\mathbb{M}^{(d)}$ once again play the role of the $\mathbb{H}$-matrices which connect solutions on dual grids.

In this section only the discretization on a single spectral element is discussed. Transformation of the domain $[-1, 1]^2$ to more general domains will be discussed in Sect. 3.4. The use of multiple elements follows the general assembly procedure from finite element methods. Results of this approach are presented in Sect. 3.5.

3.4 Transformation Rules

The basis functions used in the discretization of the different physical field quantities have only been introduced for the reference domain $\tilde{\Omega} = [-1, 1]^2$. For these basis functions to be applicable in a different domain Ω, it is fundamental to discuss how they transform under a mapping $\Phi : (\xi, \eta) \in \tilde{\Omega} \mapsto (x, y) \in \Omega \subset \mathbb{R}^2$. Within a finite element formulation this is particularly useful because the basis functions in the reference domain $\tilde{\Omega}$ can then be transformed to each of the elements Ω_e, given a mapping $\Phi_e : \tilde{\Omega} \mapsto \Omega_e$.

Consider a smooth bijective map $\Phi : (\xi, \eta) \in \tilde{\Omega} \mapsto (x, y) \in \Omega$ such that

$$x = \Phi^x(\xi, \eta) \quad \text{and} \quad y = \Phi^y(\xi, \eta),$$

and the associated rank two Jacobian tensor $\mathbf{J}$

$$\mathbf{J} := \begin{bmatrix} \dfrac{\partial \Phi^x}{\partial \xi} & \dfrac{\partial \Phi^x}{\partial \eta} \\ \dfrac{\partial \Phi^y}{\partial \xi} & \dfrac{\partial \Phi^y}{\partial \eta} \end{bmatrix}.$$

The transformation of a scalar function φ discretized by nodal values is given by

$$\tilde{\varphi}(\xi, \eta) = (\varphi \circ \Phi)(\xi, \eta) \quad \text{and} \quad \varphi(x, y) = (\tilde{\varphi} \circ \Phi^{-1})(x, y), \tag{3.39}$$

and of a scalar function ρ discretized by surface integrals is given by

$$\tilde{\rho}(\xi, \eta) = \det\mathbf{J}\,(\rho \circ \Phi)(\xi, \eta) \quad \text{and} \quad \rho(x, y) = \frac{1}{\det\mathbf{J}}(\tilde{\rho} \circ \Phi^{-1})(x, y). \tag{3.40}$$

The transformation of vector fields $\boldsymbol{v}$ discretized by line integrals is

$$\tilde{\boldsymbol{v}}(\xi, \eta) = \mathbf{J}^{\mathsf{T}}(\boldsymbol{v} \circ \Phi)(\xi, \eta) \quad \text{and} \quad \boldsymbol{v}(x, y) = (\mathbf{J}^{\mathsf{T}})^{-1}(\tilde{\boldsymbol{v}} \circ \Phi^{-1})(x, y), \tag{3.41}$$

and of vector fields $\mathbf{u}$ discretized by flux integrals is

$$\tilde{\mathbf{u}}(\xi, \eta) = \det\mathbf{J}\;\mathbf{J}^{-1}(\mathbf{u} \circ \Phi)(\xi, \eta) \quad \text{and} \quad \mathbf{u}(x, y) = \frac{1}{\det\mathbf{J}}\mathbf{J}(\tilde{\mathbf{u}} \circ \Phi^{-1})(x, y). \tag{3.42}$$

These transformations affect only the mass matrices and not the incidence matrices. This is fundamental to ensure the topological nature of the incidence matrices.

3.5 Numerical Results

In this section three test cases are presented to illustrate the accuracy of the discretization scheme developed in this work. The first test case, Sect. 3.5.1, is an analytical solution taken from [68] to assess the convergence rates of the method. The second test case, Sect. 3.5.2, is the flow through a system of sand and shale blocks with highly heterogeneous permeability in the domain, see for more details [54]. The third test case, Sect. 3.5.3, is a highly anisotropic and heterogeneous permeability tensor in the domain, see for more details, [53].

3.5.1 Manufactured Solution

We first test the method using the exact solution

$$p_{\text{exact}}(x, y) = \sin(\pi x)\sin(\pi y), \tag{3.43}$$

with the permeability tensor given by

$$\mathbb{K} = \frac{1}{\left(x^2 + y^2 + \alpha\right)} \begin{pmatrix} 10^{-3}x^2 + y^2 + \alpha & \left(10^{-3} - 1\right)xy \\ \left(10^{-3} - 1\right)xy & x^2 + 10^{-3}y^2 + \alpha \end{pmatrix}. \tag{3.44}$$

The mixed formulation (3.3) in the form of (3.38) is then solved in the domain $(x, y) \in \Omega = [0, 1]^2$ with the source term $f = -\nabla \cdot (\mathbb{K}\nabla p_{\text{exact}})$ and the Dirichlet boundary condition $p|_{\partial\Omega} = 0$. A benchmark of this test case for $\alpha = 0$ using multiple numerical schemes can be found in [68].

When $\alpha = 0$, $\mathbb{K}$ is multi-valued at the origin which makes this test case a challenging one. To see this, we can first convert the Cartesian coordinates (x, y) to polar coordinates (r, θ) by $x = r\cos\theta$, $y = r\sin\theta$. Then we have

$$\mathbb{K}|_{\alpha=0} = \begin{pmatrix} 10^{-3}\cos^2\theta + \sin^2\theta & \left(10^{-3} - 1\right)\cos\theta\sin\theta \\ \left(10^{-3} - 1\right)\cos\theta\sin\theta & \cos^2\theta + 10^{-3}\sin^2\theta \end{pmatrix}. \tag{3.45}$$

It can be seen that we get different $\mathbb{K}|_{\alpha=0}$ when we approach the origin along different angles, θ. It must be noted that inverse of $\mathbb{K}$ does not exist at the origin. The inverse of the tensor term appears in (3.35). We use Gauss integration and thus the inverse term is not evaluated at the origin.

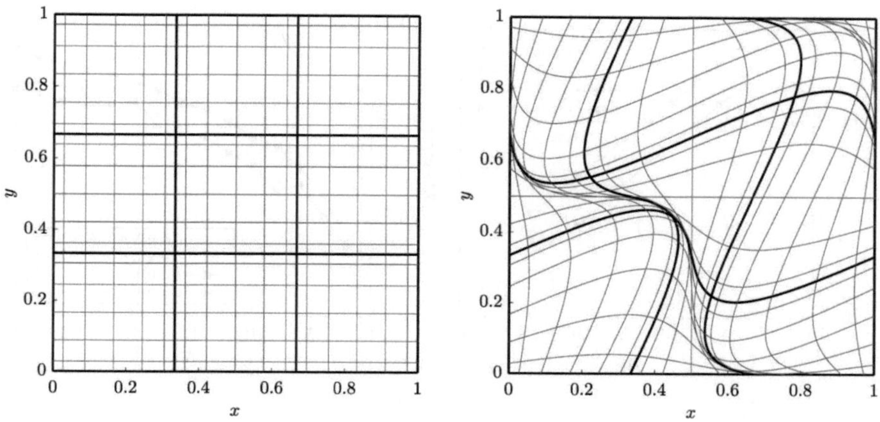

Fig. 3.5 Example meshes with 3×3 elements of polynomial degree $N = 6$. *Left:* $c = 0$ (orthogonal mesh). *Right:* $c = 0.3$ (highly deformed mesh)

The meshes we use here are obtained by deforming the GLL meshes in the reference domain $(\xi, \eta) \in \Omega_{\text{ref}} = [-1, 1]^2$ with the mapping, Φ, given as

$$\begin{cases} x = \dfrac{1}{2} + \dfrac{1}{2}\left(\xi + c\sin(\pi\xi)\sin(\pi\eta)\right) \\ y = \dfrac{1}{2} + \dfrac{1}{2}\left(\eta + c\sin(\pi\xi)\sin(\pi\eta)\right) \end{cases}, \tag{3.46}$$

where c is the deformation coefficient. The two meshes, for $c = 0.0$ and $c = 0.3$, are shown in Fig. 3.5.

The method is tested for $\alpha \in \{0, 0.01\}$ and $c \in \{0, 0.3\}$.

In Fig. 3.6, the results for $||\nabla \cdot \boldsymbol{u}_h - f_h||_{L^2}$ are presented. They show that the relation $\nabla \cdot \boldsymbol{u}_h = f_h$ is conserved to machine precision even on a highly deformed and coarse mesh i.e. of 2×2 elements with $N = 2$ and $c = 0.3$.

When $\alpha = 0.01$, $\mathbb{K}$ is no longer multi-valued at the origin. In this case the source term f is smooth over the domain, see Fig. 3.7 (bottom). For this smooth case, the method displays optimal convergence rates on both the orthogonal mesh and the deformed mesh, i.e. see Fig. 3.8 (bottom) and Fig. 3.9 (bottom).

When $\alpha = 0$, both the h-convergence rate and p-convergence rates are sub-optimal, see Fig. 3.8 (top) and Fig. 3.9 (top). This is because $\mathbb{K}$ is multi-valued and therefore f becomes singular at the origin when $\alpha = 0$, see Fig. 3.7 (top left).

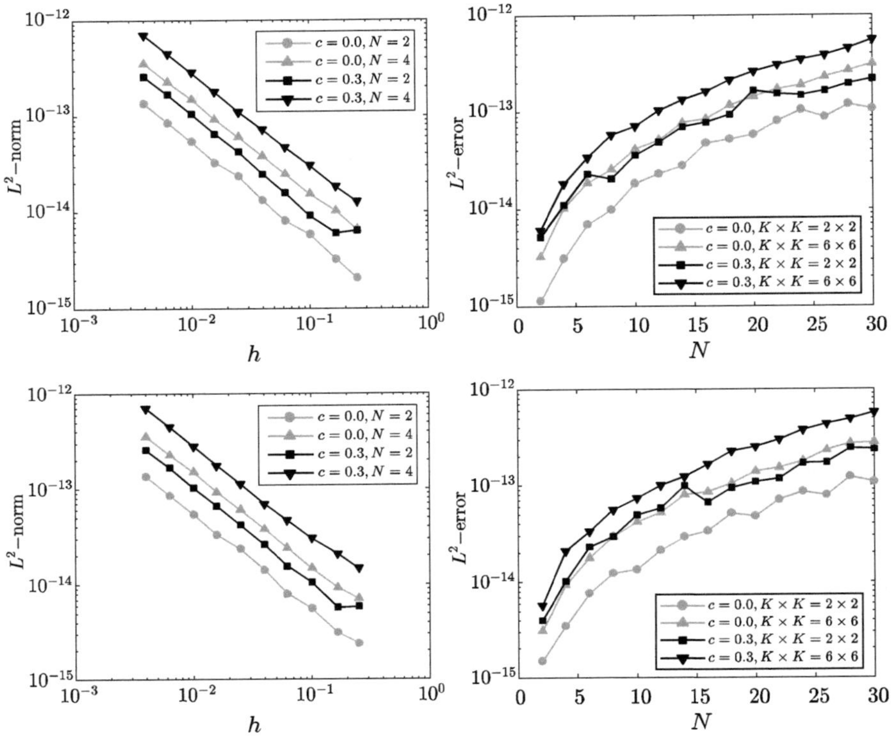

Fig. 3.6 The L^2-norm of $(\nabla \cdot \boldsymbol{u}_h - f_h)$. *Left:* $K \times K$ elements, $K = 4, \ldots, 250$, and $N = 2, 4$. *Right:* 2×2, 6×6 elements, and $N = 2, \ldots, 30$. *Top:* $\alpha = 0$. *Bottom:* $\alpha = 0.01$

3.5.2 The Sand-Shale System

This example is taken from [54, 76, 78]. The domain is a 2D unit square, $\Omega = [0, 1]^2$, with 80 shale blocks, Ω_s, placed in the domain such that the total area fraction of shale blocks is $A_{shale} = 20\%$, as shown in Fig. 3.10.

We solve the mixed formulation (3.38) with $f = 0$ in this domain. The flux across the top and the bottom boundaries is $\boldsymbol{u} \cdot \boldsymbol{n} = 0$. The flow is pressure driven with the pressure at the left boundary, $p = 1$, and the pressure at the right boundary, $p = 0$. The permeability in the domain is defined as $\mathbb{K} = k\mathbb{I}$, where k is given by:

$$k = \begin{cases} 10^{-6} & \text{in} \quad \Omega_s \\ 1 & \text{in} \quad \Omega \setminus \Omega_s \end{cases} .$$

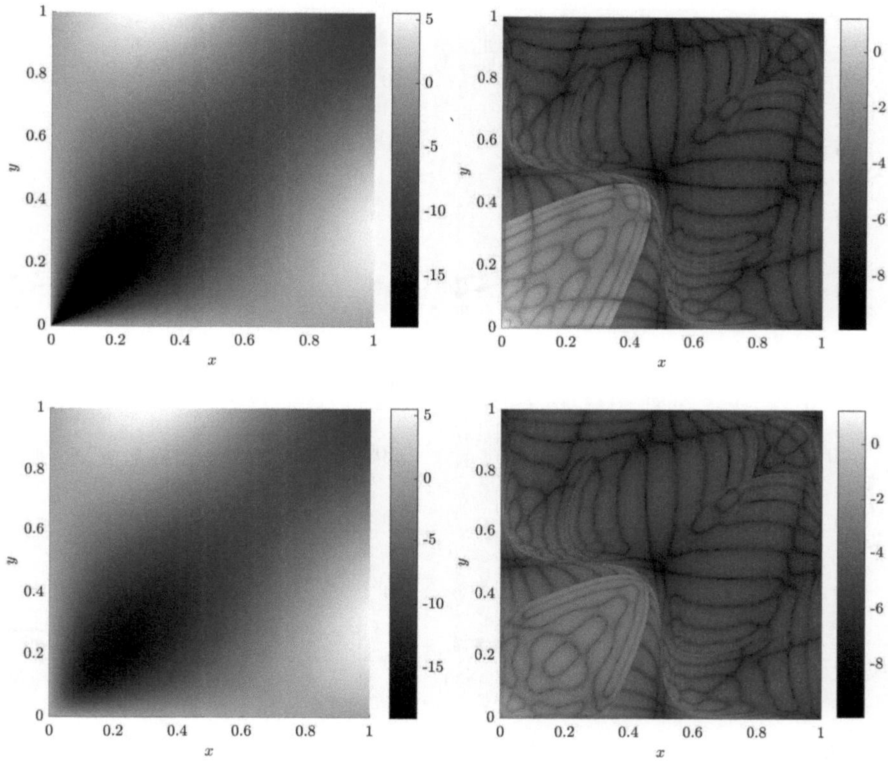

Fig. 3.7 *Left:* the source term f. *Right:* the $\log_{10}$ distribution of the projection error of f_h for 3×3 elements, $N = 10$ and $c = 0.3$. *Top:* $\alpha = 0$. *Bottom:* $\alpha = 0.01$

For this test case an orthogonal uniform grid of 20×20 elements is used. The polynomial degree is varied to achieve convergence. Streamlines through the domain for 20×20 elements and polynomial degree $N = 15$ are shown in Fig. 3.11. It can be seen that the streamlines do not pass through, but pass around the shale blocks of low permeability.

The $||\nabla \cdot \boldsymbol{u}_h||_{L^2}$ over the entire domain as a function of polynomial degree is shown in Fig. 3.12. We observe that $\nabla \cdot \boldsymbol{u}_h = 0$ is satisfied up to machine precision.

The net flux entering the domain (the same as the net flux leaving the domain) is given in Table 3.1 for varying polynomial degree. A reference value for this solution is given in [54] as 0.5205, and in [78] as 0.519269. In this work the maximum resolution corresponds to 20×20 elements and a polynomial degree $N = 19$, for which the net flux entering the domain is obtained as 0.52010.

In Fig. 3.13 we compare the net flux entering the sand-shale domain, calculated using the mixed and the direct formulation of equations, as a function of polynomial degree for different values of k in the shale blocks. The data for these figures is

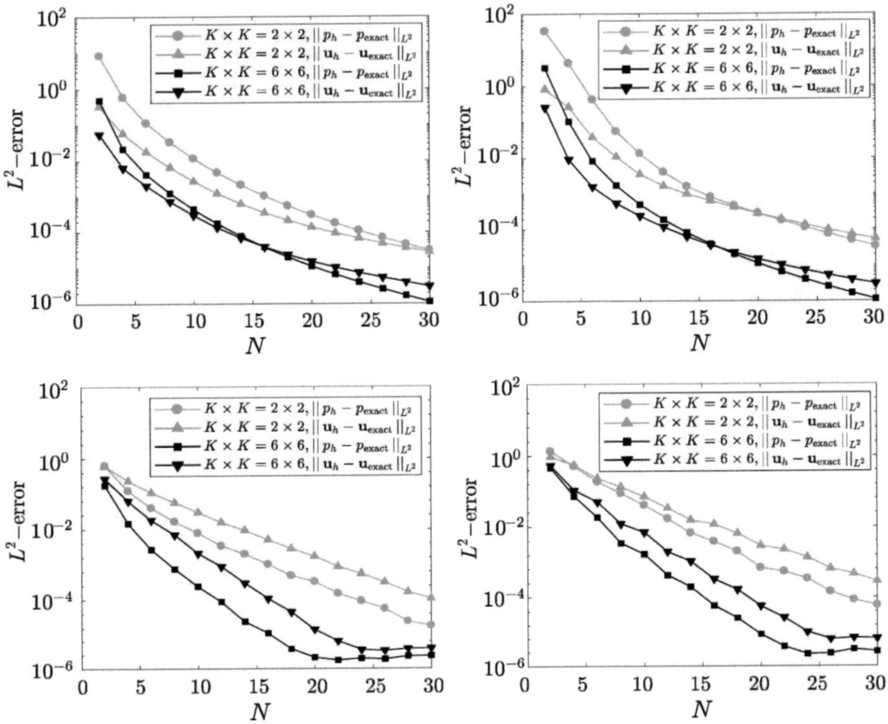

Fig. 3.8 The p-convergence for 2×2, 6×6 elements and $N = 2, \ldots, 30$. *Left:* $c = 0$. *Right:* $c = 0.3$. *Top:* $\alpha = 0$. *Bottom:* $\alpha = 0.01$

given in Table 3.2. Note that the direct formulation converges from above towards the correct inflow flux, whereas the mixed formulation converges from below.

3.5.3 *The Impermeable-Streak System*

The next example is from [53, 76, 78]. The physical domain is a 2D unit square, $\Omega = [0, 1]^2$. The domain is divided into three different regions, Ω_1, Ω_2, and Ω_3, as shown in Fig. 3.14 (left). For calculations, each region is further divided into $K \times K$ elements. Therefore, the total number of elements in the domain is given by $K \times K \times 3$. In Fig. 3.14 (right) we show the domain with each region divided into 2×2 elements.

The mixed formulation (3.38) is solved, with $f = 0$ and mixed boundary conditions, such that at the top and the bottom boundaries the net flux $\boldsymbol{u} \cdot \boldsymbol{n} = 0$, and at the left and the right boundaries, $p = 1$ and $p = 0$, respectively. Permeability in Ω_1 and Ω_3 is given by $\mathbb{K} = \mathbb{I}$. Ω_2 has a low permeability and defined such that the

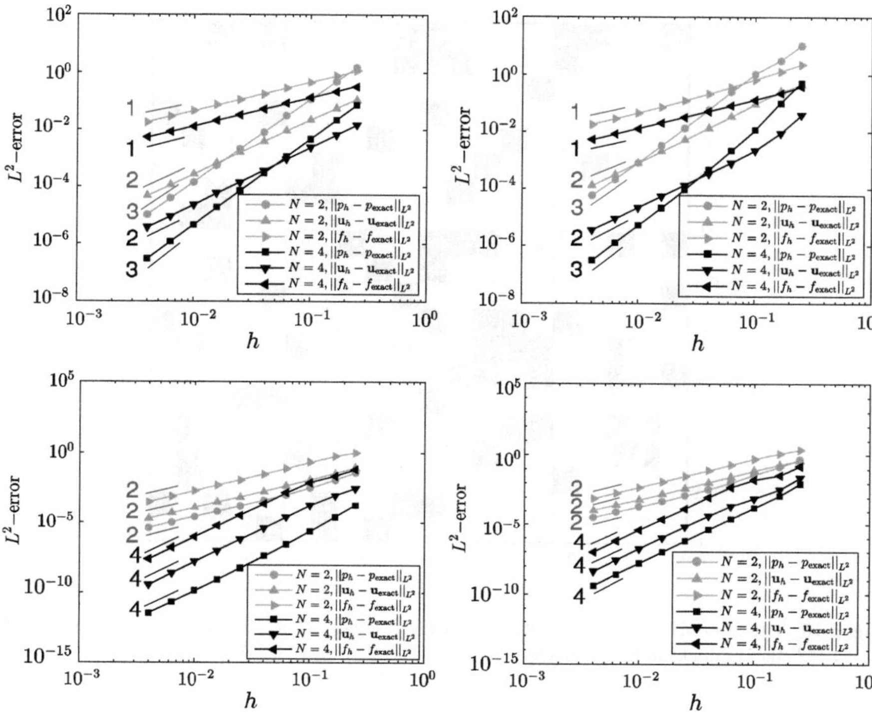

Fig. 3.9 The h-convergence of the L^2-error for $K \times K$ elements, $K = 4, \ldots, 250$ and $N = 2, 4$. *Left:* $c = 0$. *Right:* $c = 0.3$. *Top:* $\alpha = 0$. *Bottom:* $\alpha = 0.01$

component parallel to the local streak orientation is $k_{\parallel} = 10^{-1}$, and the component perpendicular to the local streak orientation is $k_{\perp} = 10^{-3}$. The analytical expression for the permeability in terms of Cartesian coordinates is given in [76] as,

$$K_{xx} = \frac{k_{\parallel}(y + 0.4)^2 + k_{\perp}(x - 0.1)^2}{(x - 0.1)^2 + (y + 0.4)^2},$$

$$K_{xy} = \frac{-(k_{\parallel} - k_{\perp})(x - 0.1)(y + 0.4)}{(x - 0.1)^2 + (y + 0.4)^2},$$

$$K_{yy} = \frac{k_{\parallel}(x - 0.1)^2 + k_{\perp}(y + 0.4)^2}{(x - 0.1)^2 + (y + 0.4)^2}.$$

The flow field in the domain is shown in Fig. 3.15. The magnitude of velocity in Ω_2 is small due to low values of the permeability tensor in this region. The velocity vectors bend in the direction of the permeability streak Ω_2. The L^2-norm of $\nabla \cdot \boldsymbol{u}$ over the entire domain as a function of polynomial degree, N, is shown in Fig. 3.16.

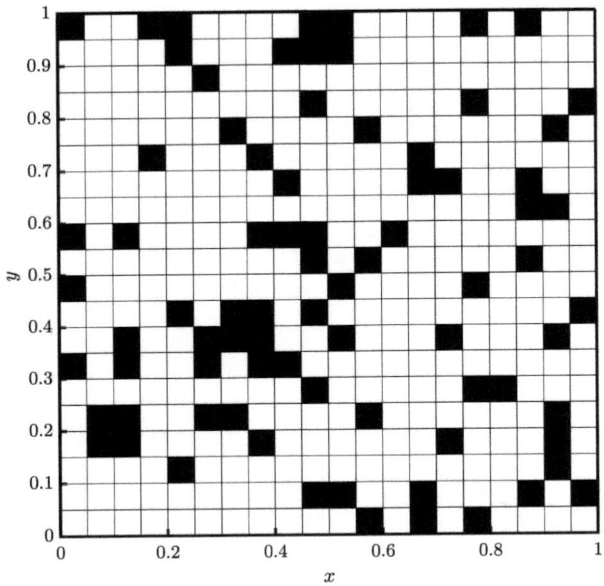

Fig. 3.10 The discretized domain for the sand-shale test case. *Black* blocks are shale blocks with $k = 10^{-6}$. *White* blocks are sand blocks with $k = 1$

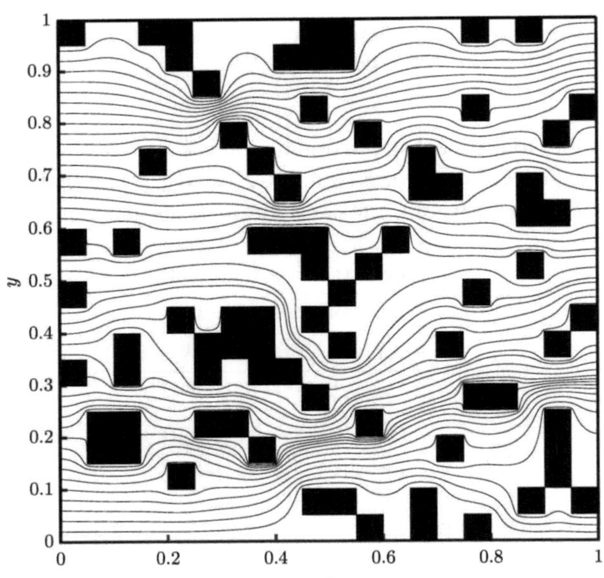

Fig. 3.11 Streamlines through the domain of sand-shale test case

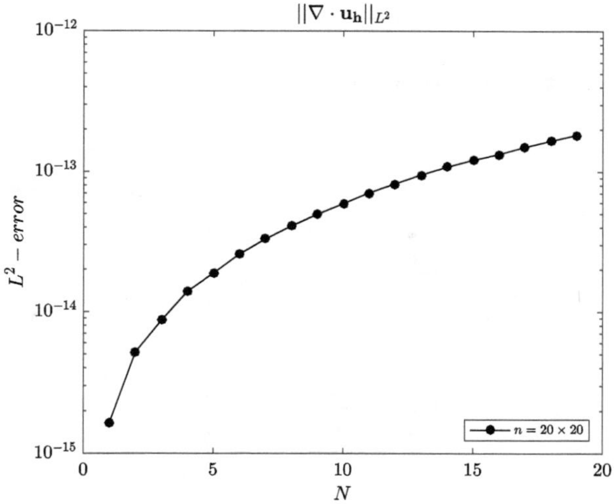

Fig. 3.12 The L^2-norm of $\nabla \cdot \boldsymbol{u}_h$ for 20×20 elements for a polynomial approximation of $N = 1, \ldots, 19$

Table 3.1 Net flux through the left boundary of the sand-shale domain for $k = 10^{-6}$, 20×20 elements, $N = 1, \ldots, 19$

N	Net flux	No. of unknowns
1	0.49041	1240
2	0.51247	4880
3	0.51744	10,920
4	0.51863	19,360
5	0.51931	30,200
6	0.51957	43,440
7	0.51977	59,080
8	0.51985	77,120
9	0.51993	97,560
10	0.51997	120,400
11	0.52001	145,640
12	0.52003	173,280
13	0.52005	203,320
14	0.52007	235,760
15	0.52008	270,600
16	0.52009	307,840
17	0.52009	347,480
18	0.52010	389,520
19	0.52010	433,960

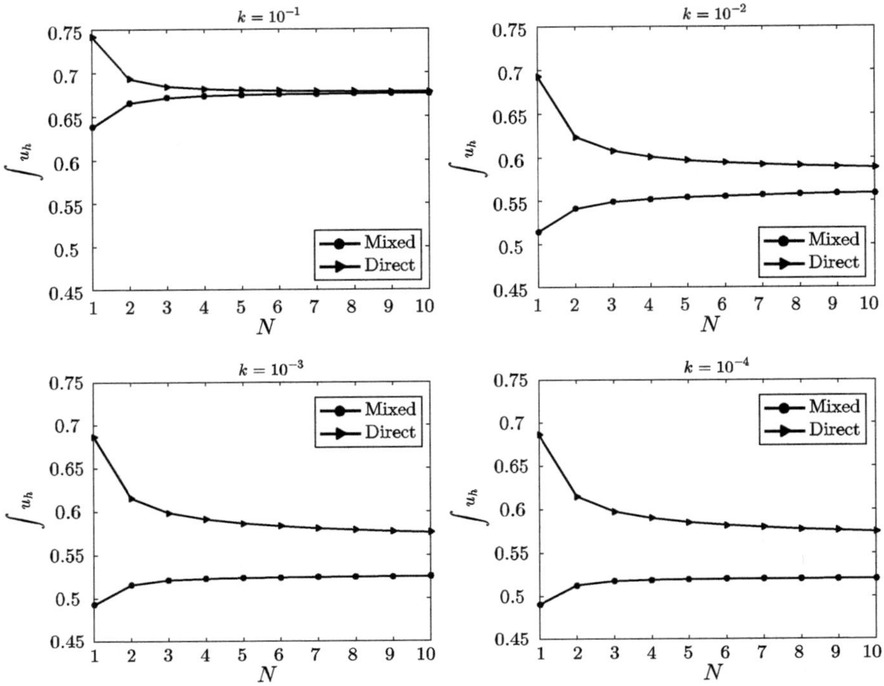

Fig. 3.13 Convergence of the net flux through the left boundary of the sand-shale domain using the mixed formulation and the direct formulation for 20×20 elements, $N = 1, \ldots, 10$. *Top left:* $k = 10^{-1}$. *Top right:* $k = 10^{-2}$. *Bottom left:* $k = 10^{-3}$. *Bottom right:* $k = 10^{-4}$

We can see that the flow field is divergence free up to machine precision because $f = 0$.

The net flux through the system for varying number of elements and polynomial degree is given in Table 3.3. In this work the finest resolution corresponds to $12 \times 12 \times 3$ elements and $N = 15$. For this case the net influx at the left boundary is 0.75668. The net influx and outflux from the region Ω_1, Ω_2 and Ω_3 is given in Tables 3.4, 3.5, and 3.6, respectively. The net influx for Ω_1 is larger than the net outflux. And the net outflux for Ω_2 and Ω_3 is larger than the net influx.

Table 3.2 Data of net flux through the left boundary of the sand-shale domain using mixed formulation and direct formulation for 20 × 20 elements, $N = 1, \ldots, 10$, $k = 10^{-1}$ (top-left), 10^{-2} (top-right), 10^{-3} (bottom-left) and 10^{-4} (bottom-right)

	$k = 10^{-1}$		$k = 10^{-2}$		$k = 10^{-3}$		$k = 10^{-4}$	
N	Mixed	Direct	Mixed	Direct	Mixed	Direct	Mixed	Direct
1	0.63805	0.74149	0.51384	0.69273	0.49296	0.68699	0.49066	0.68641
2	0.66541	0.69316	0.54101	0.62399	0.51573	0.61572	0.51279	0.61488
3	0.67131	0.68423	0.54906	0.60794	0.52121	0.59856	0.51782	0.59760
4	0.67339	0.68139	0.55208	0.60113	0.52272	0.59099	0.51904	0.58995
5	0.67450	0.68003	0.55436	0.59711	0.52371	0.58639	0.51975	0.58528
6	0.67512	0.67926	0.55568	0.59439	0.52417	0.58320	0.52003	0.58203
7	0.67555	0.67877	0.55690	0.59239	0.52459	0.58079	0.52026	0.57958
8	0.67582	0.67844	0.55772	0.59085	0.52483	0.57890	0.52036	0.57765
9	0.67604	0.67821	0.55852	0.58960	0.52508	0.57734	0.52046	0.57605
10	0.67619	0.67803	0.55910	0.58857	0.52524	0.57603	0.52051	0.57471

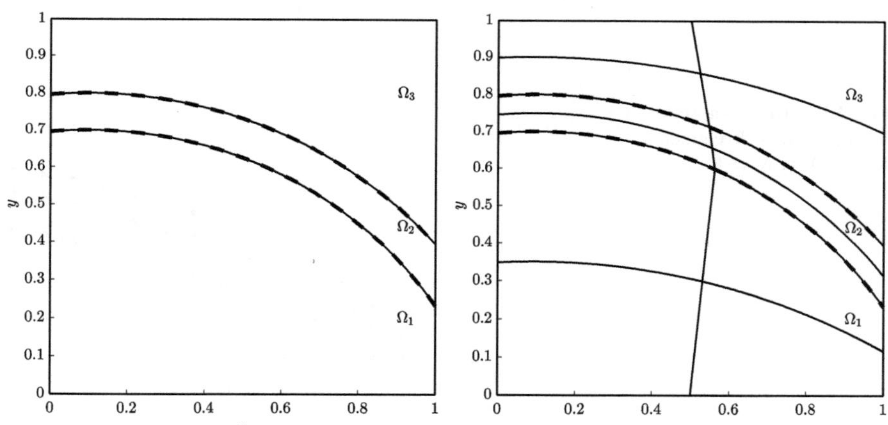

Fig. 3.14 Three regions of the domain for the impermeable streak test case. The regions are separated by the dashed lines. The solid lines indicate the element boundaries. *Left:* 1 × 1 element in each region. *Right:* 2 × 2 elements in each region

3.6 Future Work

In the above sections, mixed and direct formulations of mimetic spectral element method are discussed. The next step is to explore this framework in the direction of hybrid formulations [21, 32, 43]. Additionally, the focus will be on developing multiscale methods [118], using these formulations, for reservoir modelling applications.

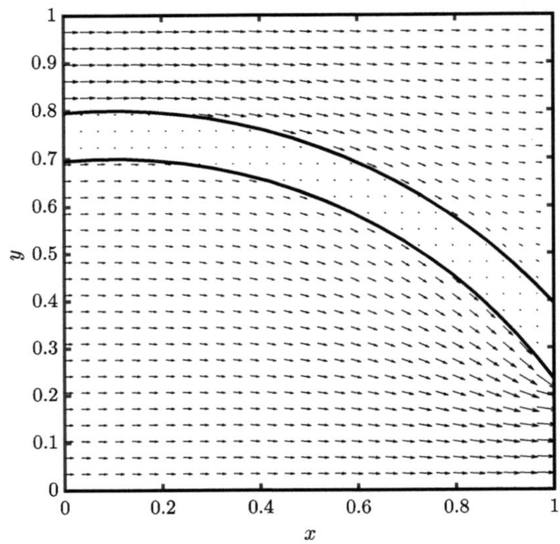

Fig. 3.15 Velocity vectors through the domain of permeability streak test case for 12×12 elements, $N = 15$

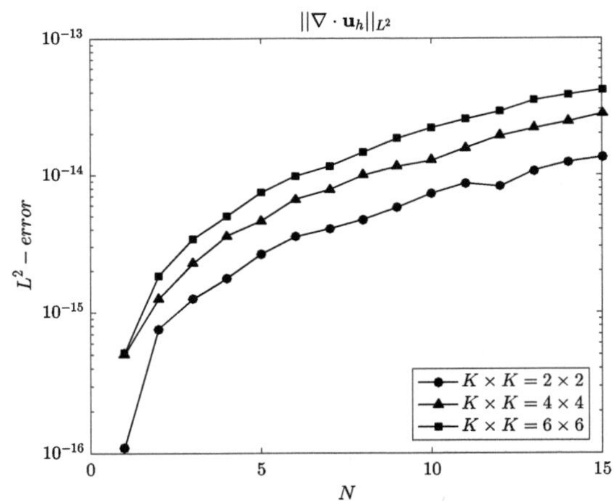

Fig. 3.16 The L^2-norm of $\nabla \cdot \boldsymbol{u}_h$ for $K \times K$ elements, $K = 2, 4, 6$, $N = 1, \ldots, 15$

Table 3.3 Net flux through the left boundary of the permeability streak test case domain for $K \times K$ elements, $K = 4, 6, 8, 10, 12$ and $N = 1, \ldots, 15$

	Elements division ($K \times K$)				
N	4×4	6×6	8×8	10×10	12×12
1	0.74689	0.74908	0.75061	0.75169	0.75247
2	0.75268	0.75407	0.75479	0.75522	0.75550
3	0.75479	0.75548	0.75582	0.75602	0.75615
4	0.75561	0.75600	0.75620	0.75631	0.75639
5	0.75600	0.75625	0.75638	0.75645	0.75650
6	0.75621	0.75639	0.75648	0.75653	0.75657
7	0.75635	0.75648	0.75654	0.75658	0.75660
8	0.75643	0.75653	0.75658	0.75661	0.75663
9	0.75649	0.75657	0.75661	0.75663	0.75665
10	0.75654	0.75660	0.75663	0.75665	0.75666
11	0.75657	0.75662	0.75664	0.75666	0.75667
12	0.75659	0.75663	0.75665	0.75666	0.75667
13	0.75661	0.75664	0.75666	0.75667	0.75668
14	0.75662	0.75665	0.75667	0.75668	0.75668
15	0.75663	0.75666	0.75667	0.75668	0.75668

Table 3.4 Net flux through the left boundary of the region Ω_1 for $K \times K$ elements, $K = 4, 6, 8, 10, 12$ and $N = 1, \ldots, 15$

	Elements division ($K \times K$)									
	4×4		6×6		8×8		10×10		12×12	
N	In flux	Out flux	In flux	Out flux	In flux	Out flux	In flux	Out flux	In flux	Out flux
1	0.47155	0.46497	0.47342	0.46657	0.47483	0.46786	0.47585	0.46881	0.47659	0.46952
2	0.47674	0.46965	0.47812	0.47098	0.47883	0.47168	0.47926	0.47211	0.47954	0.47239
3	0.47883	0.47168	0.47952	0.47236	0.47986	0.47270	0.48005	0.47291	0.48018	0.47304
4	0.47964	0.47249	0.48004	0.47289	0.48023	0.47309	0.48035	0.47321	0.48042	0.47329
5	0.48003	0.47288	0.48029	0.47315	0.48041	0.47328	0.48049	0.47336	0.48053	0.47341
6	0.48025	0.47311	0.48043	0.47330	0.48051	0.47339	0.48056	0.47345	0.48060	0.47348
7	0.48038	0.47325	0.48051	0.47339	0.48057	0.47346	0.48061	0.47350	0.48063	0.47353
8	0.48047	0.47334	0.48057	0.47345	0.48061	0.47350	0.48064	0.47354	0.48066	0.47356
9	0.48053	0.47340	0.48060	0.47349	0.48064	0.47354	0.48066	0.47356	0.48068	0.47358
10	0.48057	0.47345	0.48063	0.47352	0.48066	0.47356	0.48068	0.47358	0.48069	0.47360
11	0.48060	0.47349	0.48065	0.47355	0.48067	0.47358	0.48069	0.47360	0.48070	0.47361
12	0.48062	0.47352	0.48066	0.47357	0.48068	0.47359	0.48070	0.47361	0.48070	0.47362
13	0.48064	0.47354	0.48067	0.47358	0.48069	0.47360	0.48070	0.47362	0.48071	0.47363
14	0.48065	0.47355	0.48068	0.47359	0.48070	0.47361	0.48071	0.47362	0.48071	0.47363
15	0.48067	0.47357	0.48069	0.47360	0.48070	0.47362	0.48071	0.47363	0.48071	0.47364

Table 3.5 Net flux through the left boundary of the region Ω_2 for $K \times K$ elements, $K = 4, 6, 8, 10, 12$ and $N = 1, \ldots, 15$

	Elements division ($K \times K$)									
	4×4		6×6		8×8		10×10		12×12	
N	In flux	Out flux	In flux	Out flux	In flux	Out flux	In flux	Out flux	In flux	Out flux
1	0.00930	0.01080	0.00931	0.01106	0.00931	0.01119	0.00932	0.01130	0.00932	0.01130
2	0.00932	0.01132	0.00933	0.01138	0.00933	0.01139	0.00933	0.01140	0.00933	0.01140
3	0.00933	0.01139	0.00933	0.01140	0.00933	0.01140	0.00933	0.01140	0.00933	0.01139
4	0.00933	0.01140	0.00933	0.01140	0.00934	0.01140	0.00934	0.01139	0.00934	0.01139
5	0.00933	0.01140	0.00934	0.01139	0.00934	0.01139	0.00934	0.01138	0.00934	0.01138
6	0.00934	0.01139	0.00934	0.01139	0.00934	0.01138	0.00934	0.01137	0.00934	0.01137
7	0.00934	0.01139	0.00934	0.01138	0.00934	0.01137	0.00934	0.01137	0.00934	0.01136
8	0.00934	0.01138	0.00934	0.01137	0.00934	0.01137	0.00934	0.01136	0.00934	0.01136
9	0.00934	0.01138	0.00934	0.01137	0.00934	0.01136	0.00934	0.01136	0.00934	0.01135
10	0.00934	0.01137	0.00934	0.01136	0.00934	0.01136	0.00934	0.01135	0.00934	0.01135
11	0.00934	0.01137	0.00934	0.01136	0.00934	0.01135	0.00934	0.01135	0.00934	0.01135
12	0.00934	0.01136	0.00934	0.01136	0.00934	0.01135	0.00934	0.01135	0.00934	0.01134
13	0.00934	0.01136	0.00934	0.01135	0.00934	0.01135	0.00934	0.01134	0.00934	0.01134
14	0.00934	0.01136	0.00934	0.01135	0.00934	0.01134	0.00934	0.01134	0.00934	0.01134
15	0.00934	0.01135	0.00934	0.01135	0.00934	0.01134	0.00934	0.01134	0.00934	0.01134

Table 3.6 Net flux through the left boundary of the region Ω_3 for $K \times K$ elements, $K = 4, 6, 8, 10, 12$ and $N = 1, \ldots, 15$

	Elements division ($K \times K$)									
	4×4		6×6		8×8		10×10		12×12	
N	In flux	Out flux	In flux	Out flux	In flux	Out flux	In flux	Out flux	In flux	Out flux
1	0.26604	0.27112	0.26636	0.27146	0.26647	0.27157	0.26653	0.27163	0.26656	0.27166
2	0.26662	0.27172	0.26663	0.27172	0.26663	0.27172	0.26663	0.27172	0.26663	0.27172
3	0.26663	0.27172	0.26663	0.27172	0.26664	0.27172	0.26664	0.27172	0.26664	0.27172
4–15	0.26664	0.27172	0.26664	0.27172	0.26664	0.27172	0.26664	0.27172	0.26664	0.27172

References

1. Aarnes, J.E., Krogstad, S., Lie, K.A.: Multiscale mixed/mimetic methods on corner-point grids. Comput. Geosci. **12**, 297–315 (2008). https://doi.org/10.1007/s10596-007-9072-8
2. Aavatsmark, I.: An introduction to multipoint flux approximations for quadrilateral grids. Comput. Geosci. **6**, 405–432 (2002). https://doi.org/10.1023/A:1021291114475
3. Aavatsmark, I.: Interpretation of a two-point flux stencil for skew parallelogram grids. Comput. Geosci. **11**, 199–206 (2007). https://doi.org/10.1007/s10596-007-9042-1
4. Aavatsmark, I., Barkve, T., Bøe, O., Mannseth, T.: Discretization on unstructured grids for inhomogeneous, anisotropic media. Part I: derivation of the methods. SIAM J. Sci. Comput. **19**, 1700–1716 (1998). https://doi.org/10.1137/S1064827595293582
5. Aavatsmark, I., Barkve, T., Bøe, O., Mannseth, T.: Discretization on unstructured grids for inhomogeneous, anisotropic media. Part II: discussion and numerical results. SIAM J. Sci. Comput. **19**, 1717–1736 (1998). https://doi.org/10.1137/S1064827595293594
6. Alpak, F.O.: A mimetic finite volume discretization operator for reservoir simulation. In: SPE Reservoir Simulation Symposium. Society of Petroleum Engineers, Richardson (2007). https://doi.org/10.2118/106445-MS
7. Alpak, F.O.: A mimetic finite volume discretization method for reservoir simulation. SPE J. **15**, 436–453 (2010). https://doi.org/10.2118/106445-PA
8. Arnold, D.N., Boffi, D., Falk, R.S.: Quadrilateral H(div) finite elements. SIAM J. Numer. Anal. **42**, 2429–2451 (2005). https://doi.org/10.1137/S0036142903431924
9. Arnold, D.N., Falk, R.S., Winther, R.: Finite element exterior calculus, homological techniques, and applications. Acta Numer. **15**, 1–155 (2006)
10. Arnold, D.N., Falk, R.S., Winther, R.: Finite element exterior calculus: from Hodge theory to numerical stability. Bull. Am. Math. Soc. **47**, 281–354 (2010)
11. Aziz, K.: Reservoir simulation grids: opportunities and problems. J. Pet. Technol. **45**, 658–663 (1993). https://doi.org/10.2118/25233-PA
12. Babuska, I., Suri, M.: On locking and robustness in the finite element method. SIAM J. Numer. Anal. **29**, 1261–1293 (1992). https://doi.org/10.1137/0729075
13. Bastian, P., Ippisch, O., Marnach, S.: Benchmark 3D: a mimetic finite difference method. In: Finite Volumes for Complex Applications VI: Problems and Perspectives, pp. 961–968. Springer, Berlin (2011). https://doi.org/10.1007/978-3-642-20671-9_93
14. Bauer, W., Gay-Balmaz, F.: Variational integrators for an elastic and pseudo-incompressible flows (2017). http://arxiv.org/abs/1701.06448. ArXiv preprint n.1701.06448
15. Bergman, T.L., Incropera, F.P.: Fundamentals of Heat and Mass Transfer. Wiley, Hoboken (2011)
16. Bochev, P.B., Gerritsma, M.: A spectral mimetic least-squares method. Comput. Math. Appl. **68**, 1480–1502 (2014). https://doi.org/10.1016/j.camwa.2014.09.014
17. Bochev, P.B., Gunzburger, M.D.: Least-Squares Finite Element Methods. Springer Series in Applied Mathematical Sciences. Springer, New York (2009)
18. Bochev, P.B., Hyman, J.M.: Principles of mimetic discretizations of differential operators. IMA Vol. Math. Appl. **142**, 89 (2006)
19. Bochev, P.B., Ridzal, D.: Rehabilitation of the lowest-order Raviart–Thomas element on quadrilateral grids. SIAM J. Numer. Anal. **47**, 487–507 (2008). https://doi.org/10.1137/070704265
20. Boffi, D., Gastaldi, L.: Some remarks on quadrilateral mixed finite elements. Comput. Struct. **87**, 751–757 (2009)
21. Boffi, D., Brezzi, F., Fortin, M.: Mixed Finite Element Methods and Applications. Springer Series in Computational Mathematics. Springer, Berlin (2013)
22. Bonelle, J., Ern, A.: Analysis of compatible discrete operator schemes for elliptic problems on polyhedral meshes. ESAIM: Math. Model. Numer. Anal. **48**, 553–581 (2014). https://doi.org/10.1051/m2an/2013104

23. Bonelle, J., Ern, A.: Analysis of compatible discrete operator schemes for the Stokes equations on polyhedral meshes. IMA J. Numer. Anal. **35**, 1672–1697 (2015). https://doi.org/10.1093/imanum/dru051
24. Bonelle, J., Di Pietro, D.A., Ern, A.: Low-order reconstruction operators on polyhedral meshes: application to compatible discrete operator schemes. Comput. Aided Geom. Des. **35–36**, 27–41 (2015). https://doi.org/10.1016/j.cagd.2015.03.015
25. Bossavit, A.: Computational electromagnetism and geometry: (1) network equations. J. Jpn. Soc. Appl. Electromagn. **7**, 150–159 (1999)
26. Bossavit, A.: Computational electromagnetism and geometry: (2) network constitutive laws. J. Jpn. Soc. Appl. Electromagn. **7**, 294–301 (1999)
27. Bossavit, A.: Computational electromagnetism and geometry: (3) convergence. J. Jpn. Soc. Appl. Electromagn. **7**, 401–408 (1999)
28. Bossavit, A.: Computational electromagnetism and geometry: (4) from degrees of freedom to fields. J. Jpn. Soc. Appl. Electromagn. **8**, 102–109 (2000)
29. Bossavit, A.: Computational electromagnetism and geometry: (5) the "Galerkin Hodge". J. Jpn. Soc. Appl. Electromagn. **8**, 203–209 (2000)
30. Bouman, M., Palha, A., Kreeft, J., Gerritsma, M.: A conservative spectral element method for curvilinear domains. In: Spectral and High Order Methods for Partial Differential Equations. Lecture Notes in Computational Science and Engineering, vol. 76, pp. 111–119. Springer, Berlin (2011). https://doi.org/10.1007/978-3-642-15337-2_8
31. Brezzi, F., Buffa, A.: Innovative mimetic discretizations for electromagnetic problems. J. Comput. Appl. Math. **234**, 1980–1987 (2010)
32. Brezzi, F., Fortin, M.: Mixed and Hybrid Finite Element Methods. Springer Series in Computational Mathematics, vol. 15. Springer, New York (1991)
33. Brezzi, F., Lipnikov, K., Simoncini, V.: A family of mimetic finite difference methods on polygonal and polyhedral meshes. Math. Models Methods Appl. Sci. **15**, 1533–1551 (2005). https://doi.org/10.1142/S0218202505000832
34. Brezzi, F., Lipnikov, K., Shashkov, M.: Convergence of mimetic finite difference method for diffusion problems on polyhedral meshes with curved faces. Math. Models Methods Appl. Sci. **16**, 275–297 (2006). https://doi.org/10.1142/S0218202506001157
35. Brezzi, F., Lipnikov, K., Shashkov, M., Simoncini, V.: A new discretization methodology for diffusion problems on generalized polyhedral meshes. Comput. Methods Appl. Mech. Eng. **196**, 3682–3692 (2007). https://doi.org/10.1016/j.cma.2006.10.028
36. Brezzi, F., Buffa, A., Lipnikov, K.: Mimetic finite differences for elliptic problems. Math. Model. Numer. Anal. **43**, 277–296 (2009)
37. Brezzi, F., Falk, R.S., Donatella Marini, L.: Basic principles of mixed virtual element methods. ESAIM: Math. Model. Numer. Anal. **48**, 1227–1240 (2014). https://doi.org/10.1051/m2an/2013138
38. Budd, C., Piggott, M.: Geometric Integration and its Applications. In: Handbook of Numerical Analysis, vol. 11, pp. 35–139. North-Holland, Amsterdam (2003). https://doi.org/10.1016/S1570-8659(02)11002-7
39. Buffa, A., de Falco, C., Sangalli, G.: Isogeometric analysis: stable elements for the 2D Stokes equation. Int. J. Numer. Methods Fluids **65**, 1407–1422 (2011). https://doi.org/10.1002/fld.2337
40. Cantin, P., Bonelle, J., Burman, E., Ern, A.: A vertex-based scheme on polyhedral meshes for advection-reaction equations with sub-mesh stabilization. Comput. Math. Appl. **72**(9), 2057–2071 (2016)
41. Canuto, C., Hussaini, M.Y., Quarteroni, A., Zang, T.A.: Spectral Methods in Fluid Dynamics. Springer, Berlin (1988)
42. Christiansen, S.H., Munthe-Kaas, H.Z., Owren, B.: Topics in structure-preserving discretization. Acta Numer. **20**, 1–119 (2011). https://doi.org/10.1017/S096249291100002X
43. Cockburn, B.: Static Condensation, Hybridization, and the Devising of the HDG Methods, pp. 129–177. Springer, Cham (2016). https://doi.org/10.1007/978-3-319-41640-3_5

44. da Veiga, L.B., Brezzi, F., Marini, L.D., Russo, A.: The Hitchhiker's guide to the virtual element method. Math. Models Methods Appl. Sci. **24**, 1541–1573 (2014). https://doi.org/10.1142/S021820251440003X
45. da Veiga, L.B., Lipnikov, K., Manzini, G.: The Mimetic Finite Difference Method for Elliptic Problems. Springer, Basel (2014). https://doi.org/10.1007/978-3-319-02663-3
46. da Veiga, L.B., Lovadina, C., Vacca, G.: Divergence free virtual elements for the Stokes problem on polygonal meshes (2015). arXiv:1510.01655v1
47. da Veiga, L.B., Brezzi, F., Marini, L.D., Russo, A.: $H(\mathrm{div})$ and $H(\mathrm{curl})$ conforming virtual element methods. Numer. Math., 1–30 (2015). https://doi.org/10.1007/s00211-015-0746-1
48. da Veiga, L.B., Brezzi, F., Marini, L.D., Russo, A.: Virtual element method for general second-order elliptic problems on polygonal meshes. Math. Models Methods Appl. Sci. **26**, 729–750 (2016). https://doi.org/10.1142/S0218202516500160
49. Desbrun, M., Hirani, A.N., Leok, M., Marsden, J.E.: Discrete exterior calculus (2005). arXiv:math/0508341v2
50. Di Pietro, D.A., Ern, A.: Hybrid high-order methods for variable-diffusion problems on general meshes. C.R. Math. **353**, 31–34 (2015). https://doi.org/10.1016/j.crma.2014.10.013
51. Di Pietro, D.A., Ern, A., Lemaire, S.: An arbitrary-order and compact-stencil discretization of diffusion on general meshes based on local reconstruction operators. Comput. Methods Appl. Math. **14** (2014). https://doi.org/10.1515/cmam-2014-0018
52. Dodziuk, J.: Finite difference approach to the Hodge theory of harmonic functions. Am. J. Math. **98**, 79–104 (1976)
53. Durlofsky, L.J.: A triangle based mixed finite element finite volume technique for modeling two phase flow through porous media. J. Comput. Phys. **105**, 252–266 (1993). https://doi.org/10.1006/jcph.1993.1072
54. Durlofsky, L.J.: Accuracy of mixed and control volume finite element approximations to Darcy velocity and related quantities. Water Resour. Res. **30**, 965–973 (1994). https://doi.org/10.1029/94WR00061
55. Dziubek, A., Guidoboni, G., Harris, A., Hirani, A.N., Rusjan, E., Thistleton, W.: Effect of ocular shape and vascular geometry on retinal hemodynamics: a computational model. Biomech. Model. Mechanobiol. **15**, 893–907 (2016). https://doi.org/10.1007/s10237-015-0731-8
56. Edwards, M.G.: Unstructured, control-volume distributed, full-tensor finite-volume schemes with flow based grids. Comput. Geosci. **6**, 433–452 (2002). https://doi.org/10.1023/A:1021243231313
57. Edwards, M.G., Rogers, C.F.: Finite volume discretization with imposed flux continuity for the general tensor pressure equation. Comput. Geosci. **2**, 259–290 (1998). https://doi.org/10.1023/A:1011510505406
58. Elcott, S., Tong, Y., Kanso, E., Schröder, P., Desbrun, M.: Stable, circulation-preserving, simplicial fluids. ACM Trans. Graph. **26**(1), 4 (2007). https://doi.org/10.1145/1189762.1189766
59. Evans, J.A., Hughes, T.J.: Isogeometric divergence-conforming B-splines for the unsteady Navier-Stokes equations. J. Comput. Phys. **241**, 141–167 (2013). https://doi.org/10.1016/j.jcp.2013.01.006
60. Forsyth, P.A.: A control-volume, finite-element method for local mesh refinement in thermal reservoir simulation. SPE Reserv. Eng. **5**, 561–566 (1990). https://doi.org/10.2118/18415-PA
61. Gerritsma, M.: Edge functions for spectral element methods. In: Spectral and High Order Methods for Partial Differential Equations. Lecture Notes in Computational Science and Engineering, vol. 76, pp. 199–207. Springer, Berlin (2011). https://doi.org/10.1007/978-3-642-15337-2_17
62. Gerritsma, M., Bochev, P.B.: A spectral mimetic least-squares method for the Stokes equations with no-slip boundary condition. Comput. Math. Appl. **71**, 2285–2300 (2016). https://doi.org/10.1016/j.camwa.2016.01.033
63. Gerritsma, M., Bouman, M., Palha, A.: Least-squares spectral element method on a staggered grid. In: Large-Scale Scientific Computing. Lecture Notes in Computer Science, vol. 5910, pp. 653–661. Springer, Berlin (2010)

64. Gerritsma, M., Hiemstra, R., Kreeft, J., Palha, A., Rebelo, P.P., Toshniwal, D.: The geometric basis of numerical methods. In: Spectral and High Order Methods for Partial Differential Equations. Lecture Notes in Computational Science and Engineering, vol. 95, pp. 17–35. Springer, Cham (2013)
65. Gunasekera, D., Cox, J., Lindsey, P.: The generation and application of K-orthogonal grid systems. In: SPE Reservoir Simulation Symposium, pp. 199–214. Society of Petroleum Engineers, Richardson (1997). https://doi.org/10.2118/37998-MS
66. Hairer, E., Lubich, C., Wanner, G.: Geometric Numerical Integration. Springer, Berlin (2006)
67. Heinemann, Z.E., Brand, C.W., Munka, M., Chen, Y.M.: Modeling reservoir geometry with irregular grids. SPE Reserv. Eng. **6**, 225–232 (1991). https://doi.org/10.2118/18412-PA
68. Herbin, R., Hubert, F.: Benchmark on discretization schemes for anisotropic diffusion problems on general grids. In: Finite Volumes for Complex Applications V: Problems and Perspectives, pp. 659–692. Wiley, Hoboken (2008)
69. Hermeline, F.: A finite volume method for the approximation of diffusion operators on distorted meshes. J. Comput. Phys. **160**, 481–499 (2000). https://doi.org/10.1006/jcph.2000.6466
70. Hiemstra, R., Toshniwal, D., Huijsmans, R., Gerritsma, M.: High order geometric methods with exact conservation properties. J. Comput. Phys. **257**, 1444–1471 (2014). https://doi.org/10.1016/j.jcp.2013.09.027
71. Hiptmair, R.: PIER. In: Geometric Methods for Computational Electromagnetics, vol. 42, pp. 271–299. EMW Publishing, Cambridge (2001)
72. Hirani, A.: Discrete exterior calculus. Ph.D. thesis, California Institute of Technology (2003)
73. Hirani, A.N., Nakshatrala, K.B., Chaudhry, J.H.: Numerical method for Darcy flow derived using discrete exterior calculus. Int. J. Comput. Methods Eng. Sci. Mech. **16**, 151–169 (2015). https://doi.org/10.1080/15502287.2014.977500
74. Hyman, J.M., Scovel, J.C.: Deriving mimetic difference approximations to differential operators using algebraic topology. Technical report, Los Alamos National Laboratory (1990)
75. Hyman, J.M., Steinberg, S.: The convergence of mimetic methods for rough grids. Comput. Math. Appl. **47**, 1565–1610 (2004)
76. Hyman, J.M., Shashkov, M., Steinberg, S.: The numerical solution of diffusion problems in strongly heterogeneous non-isotropic materials. J. Comput. Phys. **132**, 130–148 (1997)
77. Hyman, J.M., Morel, J., Shashkov, M., Steinberg, S.: Mimetic finite difference methods for diffusion equations. Comput. Geosci. **6**, 333–352 (2002)
78. Kikinzon, E., Kuznetsov, Y., Lipnikov, K., Shashkov, M.: Approximate static condensation algorithm for solving multi-material diffusion problems on meshes non-aligned with material interfaces. J. Comput. Phys. (2017). https://doi.org/10.1016/j.jcp.2017.06.048
79. Kouranbaeva, S., Shkoller, S.: A variational approach to second-order multisymplectic field theory. J. Geom. Phys. **35**, 333–366 (2000). https://doi.org/10.1016/S0393-0440(00)00012-7
80. Kraus, M., Maj, O.: Variational integrators for nonvariational partial differential equations. Physica D **310**, 37–71 (2015). https://doi.org/10.1016/j.physd.2015.08.002
81. Kreeft, J., Gerritsma, M.: Mixed mimetic spectral element method for Stokes flow: a pointwise divergence-free solution. J. Comput. Phys. **240**, 284–309 (2013). https://doi.org/10.1016/j.jcp.2012.10.043
82. Kreeft, J., Palha, A., Gerritsma, M.: Mimetic framework on curvilinear quadrilaterals of arbitrary order, p. 69. arXiv:1111.4304 (2011)
83. Lie, K., Krogstad, S., Ligaarden, I.S., Natvig, J.R., Nilsen, H.M., Skaflestad, B.: Open-source MATLAB implementation of consistent discretisations on complex grids. Comput. Geosci. **16**, 297–322 (2012). https://doi.org/10.1007/s10596-011-9244-4
84. Manzini, G., Putti, M.: Mesh locking effects in the finite volume solution of 2-D anisotropic diffusion equations. J. Comput. Phys. **220**, 751–771 (2007). https://doi.org/10.1016/j.jcp.2006.05.026
85. Marsden, J.E., West, M.: Discrete mechanics and variational integrators. Acta Numer. **10**, 357–514 (2001). https://doi.org/10.1017/S096249290100006X, published online:2003

86. Mullen, P., Crane, K., Pavlov, D., Tong, Y., Desbrun, M.: Energy-preserving integrators for fluid animation. ACM Trans. Graph. **28**(3), 38 (2009). https://doi.org/10.1145/1531326.1531344
87. Neuman, S.P.: Theoretical derivation of Darcy's law. Acta Mech. **25**, 153–170 (1977). https://doi.org/10.1007/BF01376989
88. Nicolaides, R.: Discrete discretization of planar div-curl problems. SIAM J. Numer. Anal. **29**, 32–56 (1992)
89. Nilsen, H.M., Natvig, J.R., Lie, K.A.: Accurate modeling of faults by multipoint, mimetic, and mixed methods. SPE J., 568–579 (2012). https://doi.org/10.2118/149690-pa
90. Palagi, C.L., Aziz, K.: Use of Voronoi grid in reservoir simulation. SPE Adv. Technol. Ser. **2**, 69–77 (1994). https://doi.org/10.2118/22889-PA
91. Palha, A., Gerritsma, M.: Mimetic least-squares spectral/*hp* finite element method for the Poisson equation. In: Large-Scale Scientific Computing. Lecture Notes in Computer Science, vol. 5910, pp. 662–670. Springer, Berlin (2010)
92. Palha, A., Gerritsma, M.: Spectral element approximation of the Hodge-$\star$ operator in curved elements. In: Spectral and High Order Methods for Partial Differential Equations. Lecture Notes in Computational Science and Engineering, vol. 76, pp. 283–291. Springer, Berlin (2010)
93. Palha, A., Gerritsma, M.: A mass, energy, enstrophy and vorticity conserving (MEEVC) mimetic spectral element discretization for the 2D incompressible Navier-Stokes equations. J. Comput. Phys. **328**, 200–220 (2017). https://doi.org/10.1016/j.jcp.2016.10.009
94. Palha, A., Rebelo, P.P., Hiemstra, R., Kreeft, J., Gerritsma, M.: Physics-compatible discretization techniques on single and dual grids, with application to the Poisson equation of volume forms. J. Comput. Phys. **257**, 1394–1422 (2014). https://doi.org/10.1016/j.jcp.2013.08.005
95. Palha, A., Rebelo, P.P., Gerritsma, M.: Mimetic spectral element advection. In: Spectral and High Order Methods for Partial Differential Equations - ICOSAHOM 2012. Lecture Notes in Computational Science and Engineering, vol. 95, pp. 325–335. Springer, Cham (2014). https://doi.org/10.1007/978-3-319-01601-6
96. Palha, A., Koren, B., Felici, F.: A mimetic spectral element solver for the Grad–Shafranov equation. J. Comput. Phys. **316**, 63–93 (2016). https://doi.org/10.1016/j.jcp.2016.04.002
97. Pavlov, D., Mullen, P., Tong, Y., Kanso, E., Marsden, J.E., Desbrun, M.: Structure preserving discretization of incompressible fluids. Physica D **240**, 443–458 (2011)
98. Perona, P., Malik, J.: Scale-space and edge detection using anisotropic diffusion. IEEE Trans. Pattern Anal. Mach. Intell. **12**, 629–639 (1990). https://doi.org/10.1109/34.56205
99. Perot, J.B.: Conservation properties of unstructured staggered mesh schemes. J. Comput. Phys. **159**, 58–89 (2000)
100. Perot, J.B.: Discrete conservation properties of unstructured mesh schemes. Annu. Rev. Fluid Mech. **43**, 299–318 (2011)
101. Perot, J.B., Subramanian, V.: A discrete calculus analysis of the Keller Box scheme and a generalization of the method to arbitrary meshes. J. Comput. Phys. **226**, 494–508 (2007)
102. Perot, J.B., Subramanian, V.: Discrete calculus methods for diffusion. J. Comput. Phys. **224**, 59–81 (2007)
103. Perot, J.B., Vidovic, D., Wesseling, P.: Mimetic reconstruction of vectors. IMA Vol. Math. Appl. **142**, 173 (2006)
104. Rapetti, F.: High order edge elements on simplicial meshes. ESAIM Math. Model. Numer. Anal. **41**, 1001–1020 (2007)
105. Rapetti, F.: Whitney forms of higher order. SIAM J. Numer. Anal. **47**, 2369–2386 (2009)
106. Rebelo, P.P., Palha, A., Gerritsma, M.: Mixed mimetic spectral element method applied to Darcy's problem. In: Spectral and High Order Methods for Partial Differential Equations - ICOSAHOM 2012. Lecture Notes in Computational Science and Engineering, vol. 95, pp. 373–382. Springer, Cham (2014). https://doi.org/10.1007/978-3-319-01601-6__30
107. Robidoux, N.: A new method of construction of adjoint gradients and divergences on logically rectangular smooth grids. In: Finite Volumes for Complex Applications: Problems and Perspectives, pp. 261–272. Éditions Hermès, Rouen (1996)

108. Robidoux, N.: Numerical solution of the steady diffusion equation with discontinuous coefficients. Ph.D. thesis, University of New Mexico, Albuquerque (2002)
109. Robidoux, N.: Polynomial histopolation, superconvergent degrees of freedom, and pseudospectral discrete Hodge operators (2008). Unpublished: http://people.math.sfu.ca/~nrobidou/public_html/prints/histogram/histogram.pdf
110. Robidoux, N., Steinberg, S.: A discrete vector calculus in tensor grids. Comput. Methods Appl. Math. **11**, 23–66 (2011). https://doi.org/10.2478/cmam-2011-0002
111. Shashkov, M.: Conservative finite-difference methods on general grids. CRC Press, Boca Raton (1996)
112. Sovinec, C., Glasser, A., Gianakon, T., Barnes, D., Nebel, R., Kruger, S., Schnack, D., Plimpton, S., Tarditi, A., Chu, M., Team, N.: Nonlinear magnetohydrodynamics simulation using high-order finite elements. J. Comput. Phys. **195**, 355–386 (2004). https://doi.org/10.1016/j.jcp.2003.10.004
113. Steinberg, S.: A discrete calculus with applications of higher-order discretizations to boundary-value problems. Comput. Methods Appl. Math. **42**, 228–261 (2004)
114. Steinberg, S., Zingano, J.P.: Error estimates on arbitrary grids for 2nd-order mimetic discretization of Sturm-Liouville problems. Comput. Methods Appl. Math. **9**, 192–202 (2009)
115. Tarhasaari, T., Kettunen, L., Bossavit, A.: Some realizations of a discrete Hodge operator: a reinterpretation of finite element techniques. IEEE Trans. Magn. **35**, 1494–1497 (1999)
116. Taylor, G.I.: Production and dissipation of vorticity in a turbulent fluid. Proc. R. Soc. A Math. Phys. Eng. Sci. **164**, 15–23 (1938). https://doi.org/10.1098/rspa.1938.0002
117. Tonti, E.: On the formal structure of physical theories. Technical report, Italian National Research Council (1975)
118. Wang, Y., Hajibeygi, H., Tchelepi, H.A.: Algebraic multiscale solver for flow in heterogeneous porous media. J. Comput. Phys. **259**, 284–303 (2014). https://doi.org/10.1016/j.jcp.2013.11.024
119. Whitney, H.: Geometric Integration Theory. Dover Publications, Mineola (1957)
120. Wu, X.H., Parashkevov, R.: Effect of grid deviation on flow solutions. SPE J. **14**, 67–77 (2009). https://doi.org/10.2118/92868-PA
121. Younes, A., Ackerer, P., Delay, F.: Mixed finite elements for solving 2-D diffusion-type equations. Rev. Geophys. **48**, RG1004 (2010). https://doi.org/10.1029/2008RG000277
122. Young, L.C.: Rigorous treatment of distorted grids in 3D. In: SPE Reservoir Simulation Symposium. Society of Petroleum Engineers, Richardson (1999). https://doi.org/10.2118/51899-MS
123. Zhang, X., Schmidt, D., Perot, J.B.: Accuracy and conservation properties of a three-dimensional unstructured staggered mesh scheme for fluid dynamics. J. Comput. Phys. **175**, 764–791 (2002)

Chapter 4
An Introduction to Hybrid High-Order Methods

Daniele Antonio Di Pietro and Roberta Tittarelli

Abstract This chapter provides an introduction to Hybrid High-Order (HHO) methods. These are new generation numerical methods for PDEs with several advantageous features: the support of arbitrary approximation orders on general polyhedral meshes, the reproduction at the discrete level of relevant continuous properties, and a reduced computational cost thanks to static condensation and compact stencil. After establishing the discrete setting, we introduce the basics of HHO methods using as a model problem the Poisson equation. We describe in detail the construction, and prove a priori convergence results for various norms of the error as well as a posteriori estimates for the energy norm. We then consider two applications: the discretization of the nonlinear p-Laplace equation and of scalar diffusion-advection-reaction problems. The former application is used to introduce compactness analysis techniques to study the convergence to minimal regularity solution. The latter is used to introduce the discretization of first-order operators and the weak enforcement of boundary conditions. Numerical examples accompany the exposition.

4.1 Introduction

This chapter provides an introduction to Hybrid High-Order (HHO) methods. The material is closely inspired by a series of lectures given by the first author at Institut Henri Poincaré in September 2016 within the thematic quarter *Numerical Methods for PDEs* (see http://imag.edu.umontpellier.fr/event/ihp-nmpdes).

HHO methods, introduced in [27, 33], are discretization methods for Partial Differential Equations (PDEs) with relevant features that set them apart from

---

D. A. Di Pietro (✉) · R. Tittarelli
Institut Montpelliérain Alexander Grothendieck, CNRS, Université Montpellier, Montpellier, France
e-mail: daniele.di-pietro@umontpellier.fr; roberta.tittarelli@umontpellier.fr

D. A. Di Pietro et al. (eds.), *Numerical Methods for PDEs*,
SEMA SIMAI Springer Series 15, https://doi.org/10.1007/978-3-319-94676-4_4

classical techniques such as finite elements or finite volumes. These include, in particular:

1. The support of general polytopal meshes in arbitrary space dimension, paving the way to a seamless treatment of complex geometric features and unified 1d-2d-3d implementations;
2. The possibility to select the approximation order which, possibly combined with adaptivity, leads to a reduction of the simulation cost for a given precision or better precision for a given cost;
3. The compliance with the physics, including robustness with respect to the variations of physical coefficients and reproduction at the discrete level of key continuous properties such as local balances and flux continuity;
4. A reduced computational cost thanks to their compact stencil along with the possibility to perform static condensation.

As of today, HHO methods have been successfully applied to the discretization of several linear and nonlinear problems of engineering interest including: variable diffusion [28, 33, 35], quasi incompressible linear elasticity [26, 27], locally degenerate diffusion-advection-reaction [34], poroelasticity [9], creeping flows [1] possibly driven by volumetric forces with large irrotational part [36], electrostatics [31], phase separation problems governed by the Cahn–Hilliard equation [14], Leray–Lions type elliptic problems [22, 23]. More recent applications also include steady incompressible flows governed by the Navier–Stokes equations [29] and nonlinear elasticity [11]. Generalizations of HHO methods and comparisons with other (new generation or classical) discretization methods for PDEs can be found in [8, 18]. Implementation tools based on advanced programming techniques have been recently discussed in [15].

Discretization methods that support polytopal meshes and, possibly, arbitrary approximation orders have experienced a vigorous development over the last decade. Novel approaches to the analysis and the design have been developed borrowing ideas from other branches of mathematics (such as topology and geometry), or expanding past their initial limits the original ideas underlying finite element or finite volume methods. A brief state-of-the-art is provided in what follows.

Several lowest-order methods for diffusive problems have been proposed to circumvent the strict conditions of mesh-data compliance required for the consistency of classical (two-points) finite volume schemes; see [38] for a comprehensive review. We mention here, in particular, the Mixed and Hybrid Finite Volume methods of [39, 44]. These methods possess local conservation properties on the primal mesh, and enable an explicit identification of equilibrated numerical fluxes. Their relation with the lowest-order version of HHO methods has been studied in [33, Section 2.5] for pure diffusion and in [34, Section 5.4] for advection-diffusion-reaction. Other families of lowest-order methods have been obtained by reproducing at the discrete level salient features of the continuous problem. Mimetic Finite Difference methods are derived by emulating the Stokes theorems to formulate counterparts of differential operators and of L^2-products; cf. [12] and [40] for a study of their relation with Mixed and Hybrid Finite Volume methods.

In the Discrete Geometric Approach of [19] as well as in Compatible Discrete Operators [10], formal links with the continuous operators are expressed in terms of Tonti diagrams. To different extents, the aforementioned methods owe to the seminal ideas of Whitney on geometric integration [55]. A different approach to lowest-order schemes on general meshes consists in extending classical properties of nonconforming and penalized finite elements as in the Cell Centered Galerkin [21] and generalized Crouzeix–Raviart [30] methods. We also cite here [54] concerning the use of classical mixed finite elements on polyhedral meshes (see, in particular, Section 7 therein). Further investigations have recently lead to unifying frameworks that encompass the above (and other) methods. We mention, in particular, the Gradient Schemes discretizations of [41]. Finally, the methods discussed here can often be regarded as lowest-order versions of more recent technologies.

Methods that support the possibility to increase the approximation order have received a considerable amount of attention over the last few years. High-order discretizations on general meshes that are possibly physics-compliant can be obtained by the discontinuous Galerkin approach; cf., e.g., [2, 25] and also [3]. Discontinuous Galerkin methods, however, have some practical limitations. For problems in incompressible fluid mechanics, e.g., a key ingredient for inf-sup stability is a reduction map that can play the role of a Fortin interpolator. Unfortunately, such an interpolator is often not available for discontinuous Galerkin methods on non-standard elements. Additionally, in particular for modal implementations on general meshes, the number of unknowns can become unbearably large. This has motivated the introduction of Hybridizable Discontinuous Galerkin methods [13, 17], which mainly focus on standard meshes (the extension to general meshes is possible in some cases); see also the very recent M-decomposition techniques [16]. High-order discretization methods that support general meshes also include Virtual Element methods; cf. [7] for an introduction. In short, Virtual Element methods are finite element methods where explicit expressions for the basis functions are not available at each point, and computable approximations thereof are used instead. This provides the extra flexibility required, e.g., to handle polyhedral elements. Links between HHO and the nonconforming Virtual Element method have been pointed out in [18, Section 2.4]; see also [8] and [37] concerning the links among HHO, Virtual Element methods, and Gradient Schemes.

We next describe in detail the content of this chapter. We start in Sect. 4.2 by presenting the discrete setting: we introduce the notion of polytopal mesh (Sect. 4.2.1), formulate assumptions on the way meshes are refined that are suitable to carry out a h-convergence analysis (Sect. 4.2.2), introduce the local polynomial spaces (Sect. 4.2.3) and projectors (Sect. 4.2.4) that lie at the heart of the HHO construction.

In Sect. 4.3 we present the basic principles of HHO methods using as a model problem the Poisson equation. While the material in this section is mainly adapted from [33], some results are new and the arguments have been shortened or made more elegant. In Sect. 4.3.1 we introduce the local space of degrees of freedom (DOFs) and discuss the main ingredients upon which HHO methods rely, namely:

1. Reconstructions of relevant quantities obtained by solving small, embarrassingly parallel problems on each element;
2. High-order stabilization terms obtained by penalizing cleverly designed residuals.

In Sect. 4.3.2 we show how to combine these ingredients to formulate local contributions, which are then assembled element-by-element as in standard finite elements. The construction is conceived so that only face-based DOFs are globally coupled, which paves the way to efficient practical implementations where element-based DOFs are statically condensed in a preliminary step. In Sects. 4.3.3 and 4.3.4 we discuss, respectively, optimal a priori estimates for various norms and seminorms of the error, and residual-based a posteriori estimates for the energy-norm of the error. Finally, some numerical examples are provided in Sect. 4.3.5 to demonstrate the theoretical results.

In Sect. 4.4 we consider the HHO discretization of the p-Laplace equation. The material is inspired by [22, 23], where more general Leray–Lions operators are considered. When dealing with nonlinear problems, regularity for the exact solution is often difficult to prove and can entail stringent assumptions on the data. For this reason, the h-convergence analysis can be carried out in two steps: in a first step, convergence to minimal regularity solutions is proved by a compactness argument; in a second step, convergence rates are estimated for smooth solutions (and smooth data). Convergence by compactness typically requires discrete counterparts of functional analysis results relevant for the study of the continuous problem. In our case, two sets of discrete functional analysis results are needed: discrete Sobolev embeddings (Sect. 4.4.1) and compactness for sequences of HHO functions uniformly bounded in a $W^{1,p}$-like seminorm (Sect. 4.4.2). The interest of both results goes beyond the specific method and problem considered here. As an example, in [29] they are used for the analysis of a HHO discretization of the steady incompressible Navier–Stokes equations. The HHO method for the p-Laplacian stated in Sect. 4.4.3 is designed according to similar principles as for the Poisson problem. Convergence results are stated in Sect. 4.4.4, and numerical examples are provided in Sect. 4.4.5.

Following [34], in Sect. 4.5 we extend the HHO method to diffusion-advection-reaction problems. In this context, a crucial property from the numerical point of view is robustness in the advection-dominated regime. In Sect. 4.5.1 we modify the diffusive bilinear form introduced in Sect. 4.3.2 to incorporate weakly enforced boundary conditions. The weak enforcement of boundary conditions typically improves the behaviour of the method in the presence of boundary layers, since the discrete solution is not constrained to a fixed value on the boundary. In Sect. 4.5.2 we introduce the HHO discretization of first-order terms based on two novel ingredients: a local advective derivative reconstruction and an upwind penalty term. The former is used to formulate the consistency terms, while the role of the latter is to confer suitable stability properties to the advective-reactive bilinear form.

The HHO discretization is finally obtained in Sect. 4.5.3 combining the diffusive and advective-reactive contributions, and its stability with respect to an energy-like norm including an advective derivative contribution is studied. In Sect. 4.5.4 we state an energy-norm error estimate which accounts for the dependence of the error contribution of each mesh element on a local Péclet number. A numerical illustration is provided in Sect. 4.5.5.

4.2 Discrete Setting

Let $\Omega \subset \mathbb{R}^d$, $d \in \mathbb{N}^*$, denote a bounded connected open polyhedral domain with Lipschitz boundary and outward normal $\mathbf{n}$. We assume that Ω does not have cracks, i.e., it lies on one side of its boundary. In what follows, we introduce the notion of polyhedral mesh of Ω, formulate assumptions on the way meshes are refined that enable to prove useful geometric and functional results, and introduce functional spaces and projectors that will be used in the construction and analysis of HHO methods.

4.2.1 Polytopal Mesh

The following definition enables the treatment of meshes as general as the ones depicted in Fig. 4.1.

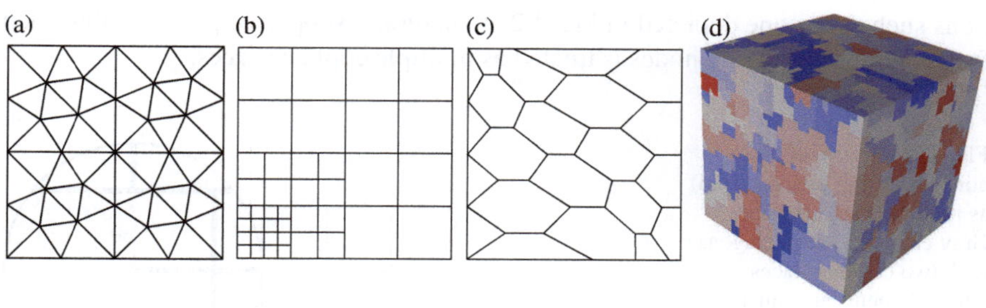

Fig. 4.1 Examples of polytopal meshes in two and three space dimensions. The triangular and nonconforming meshes are taken from the FVCA5 benchmark [47], the polygonal mesh family from [30, Section 4.2.3], and the agglomerated polyhedral mesh from [31]. (**a**) Matching triangular, (**b**) nonconforming, (**c**) polygonal, (**d**) agglomerated

Definition 4.1 (Polytopal Mesh) A polytopal mesh of Ω is a couple $\mathcal{M}_h = (\mathcal{T}_h, \mathcal{F}_h)$ where:

(i) The set of *mesh elements* $\mathcal{T}_h$ is a finite collection of nonempty disjoint open polytopes T with boundary ∂T and diameter h_T such that the *meshsize* h satisfies $h = \max_{T\in\mathcal{T}_h} h_T$ and it holds that $\overline{\Omega} = \bigcup_{T\in\mathcal{T}_h} \overline{T}$.

(ii) The set of *mesh faces* $\mathcal{F}_h$ is a finite collection of disjoint subsets of $\overline{\Omega}$ such that, for any $F \in \mathcal{F}_h$, F is an open subset of a hyperplane of $\mathbb{R}^d$, the $(d-1)$-dimensional Hausdorff measure of F is strictly positive, and the $(d-1)$-dimensional Hausdorff measure of its relative interior $\overline{F}\backslash F$ is zero. Moreover, (a) for each $F \in \mathcal{F}_h$, either there exist two distinct mesh elements $T_1, T_2 \in \mathcal{T}_h$ such that $F \subset \partial T_1 \cap \partial T_2$ and F is called an *interface* or there exists one mesh element $T \in \mathcal{T}_h$ such that $F \subset \partial T \cap \partial\Omega$ and F is called a *boundary face*; (b) the set of faces is a partition of the mesh skeleton, i.e., $\bigcup_{T\in\mathcal{T}_h} \partial T = \bigcup_{F\in\mathcal{F}_h} \overline{F}$.

Interfaces are collected in the set $\mathcal{F}_h^{\mathrm{i}}$ and boundary faces in $\mathcal{F}_h^{\mathrm{b}}$, so that $\mathcal{F}_h = \mathcal{F}_h^{\mathrm{i}} \cup \mathcal{F}_h^{\mathrm{b}}$. For any mesh element $T \in \mathcal{T}_h$,

$$\mathcal{F}_T := \{F \in \mathcal{F}_h \mid F \subset \partial T\}$$

denotes the set of faces contained in ∂T. Similarly, for any mesh face $F \in \mathcal{F}_h$,

$$\mathcal{T}_F := \{T \in \mathcal{T}_h \mid F \subset \partial T\}$$

is the set of mesh elements sharing F. Finally, for all $F \in \mathcal{F}_T$, $\mathbf{n}_{TF}$ is the unit normal vector to F pointing out of T.

Remark 4.1 (Nonconforming Junctions) Meshes including nonconforming junctions such as the one depicted in Fig. 4.2 are naturally supported provided that each face containing hanging nodes is treated as multiple coplanar faces.

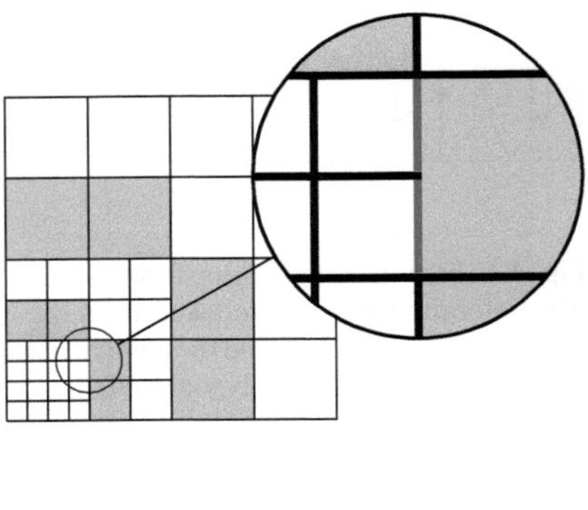

Fig. 4.2 Treatment of a nonconforming junction (red) as multiple coplanar faces. Gray elements are pentagons with two coplanar faces, white elements are squares

4.2.2 Regular Mesh Sequences

When studying the convergence of HHO methods with respect to the meshsize h, one needs to make assumptions on how the mesh is refined. The ones provided here are closely inspired by [25, Chapter 1], and refer to the case of isotropic meshes with non-degenerate faces. Isotropic means here that we do not consider the case of elements that become more and more stretched when refining. Non-degenerate faces means, on the other hand, that the diameter of each mesh face is uniformly comparable to that of the element(s) it belongs to; see (4.2) below.

Definition 4.2 (Matching Simplicial Submesh) Let $\mathcal{M}_h = (\mathcal{T}_h, \mathcal{F}_h)$ be a polytopal mesh of Ω. We say that $\mathfrak{T}_h$ is a matching simplicial submesh of $\mathcal{M}_h$ if (i) $\mathfrak{T}_h$ is a matching simplicial mesh of Ω; (ii) for all simplices $\tau \in \mathfrak{T}_h$, there is only one mesh element $T \in \mathcal{T}_h$ such that $\tau \subset T$; (iii) for all $\sigma \in \mathfrak{F}_h$, the set collecting the simplicial faces of $\mathfrak{T}_h$, there is at most one face $F \in \mathcal{F}_h$ such that $\sigma \subset F$.

If $\mathcal{T}_h$ itself is matching simplicial and $\mathcal{F}_h$ collects the corresponding simplicial faces, we can simply take $\mathfrak{T}_h = \mathcal{T}_h$, so that $\mathfrak{F}_h = \mathcal{F}_h$. The notion of regularity for refined mesh sequences is made precise by the following

Definition 4.3 (Regular Mesh Sequence) Denote by $\mathcal{H} \subset \mathbb{R}_*^+$ a countable set of meshsizes having 0 as its unique accumulation point. A sequence of refined meshes $(\mathcal{M}_h)_{h\in\mathcal{H}}$ is said to be *regular* if there exists a real number $\varrho \in (0, 1)$ such that, for all $h \in \mathcal{H}$, there exists a matching simplicial submesh $\mathfrak{T}_h$ of $\mathcal{M}_h$ and (i) for all simplices $\tau \in \mathfrak{T}_h$ of diameter h_τ and inradius r_τ, $\varrho h_\tau \leq r_\tau$; (ii) for all mesh elements $T \in \mathcal{T}_h$ and all simplices $\tau \in \mathfrak{T}_h$ such that $\tau \subset T$, $\varrho h_T \leq h_\tau$.

Remark 4.2 (Role of the Simplicial Submesh) The simplicial submesh introduced in Definition 4.3 is merely a theoretical tool, and needs not be constructed in practice.

Geometric bounds on regular mesh sequences can be proved as in [25, Section 1.4.2] (the definition of mesh face is slightly different therein since planarity is not required, but the proofs are based on the matching simplicial submesh and one can check that they carry out unchanged). We recall here, in particular, that the number of faces of one mesh element is uniformly bounded: There is $N_\partial \geq d + 1$ such that

$$\max_{h\in\mathcal{H}} \max_{T\in\mathcal{T}_h} \operatorname{card}(\mathcal{F}_T) \leq N_\partial. \tag{4.1}$$

Moreover, according to [25, Lemma 1.42], for all $h \in \mathcal{H}$, all $T \in \mathcal{T}_h$, and all $F \in \mathcal{F}_T$

$$\varrho^2 h_T \leq h_F \leq h_T. \tag{4.2}$$

Discrete functional analysis results for arbitrary-order methods on regular mesh sequences can be found in [25, Chapter 1] and [22, 23]. We also refer the reader to [43] for a first theorization of discrete functional analysis in the context of lowest-order finite volume methods, as well as to the subsequent extensions of [42, 44].

Throughout the rest of this work, it is tacitly understood that we work on regular mesh sequences.

4.2.3 Local and Broken Spaces

Throughout the rest of this chapter, for any $X \subset \overline{\Omega}$, we denote by $(\cdot,\cdot)_X$ and $\|\cdot\|_X$ the standard $L^2(X)$-product and norm, with the convention that the subscript is omitted whenever $X = \Omega$. The same notation is used for the vector-valued space $L^2(X)^d$.

Let now the set X be a mesh element or face. For an integer $l \geq 0$, we denote by $\mathbb{P}^l(X)$ the space spanned by the restriction to X of scalar-valued, d-variate polynomials of total degree l. We note the following trace inequality (see [25, Lemma 1.46]): There is a real number $C > 0$ only depending on d, ϱ, and l such that, for all $h \in \mathcal{H}$, all $T \in \mathcal{T}_h$, all $v \in \mathbb{P}^l(T)$, and all $F \in \mathcal{F}_T$,

$$\|v\|_F \leq C h_T^{-1/2} \|v\|_T. \tag{4.3}$$

At the global level, we define the broken polynomial space

$$\mathbb{P}^l(\mathcal{T}_h) := \left\{ v_h \in L^2(\Omega) \mid v_{h|T} \in \mathbb{P}^l(T) \quad \forall T \in \mathcal{T}_h \right\}.$$

Functions in $\mathbb{P}^l(\mathcal{T}_h)$ belong to the broken Sobolev space

$$W^{1,1}(\mathcal{T}_h) := \left\{ v \in L^1(\Omega) \mid v_{|T} \in W^{1,1}(T) \quad \forall T \in \mathcal{T}_h \right\}.$$

We denote by $\nabla_h : W^{1,1}(\mathcal{T}_h) \to L^1(\Omega)^d$ the usual broken gradient operator such that, for all $v \in W^{1,1}(\mathcal{T}_h)$,

$$(\nabla_h v)_{|T} = \nabla v_{|T} \qquad \forall T \in \mathcal{T}_h.$$

4.2.4 Projectors on Local Polynomial Spaces

Projectors on local polynomial spaces play a key role in the design and analysis of HHO methods.

4.2.4.1 L^2-Orthogonal Projector

Let X denote a mesh element or face. The L^2-orthogonal projector (in short, L^2-projector) $\pi_X^{0,l} : L^1(X) \to \mathbb{P}^l(X)$ is defined as follows: For all $v \in L^1(X)$, $\pi_X^{0,l}$ is the unique polynomial in $\mathbb{P}^l(X)$ that satisfies

$$(\pi_X^{0,l} v - v, w)_X = 0 \qquad \forall w \in \mathbb{P}^l(X). \tag{4.4}$$

Existence and uniqueness of $\pi_X^{0,l} v$ follow from the Riesz representation theorem in $\mathbb{P}^l(X)$ for the standard $L^2(X)$-inner product. Moreover, we have the following characterization:

$$\pi_X^{0,l} v = \arg\min_{w \in \mathbb{P}^l(X)} \|w - v\|_X^2.$$

In what follows, we will also need the vector-valued L^2-projector denoted by $\boldsymbol{\pi}_X^{0,l}$ and obtained by applying $\pi_X^{0,l}$ component-wise. The following H^s-boundedness result is a special case of [22, Corollary 3.7]: For any $s \in \{0, \ldots, l+1\}$, there exists a real number $C > 0$ depending only on d, ϱ, l, and s such that, for all $h \in \mathcal{H}$, all $T \in \mathcal{T}_h$, and all $v \in H^s(T)$,

$$|\pi_T^{0,l} v|_{H^s(T)} \le C |v|_{H^s(T)}. \tag{4.5}$$

At the global level, we denote by $\pi_h^{0,l} : L^1(\Omega) \to \mathbb{P}^l(\mathcal{T}_h)$ the L^2-projector on the broken polynomial space $\mathbb{P}^l(\mathcal{T}_h)$ such that, for all $v \in L^1(\Omega)$,

$$(\pi_h^{0,l} v)_{|T} := \pi_T^{0,l} v_{|T}.$$

4.2.4.2 Elliptic Projector

For any mesh element $T \in \mathcal{T}_h$, we also define the elliptic projector $\pi_T^{1,l} : W^{1,1}(T) \to \mathbb{P}^l(T)$ as follows: For all $v \in W^{1,1}(T)$, $\pi_T^{1,l} v$ is a polynomial in $\mathbb{P}^l(T)$ that satisfies

$$(\nabla(\pi_T^{1,l} v - v), \nabla w)_T = 0 \qquad \forall w \in \mathbb{P}^l(T). \tag{4.6a}$$

By the Riesz representation theorem in $\nabla\mathbb{P}^l(T)$ for the $L^2(T)^d$-inner product, this relation defines a unique element $\nabla\pi_T^{1,l} v$, and thus a polynomial $\pi_T^{1,l} v$ up to an additive constant. This constant is fixed by writing

$$(\pi_T^{1,l} v - v, 1)_T = 0. \tag{4.6b}$$

Observing that (4.6a) is trivially verified when $l = 0$, it follows from (4.6b) that $\pi_T^{1,0} = \pi_T^{0,0}$. Finally, the following characterization holds:

$$\pi_T^{1,l} v = \operatorname*{arg\,min}_{w \in \mathbb{P}^l(T),\ (w-v,1)_T=0} \|\nabla(w-v)\|_{L^2(T)^d}^2.$$

4.2.4.3 Approximation Properties

On regular mesh sequences, both $\pi_T^{0,l}$ and $\pi_T^{1,l}$ have optimal approximation properties in $\mathbb{P}^l(T)$, as summarized by the following result (for a proof, see Theorem 1, Theorem 2, and Lemma 13 in [22]): For any $\alpha \in \{0, 1\}$ and $s \in \{\alpha, \ldots, l+1\}$, there exists a real number $C > 0$ depending only on d, ϱ, l, α, and s such that, for all $h \in \mathcal{H}$, all $T \in \mathcal{T}_h$, and all $v \in H^s(T)$,

$$|v - \pi_T^{\alpha,l} v|_{H^m(T)} \le C h_T^{s-m} |v|_{H^s(T)} \qquad \forall m \in \{0, \ldots, s\}, \tag{4.7a}$$

and, if $s \ge 1$,

$$|v - \pi_T^{\alpha,l} v|_{H^m(\mathcal{F}_T)} \le C h_T^{s-m-\frac{1}{2}} |v|_{H^s(T)} \qquad \forall m \in \{0, \ldots, s-1\}, \tag{4.7b}$$

where $H^m(\mathcal{F}_T) := \left\{ v \in L^2(\partial T) \mid v_{|F} \in H^m(F) \quad \forall F \in \mathcal{F}_T \right\}$.

4.3 Basic Principles of Hybrid High-Order Methods

To fix the main ideas and notation, we study in this section the HHO discretization of the Poisson problem: Find $u : \Omega \to \mathbb{R}$ such that

$$-\Delta u = f \qquad \text{in } \Omega, \tag{4.8a}$$

$$u = 0 \qquad \text{on } \partial\Omega, \tag{4.8b}$$

where $f \in L^2(\Omega)$ is a given volumetric source term. More general boundary conditions can replace (4.8b), but we restrict the discussion to the homogeneous Dirichlet case for the sake of simplicity. A detailed treatment of more general boundary conditions including also variable diffusion coefficients can be found in [35].

The starting point to devise a HHO discretization is the following weak formulation of problem (4.8): Find $u \in H_0^1(\Omega)$ such that

$$a(u, v) = (f, v) \qquad \forall v \in H_0^1(\Omega), \tag{4.9}$$

where the bilinear form $a : H^1(\Omega) \times H^1(\Omega) \to \mathbb{R}$ is such that

$$a(u, v) := (\nabla u, \nabla v). \tag{4.10}$$

In what follows, the quantities u and $-\nabla u$ will be referred to, respectively, as the potential and the flux.

4.3.1 Local Construction

Throughout this section, we fix a polynomial degree $k \geq 0$ and a mesh element $T \in \mathcal{T}_h$. We introduce the local ingredients underlying the HHO construction: the DOFs, the potential reconstruction operator, and the discrete counterpart of the restriction to T of the global bilinear form a defined by (4.10).

4.3.1.1 Computing the Local Elliptic Projection from L^2-Projections

Consider a function $v \in H^1(T)$. We note the following integration by parts formula, valid for all $w \in C^\infty(\overline{T})$:

$$(\nabla v, \nabla w)_T = -(v, \Delta w)_T + \sum_{F \in \mathcal{F}_T} (v, \nabla w \cdot \mathbf{n}_{TF})_F. \tag{4.11}$$

Specializing (4.11) to $w \in \mathbb{P}^{k+1}(T)$, we obtain

$$(\nabla \pi_T^{1,k+1} v, \nabla w)_T = -(\pi_T^{0,k-1} v, \Delta w)_T + \sum_{F \in \mathcal{F}_T} (\pi_F^{0,k} v, \nabla w \cdot \mathbf{n}_{TF})_F, \tag{4.12a}$$

where we have used (4.6a) to insert $\pi_T^{1,k+1}$ into the left-hand side and (4.4) to insert $\pi_T^{0,k-1}$ and $\pi_F^{0,k}$ into the right-hand side after observing that $\Delta w \in \mathbb{P}^{k-1}(T) \subset \mathbb{P}^k(T)$ and $(\nabla w)_{|F} \cdot \mathbf{n}_{TF} \in \mathbb{P}^k(F)$ for all $F \in \mathcal{F}_T$. Moreover, recalling (4.6b) and using the definition (4.4) of the L^2-projector, we infer that

$$(v - \pi_T^{0,0} v, 1)_T = (\pi_T^{1,k+1} v - \pi_T^{0,\max(0,k-1)} v, 1)_T = 0. \tag{4.12b}$$

The relations (4.12) show that computing the elliptic projection $\pi_T^{1,k+1} v$ does not require a full knowledge of the function v. All that is required is

1. $\pi_T^{0,\max(0,k-1)} v$, the L^2-projection of v on the polynomial space $\mathbb{P}^{\max(0,k-1)}(T)$. Clearly, one could also choose $\pi_T^{0,k} v$ instead, which has the advantage of not requiring a special treatment of the case $k = 0$;

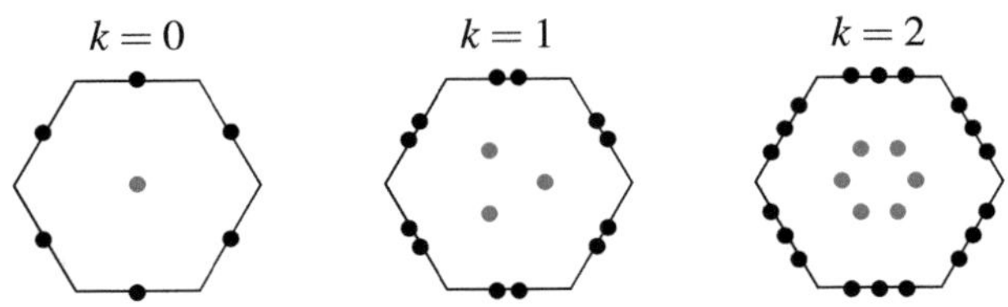

Fig. 4.3 DOFs in $\underline{U}_T^k$ for $k \in \{0, 1, 2\}$

2. for all $F \in \mathfrak{F}_T$, $\pi_F^{0,k} v_{|F}$, the L^2-projection of the trace of v on F on the polynomial space $\mathbb{P}^k(F)$.

4.3.1.2 Local Space of Degrees of Freedom

The remark at the end of the previous section motivates the introduction of the following space of DOFs (see Fig. 4.3):

$$\underline{U}_T^k := \mathbb{P}^k(T) \times \left(\bigtimes_{F \in \mathfrak{F}_T} \mathbb{P}^k(F) \right). \tag{4.13}$$

Observe that naming $\underline{U}_T^k$ space of DOFs involves a shortcut: the actual DOFs can be chosen in several equivalent ways (polynomial moments, point values, etc.), and the specific choice does not affect the following discussion. For a generic vector of DOFs in $\underline{U}_T^k$, we use the underlined notation $\underline{v}_T = (v_T, (v_F)_{F \in \mathfrak{F}_T})$. On $\underline{U}_T^k$, we define the H^1-like seminorm $\|\cdot\|_{1,T}$ such that, for all $\underline{v}_T \in \underline{U}_T^k$,

$$\|\underline{v}_T\|_{1,T}^2 := \|\nabla v_T\|_T^2 + |\underline{v}_T|_{1,\partial T}^2, \qquad |\underline{v}_T|_{1,\partial T}^2 := \sum_{F \in \mathfrak{F}_T} h_F^{-1} \|v_F - v_T\|_F^2, \tag{4.14}$$

where h_F denotes the diameter of F. The negative power of h_F in the second term ensures that both contributions have the same scaling. The DOFs corresponding to a smooth function $v \in W^{1,1}(T)$ are obtained via the reduction map $\underline{I}_T^k : W^{1,1}(T) \to \underline{U}_T^k$ such that

$$\underline{I}_T^k v := (\pi_T^{0,k} v, (\pi_F^{0,k} v_{|F})_{F \in \mathfrak{F}_T}). \tag{4.15}$$

4.3.1.3 Potential Reconstruction Operator

Inspired by formula (4.12), we introduce the potential reconstruction operator $p_T^{k+1} : \underline{U}_T^k \to \mathbb{P}^{k+1}(T)$ such that, for all $\underline{v}_T \in \underline{U}_T^k$,

$$(\nabla p_T^{k+1}\underline{v}_T, \nabla w)_T = -(v_T, \Delta w)_T + \sum_{F\in\mathfrak{F}_T} (v_F, \nabla w{\cdot}\mathbf{n}_{TF})_F \quad \forall w \in \mathbb{P}^{k+1}(T) \tag{4.16a}$$

and

$$(p_T^{k+1}\underline{v}_T - v_T, 1)_T = 0. \tag{4.16b}$$

Notice that $p_T^{k+1}\underline{v}_T$ is a polynomial function on T one degree higher than the element-based DOFs v_T. By definition, for all $v \in W^{1,1}(T)$ it holds that

$$(p_T^{k+1} \circ \underline{I}_T^k)v = \pi_T^{1,k+1}v, \tag{4.17}$$

i.e., the composition of the potential reconstruction operator with the reduction map gives the elliptic projector on $\mathbb{P}^{k+1}(T)$. An immediate consequence of (4.17) together with (4.7) is that $p_T^{k+1} \circ \underline{I}_T^k$ has optimal approximation properties in $\mathbb{P}^{k+1}(T)$.

4.3.1.4 Local Contribution

We approximate the restriction $a_{|T} : H^1(T) \times H^1(T) \to \mathbb{R}$ to T of the continuous bilinear form a defined by (4.10) by the discrete bilinear form $\mathrm{a}_T : \underline{U}_T^k \times \underline{U}_T^k \to \mathbb{R}$ such that

$$\mathrm{a}_T(\underline{u}_T, \underline{v}_T) := (\nabla p_T^{k+1}\underline{u}_T, \nabla p_T^{k+1}\underline{v}_T)_T + \mathrm{s}_T(\underline{u}_T, \underline{v}_T), \tag{4.18}$$

where the first term in the right-hand side is the usual Galerkin contribution, while the second is a stabilization contribution for which we consider the following design conditions, originally proposed in [8]:

Assumption 4.1 (Local Stabilization Bilinear Form s_T) *The local stabilization bilinear form $\mathrm{s}_T : \underline{U}_T^k \times \underline{U}_T^k \to \mathbb{R}$ satisfies the following properties:*

(S1) Symmetry and positivity. s_T *is symmetric and positive semidefinite;*
(S2) Stability. *There is a real number $\eta > 0$ independent of h and of T, but possibly depending on d, ϱ, and k, such that*

$$\eta^{-1}\|\underline{v}_T\|_{1,T}^2 \le \mathrm{a}_T(\underline{v}_T, \underline{v}_T) \le \eta\|\underline{v}_T\|_{1,T}^2 \qquad \forall \underline{v}_T \in \underline{U}_T^k; \tag{4.19}$$

(S3) Polynomial consistency. *For all* $w \in \mathbb{P}^{k+1}(T)$ *and all* $\underline{v}_T \in \underline{U}_T^k$, *it holds that*

$$\mathrm{s}_T(\underline{I}_T^k w, \underline{v}_T) = 0. \tag{4.20}$$

These requirements suggest that s_T can be obtained penalizing in a least square sense residuals that vanish for reductions of polynomial functions in $\mathbb{P}^{k+1}(T)$. Paradigmatic examples of such residuals are provided by the operators $\delta_T^k : \underline{U}_T^k \to \mathbb{P}^k(T)$ and, for all $F \in \mathcal{F}_T$, $\delta_{TF}^k : \underline{U}_T^k \to \mathbb{P}^k(F)$ such that, for all $\underline{v}_T \in \underline{U}_T^k$,

$$\delta_T^k \underline{v}_T := \pi_T^{0,k}(p_T^{k+1}\underline{v}_T - v_T), \qquad \delta_{TF}^k \underline{v}_T := \pi_F^{0,k}(p_T^{k+1}\underline{v}_T - v_F) \quad \forall F \in \mathcal{F}_T. \tag{4.21}$$

To check that δ_T^k vanishes when $\underline{v}_T = \underline{I}_T^k w$ with $w \in \mathbb{P}^{k+1}(T)$, we observe that

$$\delta_T^k \underline{I}_T^k w = \pi_T^{0,k}(p_T^{k+1}\underline{I}_T^k w - \pi_T^{0,k} w) = \pi_T^{0,k}(\pi_T^{1,k+1} w - w) = \pi_T^{0,k}(w - w) = 0,$$

where we have used the definition of δ_T^k in the first equality, the relation (4.17) to replace $p_T^{k+1}\underline{I}_T^k$ by $\pi_T^{1,k+1}$ and the fact that $\pi_T^{0,k} w \in \mathbb{P}^k(T)$ to cancel $\pi_T^{0,k}$ from the second term in parentheses, and the fact that $\pi_T^{1,k+1}$ leaves polynomials of total degree up to $(k+1)$ unaltered as a projector to conclude. A similar argument shows that $\delta_{TF}^k \underline{I}_T^k w = 0$ for all $F \in \mathcal{F}_T$ whenever $w \in \mathbb{P}^{k+1}(T)$.

Accounting for dimensional homogeneity with the Galerkin term, one possible expression for s_T is thus

$$\mathrm{s}_T(\underline{u}_T, \underline{v}_T) := h_T^{-2}(\delta_T^k \underline{u}_T, \delta_T^k \underline{v}_T)_T + \sum_{F \in \mathcal{F}_T} h_F^{-1}(\delta_{TF}^k \underline{u}_T, \delta_{TF}^k \underline{v}_T)_F. \tag{4.22}$$

This choice, inspired by the Virtual Element literature [6], differs from the original HHO stabilization of [33], where the following expression is considered instead:

$$\mathrm{s}_T(\underline{u}_T, \underline{v}_T) := \sum_{F \in \mathcal{F}_T} h_F^{-1}(\delta_{TF}^k \underline{u}_T - \delta_T^k \underline{u}_T, \delta_{TF}^k \underline{v}_T - \delta_T^k \underline{v}_T)_F. \tag{4.23}$$

In this case, only quantities at faces are penalized. Both of the above expressions match the design conditions (S1)–(S3) and are essentially equivalent in terms of implementation. A detailed proof for s_T as in (4.23) can be found in [33, Lemma 4]. Yet another example of stabilization bilinear form used in the context of HHO methods is provided by [1, Eq. (3.24)]. This expression results from the hybridization of the Mixed High-Order method of [28].

Remark 4.3 (Original HDG Stabilization) The following stabilization bilinear form is used in the original Hybridizable Discontinuous Galerkin (HDG) method of [13, 17]:

$$\mathrm{s}_T(\underline{u}_T, \underline{v}_T) = \sum_{F\in\mathcal{F}_T} h_F^{-1}(u_F - u_T, v_F - v_T)_F.$$

While this choice obviously satisfies the properties (S1)–(S2), it fails to satisfy (S3) (it is only consistent for polynomials of degree up to k). As a result, up to one order of convergence is lost with respect to the estimates of Theorems 4.1 and 4.2 below. For a discussion including fixes that restore optimal orders of convergence in HDG see [18].

4.3.1.5 Consistency Properties of the Stabilization for Smooth Functions

In the following proposition we study the consistency properties of s_T when its arguments are reductions of a smooth function. We give a detailed proof since this result is a new extension of the bound in [33, Theorem 8] (see, in particular, Eq. (45) therein) to more general stabilization bilinear forms.

Proposition 4.1 (Consistency of s_T) *Let $\{\mathrm{s}_T\}_{T\in\mathcal{T}_h}$ denote a family of stabilization bilinear forms satisfying assumptions (S1)–(S3). Then, there is a real number $C > 0$ independent of h, but possibly depending on d, ϱ, and k, such that, for all $T \in \mathcal{T}_h$ and all $v \in H^{k+2}(T)$, it holds that*

$$\mathrm{s}_T(\underline{I}_T^k v, \underline{I}_T^k v)^{1/2} \le C h_T^{k+1} \|v\|_{H^{k+2}(T)}. \tag{4.24}$$

Proof We set, for the sake of brevity, $\check{v}_T := \pi_T^{1,k+1} v$ and abridge as $A \lesssim B$ the inequality $A \le cB$ with multiplicative constant $c > 0$ having the same dependencies as C in (4.24). Using (S2) and (S3) we infer that

$$\mathrm{s}_T(\underline{I}_T^k v, \underline{I}_T^k v)^{1/2} = \mathrm{s}_T(\underline{I}_T^k (v - \check{v}_T), \underline{I}_T^k (v - \check{v}_T))^{1/2} \le \eta^{\frac{1}{2}} \|\underline{I}_T^k (v - \check{v}_T)\|_{1,T}. \tag{4.25}$$

Recalling (4.14), we have that

$$\begin{aligned}\|\underline{I}_T^k (v - \check{v}_T)\|_{1,T}^2 \\ = \|\nabla \pi_T^{0,k}(v - \check{v}_T)\|_T^2 + \sum_{F\in\mathcal{F}_T} h_F^{-1} \|\pi_F^{0,k}(v - \check{v}_T - \pi_T^{0,k}(v - \check{v}_T))\|_F^2.\end{aligned} \tag{4.26}$$

Using the $H^1(T)$-boundedness of $\pi_T^{0,k}$ resulting from (4.5) with $l = k$ and $s = 1$ followed by the optimal approximation properties (4.7a) of $\check{v}_T$ (with $\alpha = 1$, $l = k+1$, $s = k+2$, and $m = 1$), it is inferred that

$$\|\nabla\pi_T^{0,k}(v - \check{v}_T)\|_T \lesssim \|\nabla(v - \check{v}_T)\|_T \lesssim h^{k+1}\|v\|_{H^{k+2}(T)}. \tag{4.27}$$

On the other hand, for all $F \in \mathcal{F}_T$ it holds that

$$\begin{aligned} h_F^{-1/2}\|\pi_F^{0,k}(v - \check{v}_T - \pi_T^{0,k}(v - \check{v}_T))\|_F &\lesssim h_T^{-1/2}\|v - \check{v}_T - \pi_T^{0,k}(v - \check{v}_T)\|_T \\ &\lesssim \|\nabla(v - \check{v}_T)\|_T \\ &\lesssim h_T^{k+1}\|v\|_{H^{k+2}(T)}, \end{aligned} \tag{4.28}$$

where we have used the $L^2(F)$-boundedness of $\pi_F^{0,k}$ together with (4.2), the trace approximation properties (4.7b) of $\pi_T^{0,k}$ with $\alpha = 0$, $l = k$, $s = 1$, and $m = 0$ to pass to the second line, and the optimal approximation properties of $\check{v}_T$ expressed by (4.7a) with $\alpha = 1$, $l = k+1$, $s = k+2$, and $m = 1$ to conclude. Plugging (4.27) and (4.28) into (4.26), recalling that $\mathrm{card}(\mathcal{F}_T) \lesssim 1$ (see (4.1)), and using the resulting bound to estimate (4.25), (4.24) follows. □

4.3.2 *Discrete Problem*

We now show how to formulate the discrete problem from the local contributions introduced in the previous section.

4.3.2.1 Global Spaces of Degrees of Freedom

We define the following global space of DOFs with single-valued interface unknowns:

$$\underline{U}_h^k := \left(\bigtimes_{T\in\mathcal{T}_h} \mathbb{P}^k(T)\right) \times \left(\bigtimes_{F\in\mathcal{F}_h} \mathbb{P}^k(F)\right).$$

Notice that single-valued means here that interface values match from one element to the adjacent one. For a generic element $\underline{v}_h \in \underline{U}_h^k$, we use the underlined notation $\underline{v}_h = ((v_T)_{T\in\mathcal{T}_h}, (v_F)_{F\in\mathcal{F}_h})$ and, for all $T \in \mathcal{T}_h$, we denote by $\underline{v}_T = (v_T, (v_F)_{F\in\mathcal{F}_T}) \in \underline{U}_T^k$ its restriction to T. We also define the broken polynomial function $v_h \in \mathbb{P}^k(\mathcal{T}_h)$ such that

$$v_{h|T} := v_T \qquad \forall T \in \mathcal{T}_h.$$

The DOFs corresponding to a smooth function $v \in W^{1,1}(\Omega)$ are obtained via the reduction map $\underline{I}_h^k : W^{1,1}(\Omega) \to \underline{U}_h^k$ such that

$$\underline{I}_h^k v := ((\pi_T^{0,k} v_{|T})_{T\in\mathcal{T}_h}, (\pi_F^{0,k} v_{|F})_{F\in\mathcal{F}_h}).$$

We define on $\underline{U}_h^k$ the seminorm $\|\cdot\|_{1,h}$ such that, for all $\underline{v}_h \in \underline{U}_h^k$,

$$\|\underline{v}_h\|_{1,h}^2 := \sum_{T\in\mathcal{T}_h} \|\underline{v}_T\|_{1,T}^2, \tag{4.29}$$

with local seminorm $\|\cdot\|_{1,T}$ defined by (4.14). To account for the homogeneous Dirichlet boundary condition (4.8b) in a strong manner, we introduce the subspace

$$\underline{U}_{h,0}^k := \left\{\underline{v}_h \in \underline{U}_h^k \mid v_F \equiv 0 \quad \forall F \in \mathcal{F}_h^{\mathrm{b}}\right\}.$$

We recall the following discrete Poincaré inequality proved in [22, Proposition 5.4]: There exists a real number $C_{\mathrm{P}} > 0$ independent of h, but possibly depending on Ω, ϱ, and k, such that, for all $\underline{v}_h \in \underline{U}_{h,0}^k$,

$$\|v_h\| \leq C_{\mathrm{P}}\|\underline{v}_h\|_{1,h}. \tag{4.30}$$

Proposition 4.2 (Norm $\|\cdot\|_{1,h}$) *The map $\|\cdot\|_{1,h}$ defines a norm on $\underline{U}_{h,0}^k$.*

Proof The seminorm property being evident, it suffices to prove that, for all $\underline{v}_h \in \underline{U}_{h,0}^k$, $\|\underline{v}_h\|_{1,h} = 0 \implies \underline{v}_h = \underline{0}_h$. Let $\underline{v}_h \in \underline{U}_{h,0}^k$ be such that $\|\underline{v}_h\|_{1,h} = 0$. By (4.30), we have $\|v_h\| = 0$, hence $v_T \equiv 0$ for all $T \in \mathcal{T}_h$. From the definition (4.14) of the norm $\|\cdot\|_{1,T}$, we also have that $\|v_F - v_T\|_F = 0$ for all $T \in \mathcal{T}_h$ and all $F \in \mathcal{F}_T$, hence $v_F = v_{T|F} \equiv 0$. Since any mesh face belongs to the set $\mathcal{F}_T$ for at least one mesh element $T \in \mathcal{T}_h$, this concludes the proof. □

4.3.2.2 Global Bilinear Form

We define the global bilinear forms $\mathrm{a}_h : \underline{U}_h^k \times \underline{U}_h^k \to \mathbb{R}$ and $\mathrm{s}_h : \underline{U}_h^k \times \underline{U}_h^k \to \mathbb{R}$ by element-by-element assembly setting, for all $\underline{u}_h, \underline{v}_h \in \underline{U}_h^k$,

$$\mathrm{a}_h(\underline{u}_h, \underline{v}_h) := \sum_{T\in\mathcal{T}_h} \mathrm{a}_T(\underline{u}_T, \underline{v}_T), \qquad \mathrm{s}_h(\underline{u}_h, \underline{v}_h) := \sum_{T\in\mathcal{T}_h} \mathrm{s}_T(\underline{u}_T, \underline{v}_T). \tag{4.31}$$

Lemma 4.1 (Properties of a_h) *The bilinear form a_h enjoys the following properties:*

(i) Stability. *For all $\underline{v}_h \in \underline{U}_{h,0}^k$ it holds with η as in (4.19) that*

$$\eta^{-1}\|\underline{v}_h\|_{1,h}^2 \le \|\underline{v}_h\|_{\mathrm{a},h}^2 := \mathrm{a}_h(\underline{v}_h, \underline{v}_h) \le \eta\|\underline{v}_h\|_{1,h}^2. \tag{4.32}$$

(ii) Consistency. *There is a real number $C > 0$ independent of h, but possibly depending on d, ϱ, and k, such that, for all $w \in H_0^1(\Omega) \cap H^{k+2}(\Omega)$,*

$$\sup_{\underline{v}_h \in \underline{U}_{h,0}^k, \|\underline{v}_h\|_{1,h}=1} \mathcal{E}_h(w; \underline{v}_h) \le Ch^{k+1}\|w\|_{H^{k+2}(\Omega)}, \tag{4.33}$$

with linear form $\mathcal{E}_h(w; \cdot) : \underline{U}_h^k \to \mathbb{R}$ representing the conformity error such that, for all $\underline{v}_h \in \underline{U}_h^k$,

$$\mathcal{E}_h(w; \underline{v}_h) := -(\Delta w, v_h) - \mathrm{a}_h(\underline{I}_h^k w, \underline{v}_h). \tag{4.34}$$

Proof

(i) *Stability.* Summing inequalities (4.19) over $T \in \mathcal{T}_h$, (4.32) follows.

(ii) *Consistency.* Let $\underline{v}_h \in \underline{U}_{h,0}^k$ be such that $\|\underline{v}_h\|_{1,h} = 1$. Throughout the proof, we abridge as $A \lesssim B$ the inequality $A \le cB$ with multiplicative constant $c > 0$ having the same dependecies as C in (4.33). For the sake of brevity, we also let $\check{w}_T := p_T^{k+1}\underline{I}_T^k w = \pi_T^{1,k+1} w$ (cf. (4.17)) for all $T \in \mathcal{T}_h$. Integrating by parts element-by-element, we infer that

$$-(\Delta w, v_h) = \sum_{T\in\mathcal{T}_h} \left((\nabla w, \nabla v_T)_T + \sum_{F\in\mathcal{F}_T} (\nabla w{\cdot}\mathbf{n}_{TF}, v_F - v_T)_F \right). \tag{4.35}$$

To insert v_F into the second term in parentheses in (4.35), we have used the fact that $v_F \equiv 0$ for all $F \in \mathcal{F}_h^{\mathrm{b}}$ while, for all $F \in \mathcal{F}_h^{\mathrm{i}}$ such that $F \subset \partial T_1 \cap \partial T_2$ for distinct mesh elements $T_1, T_2 \in \mathcal{T}_h$, $(\nabla w)_{|T_1}{\cdot}\mathbf{n}_{T_1F} + (\nabla w)_{|T_2}{\cdot}\mathbf{n}_{T_2F} = 0$ (since $w \in H^{k+2}(\Omega)$), so that

$$\sum_{T\in\mathcal{T}_h}\sum_{F\in\mathcal{F}_T} (\nabla w{\cdot}\mathbf{n}_{TF}, v_F)_F = \sum_{F\in\mathcal{F}_h^{\mathrm{i}}} (\sum_{T\in\mathcal{T}_F} (\nabla w)_{|T}{\cdot}\mathbf{n}_{TF}, v_F)_F + \sum_{F\in\mathcal{F}_h^{\mathrm{b}}} (\nabla w{\cdot}\mathbf{n}, v_F)_F = 0.$$

On the other hand, plugging the definition (4.18) of a_T into (4.31), and expanding $p_T^{k+1}\underline{v}_T$ according to (4.16) with $w = \check{w}_T$, it is inferred that

$$\mathrm{a}_h(\underline{I}_h^k w, \underline{v}_h) = \sum_{T\in\mathcal{T}_h}\left((\nabla\check{w}_T, \nabla v_T)_T + \sum_{F\in\mathcal{F}_T}(\nabla\check{w}_T\cdot\mathbf{n}_{TF}, v_F - v_T)_F \right.$$
$$\left. + \mathrm{s}_T(\underline{I}_T^k w, \underline{v}_T)\right). \tag{4.36}$$

Subtracting (4.36) from (4.35), using the definition (4.6) of $\pi_T^{1,k+1}$ to cancel the first terms in parentheses, and taking absolute values, we get

$$|\mathcal{E}_h(w; \underline{v}_h)| = \left|\sum_{T\in\mathcal{T}_h}\left(\sum_{F\in\mathcal{F}_T}(\nabla(w-\check{w}_T)\cdot\mathbf{n}_{TF}, v_F - v_T)_F + \mathrm{s}_T(\underline{I}_T^k w, \underline{v}_T)\right)\right|$$
$$\leq \left[\sum_{T\in\mathcal{T}_h}\left(h_T\|\nabla(w-\check{w}_T)\|_{\partial T}^2 + \mathrm{s}_T(\underline{I}_T^k w, \underline{I}_T^k w)\right)\right]^{1/2}$$
$$\times\left[\sum_{T\in\mathcal{T}_h}\left(|\underline{v}_T|_{1,\partial T}^2 + \mathrm{s}_T(\underline{v}_T, \underline{v}_T)\right)\right]^{1/2}.$$

Using (4.7b) with $\alpha = 1$, $l = k+1$, $s = k+2$, and $m = 1$ together with (4.24) for the first factor, and the seminorm equivalence (4.19) together with the fact that $\|\underline{v}_h\|_{1,h} = 1$ for the second, we infer the bound

$$|\mathcal{E}_h(w; \underline{v}_h)| \lesssim h^{k+1}\|w\|_{H^{k+2}(\Omega)}.$$

Since $\underline{v}_h$ is arbitrary, this yields (4.33). □

4.3.2.3 Discrete Problem and Well-Posedness

The discrete problem reads: Find $\underline{u}_h \in \underline{U}_{h,0}^k$ such that

$$\mathrm{a}_h(\underline{u}_h, \underline{v}_h) = (f, v_h) \qquad \forall\underline{v}_h \in \underline{U}_{h,0}^k. \tag{4.37}$$

Lemma 4.2 (Well-Posedness) *Problem* (4.37) *is well-posed, and we have the following a priori bound for the unique discrete solution* $\underline{u}_h \in \underline{U}_{h,0}^k$*:*

$$\|\underline{u}_h\|_{1,h} \leq \eta C_{\mathrm{P}}\|f\|.$$

Proof We check the assumptions of the Lax–Milgram lemma [49] on the finite-dimensional space $\underline{U}_{h,0}^k$ equipped with the norm $\|\cdot\|_{1,h}$. The bilinear form a_h is coercive and continuous owing to (4.32) with coercivity constant equal to η^{-1}. The linear form $\underline{v}_h \mapsto (f, v_h)$ is continuous owing to (4.30) with continuity constant equal to C_P. □

4.3.2.4 Implementation

Let a basis $\mathcal{B}_h$ for the space $\underline{U}_{h,0}^k$ be fixed such that every basis function is supported by only one mesh element or face. For a generic element $\underline{v}_h \in \underline{U}_{h,0}^k$, denote by V_h the corresponding vector of coefficients in $\mathcal{B}_h$ partitioned as

$$\mathsf{V}_h = \begin{bmatrix} \mathsf{V}_{\mathcal{T}_h} \\ \hdashline \mathsf{V}_{\mathcal{F}_h} \end{bmatrix},$$

where the subvectors $\mathsf{V}_{\mathcal{T}_h}$ and $\mathsf{V}_{\mathcal{F}_h}$ collect the coefficients associated to element-based and face-based DOFs, respectively. Denote by A_h the matrix representation of the bilinear form a_h and by B_h the vector representation of the linear form $\underline{v}_h \mapsto (f, v_h)$, both partitioned in a similar way. The algebraic problem corresponding to (4.37) reads

$$\underbrace{\begin{bmatrix} \mathsf{A}_{\mathcal{T}_h\mathcal{T}_h} & \mathsf{A}_{\mathcal{T}_h\mathcal{F}_h} \\ \hdashline \mathsf{A}_{\mathcal{T}_h\mathcal{F}_h}{}^\mathrm{T} & \mathsf{A}_{\mathcal{F}_h\mathcal{F}_h} \end{bmatrix}}_{\mathsf{A}_h} \underbrace{\begin{bmatrix} \mathsf{U}_{\mathcal{T}_h} \\ \hdashline \mathsf{U}_{\mathcal{F}_h} \end{bmatrix}}_{\mathsf{U}_h} = \underbrace{\begin{bmatrix} \mathsf{B}_{\mathcal{T}_h} \\ \hdashline \mathsf{0}_{\mathcal{F}_h} \end{bmatrix}}_{\mathsf{B}_h}. \tag{4.38}$$

The submatrix $\mathsf{A}_{\mathcal{T}_h\mathcal{T}_h}$ is block-diagonal and symmetric positive definite, and is therefore inexpensive to invert. In the practical implementation, this remark can be exploited by solving the linear system (4.38) in two steps (see, e.g., [18, Section 2.4]):

1. First, element-based coefficients in $\mathsf{U}_{\mathcal{T}_h}$ are expressed in terms of $\mathsf{B}_{\mathcal{T}_h}$ and $\mathsf{U}_{\mathcal{F}_h}$ by the inexpensive solution of the first block equation:

$$\mathsf{U}_{\mathcal{T}_h} = \mathsf{A}_{\mathcal{T}_h\mathcal{T}_h}^{-1}\left(\mathsf{B}_{\mathcal{T}_h} - \mathsf{A}_{\mathcal{T}_h\mathcal{F}_h}\mathsf{U}_{\mathcal{F}_h}\right). \tag{4.39a}$$

   This step is referred to as *static condensation* in the finite element literature;
2. Second, face-based coefficients in $\mathsf{U}_{\mathcal{F}_h}$ are obtained solving the global skeletal (i.e., involving unknowns attached to the mesh skeleton) problem

$$\left(\mathsf{A}_{\mathcal{F}_h\mathcal{F}_h} - \mathsf{A}_{\mathcal{T}_h\mathcal{F}_h}^\mathrm{T}\mathsf{A}_{\mathcal{T}_h\mathcal{T}_h}^{-1}\mathsf{A}_{\mathcal{T}_h\mathcal{F}_h}\right)\mathsf{U}_{\mathcal{F}_h} = -\mathsf{A}_{\mathcal{T}_h\mathcal{F}_h}^\mathrm{T}\mathsf{A}_{\mathcal{T}_h\mathcal{T}_h}^{-1}\mathsf{B}_{\mathcal{T}_h}. \tag{4.39b}$$

This computationally more intensive step requires to invert the matrix in parentheses in the above expression. This symmetric positive definite matrix, whose stencil is the same as that of $\mathsf{A}_{\mathcal{F}_h\mathcal{F}_h}$ and only involves neighbours through faces, has size $N_{\mathrm{dof}} \times N_{\mathrm{dof}}$ with

$$N_{\mathrm{dof}} = \mathrm{card}(\mathcal{F}_h^{\mathrm{i}}) \times \binom{k+d-1}{k}. \tag{4.39c}$$

4.3.2.5 Local Conservation and Flux Continuity

At the continuous level, the solution of problem (4.9) satisfies the following local balance for all $T \in \mathcal{T}_h$ and all $v_T \in \mathbb{P}^k(T)$:

$$(\nabla u, \nabla v_T)_T - \sum_{F\in\mathcal{F}_T} (\nabla u{\cdot}\mathbf{n}_{TF}, v_T)_F = (f, v_T)_T, \tag{4.40a}$$

and the normal flux traces are continuous in the sense that, for all $F \in \mathcal{F}_h^{\mathrm{i}}$ such that $F \subset \partial T_1 \cap \partial T_2$ with distinct mesh elements $T_1, T_2 \in \mathcal{T}_h$, it holds (see, e.g., [25, Lemma 4.3])

$$(\nabla u)_{|T_1}{\cdot}\mathbf{n}_{T_1F} + (\nabla u)_{|T_2}{\cdot}\mathbf{n}_{T_2F} = 0. \tag{4.40b}$$

We show in this section that a discrete counterpart of the relations (4.40) holds for the discrete solution. This property is relevant both from the engineering and mathematical points of view, and it can be exploited to derive a posteriori error estimators by flux equilibration. It was originally highlighted in [26] and, using different techniques, in [18] for the stabilization bilinear form s_T defined by (4.23). Here, using yet a different approach, we extend these results to more general stabilization bilinear forms.

Let a mesh element $T \in \mathcal{T}_h$ be fixed. We define the space

$$\underline{U}_{\partial T}^k := \bigtimes_{F\in\mathcal{F}_T} \mathbb{P}^k(F), \tag{4.41}$$

as well as the boundary difference operator $\underline{\Delta}_{\partial T}^k : \underline{U}_T^k \to \underline{U}_{\partial T}^k$ such that, for all $\underline{v}_T \in \underline{U}_T^k$,

$$\underline{\Delta}_{\partial T}^k \underline{v}_T = (\Delta_{TF}^k \underline{v}_T)_{F\in\mathcal{F}_T} := (v_F - v_{T|F})_{F\in\mathcal{F}_T}. \tag{4.42}$$

A useful remark is that, for all $\underline{v}_T \in \underline{U}_T^k$, it holds

$$\underline{v}_T - \underline{I}_T^k v_T = (v_T - \pi_T^{0,k} v_T, (v_F - \pi_F^{0,k} v_{T|F})_{F\in\mathcal{F}_T}) = (0, \underline{\Delta}_{\partial T}^k \underline{v}_T), \tag{4.43}$$

where the conclusion follows observing that, for all $T \in \mathcal{T}_h$ and all $F \in \mathcal{F}_T$, $\pi_T^{0,k} v_T = v_T$ and $\pi_F^{0,k} v_{T|F} = v_{T|F}$ since $v_T \in \mathbb{P}^k(T)$ and $v_{T|F} \in \mathbb{P}^k(F)$.

We show in the next proposition that any stabilization bilinear form with a suitable dependence on its arguments can be reformulated in terms of boundary differences.

Proposition 4.3 (Reformulation of the Stabilization Bilinear Form) *Let $T \in \mathcal{T}_h$, and assume that s_T is a stabilization bilinear form that satisfies assumptions (S1)–(S3) and that depends on its arguments only through the residuals defined by* (4.21)*. Then, it holds for all $\underline{u}_T, \underline{v}_T \in \underline{U}_T^k$ that*

$$\mathrm{s}_T(\underline{u}_T, \underline{v}_T) = \mathrm{s}_T((0, \underline{\Delta}_{\partial T}^k \underline{u}_T), (0, \underline{\Delta}_{\partial T}^k \underline{v}_T)). \tag{4.44}$$

Proof It suffices to show that, for all $\underline{v}_T \in \underline{U}_T^k$,

$$\delta_T^k \underline{v}_T = \delta_T^k(0, \underline{\Delta}_{\partial T}^k \underline{v}_T), \qquad \delta_{TF}^k \underline{v}_T = \delta_{TF}^k(0, \underline{\Delta}_{\partial T}^k \underline{v}_T) \quad \forall F \in \mathcal{F}_T.$$

Let us start by δ_T^k. Since $v_T \in \mathbb{P}^k(T)$, $p_T^{k+1} \underline{I}_T^k v_T = \pi_T^{1,k+1} v_T = v_T$. Hence,

$$\begin{aligned}
\delta_T^k \underline{v}_T &= \pi_T^{0,k}(p_T^{k+1} \underline{v}_T - v_T) \\
&= \pi_T^{0,k}(p_T^{k+1} \underline{v}_T - p_T^{k+1} \underline{I}_T^k v_T) \\
&= \pi_T^{0,k} p_T^{k+1}(\underline{v}_T - \underline{I}_T^k v_T) = \delta_T^k(0, \underline{\Delta}_{\partial T}^k \underline{v}_T),
\end{aligned}$$

where we have used the linearity of p_T^{k+1} to pass to the third line and (4.43) to conclude. Let now $F \in \mathcal{F}_T$ and consider δ_{TF}^k. We have

$$\begin{aligned}
\delta_{TF}^k \underline{v}_T &= \pi_F^{0,k}(p_T^{k+1} \underline{v}_T - v_F) \\
&= \pi_F^{0,k}(p_T^{k+1} \underline{v}_T - p_T^{k+1} \underline{I}_T^k v_T + v_T - v_F) \\
&= \pi_F^{0,k}(p_T^{k+1}(0, \underline{\Delta}_{\partial T}^k \underline{v}_T) - \Delta_{TF}^k \underline{v}_T) = \delta_{TF}^k(0, \underline{\Delta}_{\partial T}^k \underline{v}_T),
\end{aligned}$$

where we have introduced $v_T - p_T^{k+1} \underline{I}_T^k v_T = 0$ in the second line, used the linearity of p_T^{k+1} together with (4.43) and the definition (4.41) of $\underline{\Delta}_{\partial T}^k$ in the third line, and concluded recalling the definition (4.21) of δ_{TF}^k. □

Define the boundary residual operator $\underline{R}_{\partial T}^k : \underline{U}_T^k \to \underline{U}_{\partial T}^k$ such that, for all $\underline{v}_T \in \underline{U}_T^k$, $\underline{R}_{\partial T}^k \underline{v}_T = (R_{TF}^k \underline{v}_T)_{F \in \mathcal{F}_T}$ satisfies for all $\underline{\alpha}_{\partial T} = (\alpha_{TF})_{F \in \mathcal{F}_T} \in \underline{U}_{\partial T}^k$

$$-\sum_{F \in \mathcal{F}_T} (R_{TF}^k \underline{v}_T, \alpha_{TF})_F = \mathrm{s}_T((0, \underline{\Delta}_{\partial T}^k \underline{v}_T), (0, \underline{\alpha}_{\partial T})). \tag{4.45}$$

Problem (4.45) is well-posed, and computing $R_{TF}^k \underline{v}_T$ requires to invert the boundary mass matrix.

Lemma 4.3 (Local Balance and Flux Continuity) *Under the assumptions of Proposition 4.3, denote by* $\underline{u}_h \in \underline{U}_{h,0}^k$ *the unique solution of problem* (4.37) *and, for all* $T \in \mathcal{T}_h$ *and all* $F \in \mathcal{F}_T$*, define the numerical normal trace of the flux*

$$S_{TF}(\underline{u}_T) := -\nabla p_T^{k+1}\underline{u}_T{\cdot}\mathbf{n}_{TF} + R_{TF}^k\underline{u}_T$$

with R_{TF}^k *defined by* (4.45)*. Then, for all* $T \in \mathcal{T}_h$ *we have the following discrete counterpart of the local balance* (4.40a)*: For all* $v_T \in \mathbb{P}^k(T)$*,*

$$(\nabla p_T^{k+1}\underline{u}_T, \nabla v_T)_T + \sum_{F\in\mathcal{F}_T}(S_{TF}(\underline{u}_T), v_T)_F = (f, v_T)_T, \tag{4.46a}$$

and, for any interface $F \in \mathcal{F}_h^{\mathrm{i}}$ *such that* $F \subset \partial T_1 \cap \partial T_2$ *with distinct mesh elements* $T_1, T_2 \in \mathcal{T}_h$*, the numerical fluxes are continuous in the sense that (compare with* (4.40b)*):*

$$S_{T_1F}(\underline{u}_{T_1}) + S_{T_2F}(\underline{u}_{T_2}) = 0. \tag{4.46b}$$

Proof Let $\underline{v}_h \in \underline{U}_{h,0}^k$. Plugging the definition (4.18) of a_T into (4.31), using for all $T \in \mathcal{T}_h$ the definition of $p_T^{k+1}\underline{v}_T$ with $w = p_T^{k+1}\underline{u}_T$, and recalling the reformulation (4.44) of s_T together with the definition (4.45) of $\underline{R}_{\partial T}^k$ to write

$$\mathrm{s}_T(\underline{u}_T, \underline{v}_T) = -\sum_{F\in\mathcal{F}_T}(R_{TF}^k\underline{u}_T, v_F - v_T)_F \qquad \forall T \in \mathcal{T}_h, \tag{4.47}$$

we infer from the discrete problem (4.37) that

$$\sum_{T\in\mathcal{T}_h}\left((\nabla p_T^{k+1}\underline{u}_T, \nabla v_T)_T + \sum_{F\in\mathcal{F}_T}(\nabla p_T^{k+1}\underline{u}_T{\cdot}\mathbf{n}_{TF} - R_{TF}^k\underline{u}_T, v_F - v_T)_F\right)$$
$$= (f, v_h).$$

Selecting $\underline{v}_h$ such that v_T spans $\mathbb{P}^k(T)$ for a selected mesh element $T \in \mathcal{T}_h$ while $v_{T'} \equiv 0$ for all $T' \in \mathcal{T}_h \setminus \{T\}$ and $v_F \equiv 0$ for all $F \in \mathcal{F}_h$, we obtain (4.46a). On the other hand, selecting $\underline{v}_h$ such that $v_T \equiv 0$ for all $T \in \mathcal{T}_h$, v_F spans $\mathbb{P}^k(F)$ for a selected interface $F \in \mathcal{F}_h^{\mathrm{i}}$ such that $F \subset \partial T_1 \cap \partial T_2$ for distinct mesh elements $T_1, T_2 \in \mathcal{T}_h$, and $v_{F'} \equiv 0$ for all $F' \in \mathcal{F}_h \setminus \{F\}$ yields (4.46b).

Remark 4.4 (Interpretation of the Discrete Problem) Lemma 4.3 and its proof provide further insight into the structure of the discrete problem (4.37), which consists of the local balances (4.46a) (corresponding to the local block equations (4.39a)) and a global transmission condition enforcing the continuity (4.46b) of numerical fluxes (corresponding to the global skeletal problem (4.39b)).

4.3.3 A Priori Error Analysis

Having proved that the discrete problem (4.37) is well-posed, it remains to determine the convergence of the discrete solution towards the exact solution, which is precisely the goal of this section.

4.3.3.1 Energy Error Estimate

We start by deriving a basic convergence result. The error is measured as the difference between the exact solution and the global reconstruction obtained from the discrete solution through the operator $p_h^{k+1} : \underline{U}_h^k \to \mathbb{P}^{k+1}(\mathcal{T}_h)$ such that, for all $\underline{v}_h \in \underline{U}_h^k$,

$$(p_h^{k+1}\underline{v}_h)_{|T} := p_T^{k+1}\underline{v}_T \qquad \forall T \in \mathcal{T}_h. \tag{4.48}$$

Theorem 4.1 (Energy Error Estimate) *Let a polynomial degree $k \geq 0$ be fixed. Let $u \in H_0^1(\Omega)$ denote the unique solution to* (4.9), *for which we assume the additional regularity $u \in H^{k+2}(\Omega)$. Let $\underline{u}_h \in \underline{U}_{h,0}^k$ denote the unique solution to* (4.37) *with stabilization bilinear form s_T in* (4.18) *satisfying assumptions (S1)–(S3) for all $T \in \mathcal{T}_h$. Then, there exists a real number $C > 0$ independent of h, but possibly depending on d, ϱ, and k, such that*

$$\|\nabla_h(p_h^{k+1}\underline{u}_h - u)\| + |\underline{u}_h|_{\mathrm{s},h} \leq Ch^{k+1}\|u\|_{H^{k+2}(\Omega)}, \tag{4.49}$$

where $|\cdot|_{\mathrm{s},h}$ is the seminorm defined by the bilinear form s_h on $\underline{U}_h^k$.

Proof Let, for the sake of brevity, $\underline{\hat{u}}_h := \underline{I}_h^k u$ and $\check{u}_h := p_h^{k+1}\underline{\hat{u}}_h$. We abridge as $A \lesssim B$ the inequality $A \leq cB$ with multiplicative constant $c > 0$ having the same dependencies as C in (4.49). Using the triangle and Cauchy–Schwarz inequalities, it is readily inferred that

$$\|\nabla_h(p_h^{k+1}\underline{u}_h - u)\| + |\underline{u}_h|_{\mathrm{s},h} \leq \underbrace{\|\underline{u}_h - \underline{\hat{u}}_h\|_{\mathrm{a},h}}_{\mathfrak{T}_1} + \underbrace{\left(\|\nabla_h(\check{u}_h - u)\|^2 + |\underline{\hat{u}}_h|_{\mathrm{s},h}^2\right)^{1/2}}_{\mathfrak{T}_2}. \tag{4.50}$$

We have that

$$\begin{aligned}\mathfrak{T}_1^2 &= \mathrm{a}_h(\underline{u}_h, \underline{u}_h - \underline{\hat{u}}_h) - \mathrm{a}_h(\underline{\hat{u}}_h, \underline{u}_h - \underline{\hat{u}}_h)\\ &= (f, u_h - \hat{u}_h) - \mathrm{a}_h(\underline{\hat{u}}_h, \underline{u}_h - \underline{\hat{u}}_h) = \mathcal{E}_h(u; \underline{u}_h - \underline{\hat{u}}_h),\end{aligned}$$

where we have used the definition (4.32) of the $\|\cdot\|_{\mathrm{a},h}$-norm together with the linearity of a_h in its first argument in the first line, the discrete problem (4.37) to pass to the second line, and the definition (4.34) of the conformity error to conclude. As a consequence, assuming $\underline{u}_h \neq \underline{\hat{u}}_h$ (the other case is trivial), we have that

$$|\mathfrak{T}_1| \leq \mathcal{E}_h\left(u; \frac{\underline{u}_h - \underline{\hat{u}}_h}{\|\underline{u}_h - \underline{\hat{u}}_h\|_{\mathrm{a},h}}\right) \leq \eta^{1/2}\mathcal{E}_h\left(u; \frac{\underline{u}_h - \underline{\hat{u}}_h}{\|\underline{u}_h - \underline{\hat{u}}_h\|_{1,h}}\right)$$
$$\leq \eta^{1/2} \sup_{\underline{v}_h \in \underline{U}_{h,0}^k, \|\underline{v}_h\|_{1,h}=1} \mathcal{E}_h(u; \underline{v}_h),$$

where we have used the linearity of $\mathcal{E}_h(u; \cdot)$, the first bound in (4.32), and a passage to the supremum to conclude. Recalling (4.33), we arrive at

$$|\mathfrak{T}_1| \lesssim h^{k+1}\|u\|_{H^{k+2}(\Omega)}. \tag{4.51}$$

On the other hand, using the approximation properties (4.7) of $\check{u}_T$ with $\alpha = 1$, $l = k+1$, $s = k+2$, and $m = 1$ together with the approximation properties (4.24) of s_T, it is inferred for the second term

$$|\mathfrak{T}_2| \lesssim h^{k+1}\|u\|_{H^{k+2}(\Omega)}. \tag{4.52}$$

Using (4.51) and (4.52) to bound the right-hand side of (4.50), (4.49) follows. □

4.3.3.2 Convergence of the Jumps

Functions in $H^1(\mathcal{T}_h) := \left\{v \in L^2(\Omega) \mid v_{|T} \in H^1(T) \quad \forall T \in \mathcal{T}_h\right\}$ are in $H_0^1(\Omega)$ if their jumps vanish a.e. at interfaces and their trace is zero a.e. on $\partial\Omega$; see, e.g., [25, Lemma 1.23]. Thus, a measure of the nonconformity is provided by the jump seminorm $|\cdot|_{\mathrm{J},h}$ such that, for all $v \in H^1(\mathcal{T}_h)$,

$$|v|_{\mathrm{J},h}^2 := \sum_{F \in \mathcal{F}_h} h_F^{-1}\|\pi_F^{0,k}[v]_F\|_F^2, \tag{4.53}$$

where $[\cdot]_F$ denotes the usual jump operator such that, for all faces $F \in \mathcal{F}_h$ and all functions $v : \bigcup_{T \in \mathcal{T}_F} T \to \mathbb{R}$ smooth enough,

$$[v]_F := \begin{cases} v_{|T_1} - v_{|T_2} & \forall F \in \mathcal{F}_{T_1} \cap \mathcal{F}_{T_2}, \\ v & \forall F \in \mathcal{F}_h^{\mathrm{b}}. \end{cases} \tag{4.54}$$

A natural question is whether the jump seminorm of $p_h^{k+1}\underline{u}_h$ converges to zero. The answer is provided by the following lemma.

Lemma 4.4 (Convergence of the Jumps) *Under the assumptions and notations of Theorem 4.1, and further supposing, for the sake of simplicity, that the local* *stabilization bilinear form* s_T *is given by* (4.22), *there is a real number* $C > 0$ *independent of h, but possibly depending on d,* ϱ*, and k, such that*

$$|p_h^{k+1}\underline{u}_h|_{J,h} \le Ch^{k+1}\|u\|_{H^{k+2}(\Omega)}. \tag{4.55}$$

Proof Inserting u_F inside the jump and using the triangle inequality for every interface $F \in \mathcal{F}_h^{\mathrm{i}}$, and recalling that $v_F = 0$ on every boundary face $F \in \mathcal{F}_h^{\mathrm{b}}$, it is inferred that

$$\begin{aligned}
\sum_{F\in\mathcal{F}_h} h_F^{-1}\|\pi_F^{0,k}[p_h^{k+1}\underline{u}_h]_F\|_F^2 &\le 2\sum_{F\in\mathcal{F}_h}\sum_{T\in\mathcal{T}_F} h_F^{-1}\|\pi_F^{0,k}(p_T^{k+1}\underline{u}_T - u_F)\|_F^2\\
&\le 2\sum_{T\in\mathcal{T}_h}\sum_{F\in\mathcal{F}_T} h_F^{-1}\|\pi_F^{0,k}(p_T^{k+1}\underline{u}_T - u_F)\|_F^2\\
&\le 2|\underline{u}_h|_{s,h}^2.
\end{aligned}$$

Using (4.49) to bound the right-hand side yields (4.55).

4.3.3.3 L^2-Error Estimate

To close this section, we state a result concerning the convergence of the error in the L^2-norm. Optimal error estimates require in this context further regularity for the continuous operator. More precisely, we assume that, for all $g \in L^2(\Omega)$, the unique solution of the problem: Find $z \in H_0^1(\Omega)$ such that

$$a(z, v) = (g, v) \qquad \forall v \in H_0^1(\Omega)$$

satisfies the a priori estimate

$$\|z\|_{H^2(\Omega)} \le C\|g\|,$$

with real number C depending only on Ω. Elliptic regularity holds when the domain Ω is convex; see, e.g., [46]. The following result, whose detailed proof is omitted, can be obtained using the arguments of [33, Theorem 10] and [1, Corollary 4.6].

Theorem 4.2 (**L^2-Error Estimate**) *Under the assumptions and notations of Theorem 4.1, and further assuming elliptic regularity and that* $f \in H^1(\Omega)$ *if* $k = 0$, $f \in H^k(\Omega)$ *if* $k \ge 1$, *there exists a real number* $C > 0$ *independent of h, but possibly depending on* Ω, *d,* ϱ*, and k, such that*

$$\|p_h^{k+1}\underline{u}_h - u\| \le \begin{cases} Ch^2\|f\|_{H^1(\Omega)} & \text{if } k = 0,\\ Ch^{k+2}\left(\|u\|_{H^{k+2}(\Omega)} + \|f\|_{H^k(\Omega)}\right) & \text{if } k \ge 1.\end{cases} \tag{4.56}$$

Remark 4.5 (Supercloseness of Element DOFs) An intermediate step in the proof of the estimate (4.56) (see [33, Theorem 10]) consists in showing that the element DOFs are superclose to the L^2-projection of the exact solution on $\mathbb{P}^k(\mathcal{T}_h)$:

$$\|\pi_h^{0,k}u - u_h\| \le \begin{cases} Ch^2\|f\|_{H^1(\Omega)} & \text{if } k = 0, \\ Ch^{k+2}\left(\|u\|_{H^{k+2}(\Omega)} + \|f\|_{H^k(\Omega)}\right) & \text{if } k \ge 1. \end{cases} \tag{4.57}$$

This is done adapting to the HHO framework the classical Aubin–Nitsche technique.

4.3.4 A Posteriori Error Analysis

For smooth enough exact solutions, it is classically expected that increasing the polynomial degree k will reduce the computational time required to achieve a desired precision; see, e.g., the numerical test in Sect. 4.3.5.2 below and, in particular, Fig. 4.6. However, when the regularity requirements detailed in Theorems 4.1 and 4.2 are not met, the order of convergence is limited by the regularity of the solution instead of the polynomial degree. To restore optimal orders of convergence, local mesh adaptation is required. This is typically done using a posteriori error estimators to mark the elements where the error is larger, and locally refine the computational mesh based on this information. Here, we present energy-norm upper and lower bounds for the HHO method (4.37) inspired by the residual-based approach of [31].

4.3.4.1 Error Upper Bound

We start by proving an upper bound of the discretization error in terms of quantities whose computation does not require the knowledge of the exact solution. We will need the following local Poincaré and Friedrichs inequalities, valid for all $T \in \mathcal{T}_h$ and all $\varphi \in H^1(T)$:

$$\|\varphi - \pi_T^{0,0}\varphi\|_T \le C_{\mathrm{P},T} h_T \|\nabla\varphi\|_T, \tag{4.58}$$

$$\|\varphi - \pi_T^{0,0}\varphi\|_{\partial T} \le C_{\mathrm{F},T}^{1/2} h_T^{1/2} \|\nabla\varphi\|_T. \tag{4.59}$$

In (4.58), $C_{\mathrm{P},T}$ is a constant equal to $d\pi^{-1}$ if T is convex [4, 52]. In (4.59), $C_{\mathrm{F},T}$ is a constant which, if T is a simplex, can be estimated as $C_{\mathrm{F},T} = C_{\mathrm{P},T}(h_T|\partial T|_{d-1}/|T|_d)(2/d + C_{\mathrm{P},T})$ (see [25, Section 5.6.2.2]).

Theorem 4.3 (A Posteriori Error Upper Bound) *Let $u \in H_0^1(\Omega)$ and $\underline{u}_h \in \underline{U}_{h,0}^k$ denote the unique solutions to problems* (4.9) *and* (4.37), *respectively, with local stabilization bilinear form s_T satisfying the assumptions of Proposition 4.3 for all*

$T \in \mathcal{T}_h$. Let u_h^* be an arbitrary function in $H_0^1(\Omega)$. Then, it holds that

$$\|\nabla_h(p_h^{k+1}\underline{u}_h - u)\| \le \left[\sum_{T\in\mathcal{T}_h}\left(\eta_{\mathrm{nc},T}^2 + (\eta_{\mathrm{res},T} + \eta_{\mathrm{sta},T})^2\right)\right]^{1/2}, \tag{4.60}$$

with local nonconformity, residual, and stabilization estimators such that, for all $T \in \mathcal{T}_h$,

$$\eta_{\mathrm{nc},T} := \|\nabla(p_T^{k+1}\underline{u}_T - u_h^*)\|_T, \tag{4.61a}$$

$$\eta_{\mathrm{res},T} := C_{\mathrm{P},T}h_T\|(f + \Delta p_T^{k+1}\underline{u}_T) - \pi_T^{0,0}(f + \Delta p_T^{k+1}\underline{u}_T)\|_T, \tag{4.61b}$$

$$\eta_{\mathrm{sta},T} := C_{\mathrm{F},T}^{1/2}h_T^{1/2}\left(\sum_{F\in\mathcal{F}_T}\|R_{TF}^k\underline{u}_T\|_F^2\right)^{1/2}, \tag{4.61c}$$

where, for all $F \in \mathcal{F}_T$, the boundary residual R_{TF}^k is defined by (4.45).

Remark 4.6 (Nonconformity Estimator) To compute the estimator $\eta_{\mathrm{nc},T}$, we can obtain a $H_0^1(\Omega)$-conforming function u_h^* by applying a node-averaging operator to $p_h^{k+1}\underline{u}_h$. Let an integer $l \ge 1$ be fixed. When $\mathcal{T}_h$ is a matching simplicial mesh and $\mathcal{F}_h$ is the corresponding set of simplicial faces, the node-averaging operator $\mathcal{I}_h^l$: $\mathbb{P}^l(\mathcal{T}_h) \to \mathbb{P}^l(\mathcal{T}_h) \cap H_0^1(\Omega)$ is defined by setting for each (Lagrange) interpolation node N

$$\mathcal{I}_h^l v_h(N) := \begin{cases} \frac{1}{\mathrm{card}(\mathcal{T}_N)}\sum_{T\in\mathcal{T}_N}(v_h)_{|T}(N) & \text{if } N \in \Omega, \\ 0 & \text{if } N \in \partial\Omega, \end{cases}$$

where the set $\mathcal{T}_N \subset \mathcal{T}_h$ collects the simplices to which N belongs. We then set

$$u_h^* := \mathcal{I}_h^{k+1}p_h^{k+1}\underline{u}_h. \tag{4.62}$$

The generalization to polytopal meshes can be realized applying the node averaging operator to $p_h^{k+1}\underline{u}_h$ on a simplicial submesh of $\mathcal{T}_h$ (whose existence is guaranteed for regular mesh sequences, see Definition 4.3).

Proof Let the equation residual $\mathcal{R} \in H^{-1}(\Omega)$ be such that, for all $\varphi \in H_0^1(\Omega)$, $\langle\mathcal{R},\varphi\rangle_{-1,1} := (f,\varphi) - (\nabla_h p_h^{k+1}\underline{u}_h, \nabla\varphi)$. The following abstract error estimate descends from [25, Lemma 5.44] and is valid for any function $u_h^* \in H_0^1(\Omega)$:

$$\|\nabla_h(p_h^{k+1}\underline{u}_h - u)\|^2 \le \|\nabla_h(p_h^{k+1}\underline{u}_h - u_h^*)\|^2 + \left(\sup_{\varphi\in H_0^1(\Omega),\|\nabla\varphi\|=1}\langle\mathcal{R},\varphi\rangle_{-1,1}\right)^2. \tag{4.63}$$

Denote by $\mathfrak{T}_1$ and $\mathfrak{T}_2$ the addends in the right-hand side of (4.63).

(i) *Bound of* $\mathfrak{T}_1$. Recalling the definition (4.61a) of the nonconformity estimator, it is readily inferred that

$$\mathfrak{T}_1 = \sum_{T \in \mathcal{T}_h} \eta_{\mathrm{nc},T}^2. \tag{4.64}$$

(ii) *Bound of* $\mathfrak{T}_2$. We bound the argument of the supremum in $\mathfrak{T}_2$ for a generic function $\varphi \in H_0^1(\Omega)$. Using an element-by-element integration by parts, we obtain

$$\langle \mathcal{R}, \varphi \rangle_{-1,1} = \sum_{T \in \mathcal{T}_h} \left((f + \Delta p_T^{k+1} \underline{u}_T, \varphi)_T - \sum_{F \in \mathcal{F}_T} (\nabla p_T^{k+1} \underline{u}_T \cdot \mathbf{n}_{TF}, \varphi)_F \right). \tag{4.65}$$

Let now $\underline{\varphi}_h \in \underline{U}_{h,0}^k$ be such that $\varphi_T = \pi_T^{0,0} \varphi$ for all $T \in \mathcal{T}_h$ and $\varphi_F = \pi_F^{0,k} \varphi_{|F}$ for all $F \in \mathcal{F}_h$. We have that

$$\begin{aligned}
&\sum_{T \in \mathcal{T}_h} (\pi_T^{0,0}(f + \Delta p_T^{k+1} \underline{u}_T), \varphi)_T \\
&\quad = \sum_{T \in \mathcal{T}_h} (f + \Delta p_T^{k+1} \underline{u}_T, \varphi_T)_T \\
&\quad = \sum_{T \in \mathcal{T}_h} \left(\mathrm{a}_T(\underline{u}_T, \underline{\varphi}_T) + \sum_{F \in \mathcal{F}_T} (\nabla p_T^{k+1} \underline{u}_T \cdot \mathbf{n}_{TF}, \varphi_T)_F \right) \\
&\quad = \sum_{T \in \mathcal{T}_h} \left(\mathrm{s}_T(\underline{u}_T, \underline{\varphi}_T) + \sum_{F \in \mathcal{F}_T} (\nabla p_T^{k+1} \underline{u}_T \cdot \mathbf{n}_{TF}, \varphi)_F \right),
\end{aligned} \tag{4.66}$$

where we have used definition (4.4) of $\pi_T^{0,0}$ in the first line, the discrete problem (4.37) with $\underline{v}_h = \underline{\varphi}_h$ and an element-by-element integration by parts together with the fact that $\nabla \varphi_T \equiv 0$ for all $T \in \mathcal{T}_h$ in the second line. In order to pass to the third line, we have expanded a_T according to its definition (4.18) and used (4.16a) with $\underline{v}_T = \underline{\varphi}_T$ and $w = p_T^{k+1} \underline{u}_T$ for the consistency term (in the boundary integral, we can write φ instead of φ_F using the definition (4.4) of $\pi_F^{0,k}$).

Summing (4.66) and (4.65), and rearranging the terms, we obtain

$$\langle \mathcal{R}, \varphi \rangle_{-1,1} = \sum_{T \in \mathcal{T}_h} \left((f + \Delta p_T^{k+1} \underline{u}_T - \pi_T^{0,0}(f + \Delta p_T^{k+1} \underline{u}_T), \varphi - \varphi_T)_T + \mathrm{s}_T(\underline{u}_T, \underline{\varphi}_T) \right), \tag{4.67}$$

where we have used the definition (4.4) of $\pi_T^{0,0}$ to insert φ_T into the first term. Let us estimate the addends inside the summation, hereafter denoted by $\mathfrak{T}_{2,1}(T)$ and $\mathfrak{T}_{2,2}(T)$. Using the Cauchy–Schwarz and local Poincaré (4.58) inequalities, and recalling the definition (4.61b) of the residual estimator, we readily infer, for all $T \in \mathcal{T}_h$, that

$$|\mathfrak{T}_{2,1}(T)| \leq \eta_{\mathrm{res},T} \|\nabla\varphi\|_T. \tag{4.68}$$

On the other hand, recalling the reformulation (4.47) of the local stabilization bilinear form s_T we have, for all $T \in \mathcal{T}_h$,

$$|\mathfrak{T}_{2,2}(T)| = \left| \sum_{F \in \mathcal{F}_T} (R_{TF}^k \underline{u}_T, \varphi - \varphi_T)_F \right| \leq \eta_{\mathrm{sta},T} \|\nabla\varphi\|_T, \tag{4.69}$$

where we have used the fact that $\varphi_F = \pi_F^{0,k}\varphi$ and $R_{TF}^k \underline{u}_T \in \mathbb{P}^k(F)$ together with the definition (4.4) of $\pi_F^{0,k}$ to write φ instead of φ_F inside the boundary term, and the Cauchy–Schwarz and local Friedrichs (4.59) inequalities followed by definition (4.61c) of the stability estimator to conclude. Using (4.68) and (4.69) to estimate the right-hand side of (4.67) followed by a Cauchy–Schwarz inequality, and plugging the resulting bound inside the supremum in $\mathfrak{T}_2$, we arrive at

$$\mathfrak{T}_2 \leq \sum_{T \in \mathcal{T}_h} (\eta_{\mathrm{res},T} + \eta_{\mathrm{sta},T})^2. \tag{4.70}$$

(iii) *Conclusion.* Plug (4.64) and (4.70) into (4.63). □

4.3.4.2 Error Lower Bound

In practice, one wants to make sure that the error estimators are able to correctly localize the error (for use, e.g., in adaptive mesh refinement) and that they do not unduly overestimate it. We prove in this section that the error estimators defined in Theorem 4.3 are *locally efficient*, i.e., they are locally controlled by the error. This shows that they are suitable to drive mesh refinement. Moreover, they are also *globally efficient*, i.e., the right-hand side of (4.60) is (uniformly) controlled by the discretization error, so that it cannot depart from it.

Let a mesh element $T \in \mathcal{T}_h$ be fixed and define the following sets of faces and elements sharing at least one node with T:

$$\mathcal{F}_{\mathcal{N},T} := \{F \in \mathcal{F}_h \mid \overline{F} \cap \partial T \neq \emptyset\}, \qquad \mathcal{T}_{\mathcal{N},T} := \{T' \in \mathcal{T}_h \mid \overline{T}' \cap \overline{T} \neq \emptyset\}.$$

Let an integer $l \geq 1$ be fixed. The following result is proved in [48] for standard meshes: There is a real number $C > 0$ independent of h, but possibly depending on

d, ϱ, and l, such that, for all $v_h \in \mathbb{P}^l(\mathcal{T}_h)$ and all $T \in \mathcal{T}_h$,

$$\|v_h - \mathcal{I}_h^l v_h\|_T^2 \leq C \sum_{F \in \mathcal{F}_{\mathcal{N},T}} h_F \|[v_h]_F\|_F^2, \tag{4.71}$$

with jump operator defined by (4.54). Following [25, Section 5.5.2], (4.71) still holds on regular polyhedral meshes when the nodal interpolator is defined on the matching simplicial submesh of Definition 4.3. We also note the following technical result:

Proposition 4.4 (Estimate of Boundary Oscillations) *Let an integer $l \geq 0$ be fixed. There is a real number $C > 0$ independent of h, but possibly depending on d, ϱ, and l, such that, for all mesh elements $T \in \mathcal{T}_h$ and all functions $\varphi \in H^1(T)$,*

$$h_F^{-1/2} \|\varphi - \pi_F^{0,l} \varphi\|_F \leq C \|\nabla \varphi\|_T. \tag{4.72}$$

Proof We abridge as $A \lesssim B$ the inequality $A \leq cB$ with multiplicative constant $c > 0$ having the same dependencies as C in (4.72). Let $F \in \mathcal{F}_T$ and observe that

$$\begin{aligned} \|\varphi - \pi_F^{0,l} \varphi\|_F &\leq \|\varphi - \pi_T^{0,l} \varphi\|_F + \|\pi_F^{0,l}(\pi_T^{0,l} \varphi - \varphi)\|_F \\ &\leq 2\|\varphi - \pi_T^{0,l} \varphi\|_F \lesssim h_T^{1/2} \|\nabla \varphi\|_T, \end{aligned} \tag{4.73}$$

where we have inserted $\pm \pi_T^{0,l} \varphi$ and used the triangle inequality to infer the first bound, we have used the $L^2(F)$-boundedness of $\pi_F^{0,l}$ to infer the second, and invoked (4.7b) with $\alpha = 0$, $m = 0$, and $s = 1$ to conclude. Using the fact that $h_T / h_F \lesssim 1$ owing to (4.2) gives the desired result. □

Theorem 4.4 (A Posteriori Error Lower Bound) *Under the assumptions of Theorem 4.3, and further assuming, for the sake of simplicity, (i) that the local stabilization bilinear form s_T is given by (4.22) for all $T \in \mathcal{T}_h$, (ii) that u_h^* is obtained applying the node-averaging operator to $p_h^{k+1} \underline{u}_h$ on $\mathcal{T}_h$ if $\mathcal{T}_h$ is matching simplicial or on the simplicial submesh of Definition 4.3 if this is not the case, and (iii) that $f \in \mathbb{P}^{k+1}(\mathcal{T}_h)$, it holds for all $T \in \mathcal{T}_h$,*

$$\eta_{\mathrm{nc},T} \leq C \left(\|\nabla_h (p_h^{k+1} \underline{u}_h - u)\|_{\mathcal{N},T} + |\underline{u}_h|_{\mathrm{s},\mathcal{N},T} \right), \tag{4.74a}$$

$$\eta_{\mathrm{res},T} \leq C \|\nabla (p_T^{k+1} \underline{u}_T - u_{|T})\|_T, \tag{4.74b}$$

$$\eta_{\mathrm{sta},T} \leq C |\underline{u}_T|_{\mathrm{s},T}, \tag{4.74c}$$

where $C > 0$ is a real number possibly depending on d, ϱ, and on k but independent of both h, T, and of the problem data. For all $T \in \mathcal{T}_h$, $\|\cdot\|_{\mathcal{N},T}$ denotes the L^2-norm *on the union of the elements in* $\mathcal{T}_{\mathcal{N},T}$ *and we have set*

$$|\underline{u}_T|_{s,T} = s_T(\underline{u}_T, \underline{u}_T)^{1/2}, \qquad |\underline{u}_h|^2_{s,\mathcal{N},T} := \sum_{T' \in \mathcal{T}_{\mathcal{N},T}} |\underline{u}_T|^2_{s,T'}.$$

Proof Let a mesh element $T \in \mathcal{T}_h$ be fixed. In the proof, we abridge as $A \lesssim B$ the inequality $A \le cB$ with multiplicative constant $c > 0$ having the same dependencies as C in (4.74).

(i) *Bound* (4.74a) *on the nonconformity estimator.* Using a local inverse inequality (see, e.g., [25, Lemma 1.44]) and the relation (4.71), we infer from (4.61a) that

$$\eta^2_{\mathrm{nc},T} \lesssim h_T^{-2} \|p_T^{k+1}\underline{u}_T - u_h^*\|_T^2 \lesssim \sum_{F \in \mathcal{F}_{\mathcal{N},T}} h_F^{-1} \|[p_h^{k+1}\underline{u}_h]_F\|_F^2, \tag{4.75}$$

where we have used the fact that, owing to mesh regularity, $h_F \lesssim h_T$ for all $F \in \mathcal{F}_{\mathcal{N},T}$. Using the fact $[u]_F = 0$ for all $F \in \mathcal{F}_h$ (see, e.g., [25, Lemma 4.3]) to write $[p_h^{k+1}\underline{u}_h - u]_F$ instead of $[p_h^{k+1}\underline{u}_h]_F$, inserting $\pi_F^{0,k}[p_h^{k+1}\underline{u}_h]_F - \pi_F^{0,k}[p_h^{k+1}\underline{u}_h - u]_F = 0$ inside the norm, and using the triangle inequality, we have for all $F \in \mathcal{F}_h[\mathcal{N}, T]$,

$$\begin{aligned}\|[p_h^{k+1}\underline{u}_h]_F\|_F &\le \|[p_h^{k+1}\underline{u}_h - u]_F - \pi_F^{0,k}[p_h^{k+1}\underline{u}_h - u]_F\|_F + \|\pi_F^{0,k}[p_h^{k+1}\underline{u}_h]_F\|_F \\ &\le \sum_{T' \in \mathcal{T}_F} \|(p_{T'}^{k-1}\underline{u}_{T'} - u) - \pi_F^{0,k}(p_{T'}^{k-1}\underline{u}_{T'} - u)\|_F \\ &\quad + \|\pi_F^{0,k}[p_h^{k+1}\underline{u}_h]_F\|_F,\end{aligned}$$

where we have expanded the jump according to its definition (4.54) and used a triangle inequality to pass to the second line. Plugging the above bound into (4.75), and using multiple times (4.72) with $\varphi = (p_{T'}^{k-1}\underline{u}_{T'} - u)$ for $T' \in \mathcal{T}_{\mathcal{N},T}$, we arrive at

$$\eta^2_{\mathrm{nc},T} \lesssim \|\nabla_h(p_h^{k+1}\underline{u}_h - u)\|^2_{\mathcal{N},T} + \sum_{F \in \mathcal{F}_{\mathcal{N},T}} h_F^{-1} \|\pi_F^{0,k}[p_h^{k+1}\underline{u}_h]_F\|_F^2.$$

To conclude, we proceed as in Lemma 4.4 to prove that the last term is bounded by $|\underline{u}_h|^2_{s,\mathcal{N},T}$ up to a constant independent of h and of the problem data.

(ii) *Bound* (4.74b) *on the residual estimator.* We use classical bubble function techniques, see e.g. [53]. For the sake of brevity, we let $r_T := f_{|T} + \Delta p_T^{k+1}\underline{u}_T$. Denote by $\mathfrak{T}_h$ the simplicial submesh of $\mathcal{T}_h$ introduced in Definition 4.3, and let $\mathfrak{T}_T := \{\tau \in \mathfrak{T}_h \mid \tau \subset T\}$, the set of simplices contained in T. For all $\tau \in \mathfrak{T}_T$, we denote by $b_\tau \in H_0^1(\tau)$ the element bubble function equal to the product of barycentric coordinates of τ and rescaled so as to take the value 1 at the center of gravity of τ. Letting $\psi_\tau := b_\tau r_T$ for all $\tau \in \mathfrak{T}_T$, the following properties hold [53]:

$$\psi_\tau = 0 \text{ on } \partial\tau, \tag{4.76a}$$

$$\|r_T\|_\tau{}^2 \lesssim (r_T, \psi_\tau)_\tau, \tag{4.76b}$$

$$\|\psi_\tau\|_\tau \lesssim \|r_T\|_\tau. \tag{4.76c}$$

We have that

$$\begin{aligned}
\|r_T\|_T^2 &= \sum_{\tau\in\mathfrak{T}_T} \|r_T\|_\tau^2 \lesssim \sum_{\tau\in\mathfrak{T}_T} (r_T, \psi_\tau)_\tau \\
&= \sum_{\tau\in\mathfrak{T}_T} (\nabla(u - p_T^{k+1}\underline{u}_T), \nabla\psi_\tau)_\tau \\
&\le \|\nabla(u - p_T^{k+1}\underline{u}_T)\|_T \left(\sum_{\tau\in\mathfrak{T}_T} h_\tau^{-2}\|\psi_\tau\|_\tau^2\right)^{1/2} \\
&\lesssim h_T^{-1}\|\nabla(u - p_T^{k+1}\underline{u}_T)\|_T \|r_T\|_T,
\end{aligned} \tag{4.77}$$

where we have used property (4.76b) in the first line, the fact that $f = -\Delta u$ together with an integration by parts and property (4.76a) to pass to the second line, the Cauchy–Schwarz inequality together with a local inverse inequality (see, e.g., [25, Lemma 1.44]) to pass to the third line, and (4.76c) together with the fact that $h_\tau^{-1} \le (\varrho h_T)^{-1}$ for all $\tau \in \mathfrak{T}_T$ (see Definition 4.3) to conclude. Recalling the definition (4.61b) of the residual estimator, observing that $\|r_T - \pi_T^{0,0} r_T\|_T \le \|r_T\|_T$ as a result of the triangle inequality followed by the $L^2(T)$-boundedness of $\pi_T^{0,0}$, and using (4.77), the bound (4.74b) follows.

(iii) *Bound* (4.74c) *on the stabilization estimator.* Using the definition (4.45) of the boundary residual operator $\underline{R}_{\partial T}^k$ with $\underline{v}_T = \underline{u}_T$ and $\underline{\alpha}_{\partial T} = -h_T \underline{R}_{\partial T}^k \underline{u}_T = (-h_T R_{TF}^k \underline{u}_T)_{F\in\mathfrak{F}_T}$, the stabilization estimator (4.61c) can be bounded as follows:

$$\eta_{\mathrm{sta},T}^2 = C_{\mathrm{F},T} s_T(\underline{u}_T, (0, -h_T \underline{R}_{\partial T}^k \underline{u}_T)) \lesssim |\underline{u}_T|_{\mathrm{s},T} |(0, -h_T \underline{R}_{\partial T}^k \underline{u}_T)|_{\mathrm{s},T}. \tag{4.78}$$

On the other hand, from property (S2) in Assumption 4.1, the relation (4.2), and the definition (4.61c) of $\eta_{\mathrm{sta},T}$, it is inferred that

$$|(0, -h_T \underline{R}_{\partial T}^k \underline{u}_T)|_{\mathrm{s},T} \le \eta^{1/2} \left(\sum_{F\in\mathfrak{F}_T} h_F^{-1} \|h_T R_{TF}^k \underline{u}_T\|_F^2\right)^{1/2} \le \left(\frac{\eta}{\varrho}\right)^{1/2} C_{\mathrm{F},T}^{-1/2} \eta_{\mathrm{sta},T}.$$

Using this estimate to bound the right-hand side of (4.78), (4.74c) follows. □

Corollary 4.1 (Global Lower Bound) *Under the assumptions of Theorem 4.4, there exists a constant C independent of h, but possibly depending on d, ϱ and k, such that*

$$\left[\sum_{T\in\mathcal{T}_h}\left(\eta_{\mathrm{nc},T}^2+(\eta_{\mathrm{res},T}+\eta_{\mathrm{sta},T})^2\right)\right]^{1/2}\leq C\left(\|\nabla_h(p_h^{k+1}\underline{u}_h-u)\|+|\underline{u}_h|_{\mathrm{s},h}\right).$$

4.3.5 *Numerical Examples*

We illustrate the numerical performance of the HHO method on a set of model problems.

4.3.5.1 Two-Dimensional Test Case

The first test case, taken from [33], aims at demonstrating the estimated orders of convergence in two space dimensions. We solve the Dirichlet problem in the unit square $\Omega=(0,1)^2$ with

$$u(\mathbf{x})=\sin(\pi x_1)\sin(\pi x_2), \tag{4.79}$$

and corresponding right-hand side $f(\mathbf{x})=2\pi^2\sin(\pi x_1)\sin(\pi x_2)$ on the triangular and polygonal meshes of Fig. 4.1a, c. Figure 4.4 displays convergence results for both mesh families and polynomial degrees up to 4. Recalling (4.51) and (4.57), we measure the energy- and L^2-errors by the quantities $\|\underline{I}_h^k u-\underline{u}_h\|_{\mathrm{a},h}$ and $\|\pi_h^{0,k}u-u_h\|$, respectively. In all cases, the numerical results show asymptotic convergence rates that match those predicted by the theory.

4.3.5.2 Three-Dimensional Test Case

The second test case, taken from [31], demonstrates the orders of convergence in three space dimensions. We solve the Dirichlet problem in the unit cube $\Omega=(0,1)^3$ with

$$u(\mathbf{x})=\sin(\pi x_1)\sin(\pi x_2)\sin(\pi x_3),$$

and corresponding right-hand side $f(\mathbf{x})=3\pi^2\sin(\pi x_1)\sin(\pi x_2)\sin(\pi x_3)$ on a matching simplicial mesh family for polynomial degrees up to 3. The numerical results displayed in Fig. 4.5 show asymptotic convergence rates that match those predicted by (4.49) and (4.56). In Fig. 4.6 we display the error versus the total computational time t_{tot} (including the pre-processing, solution, and post-processing). It

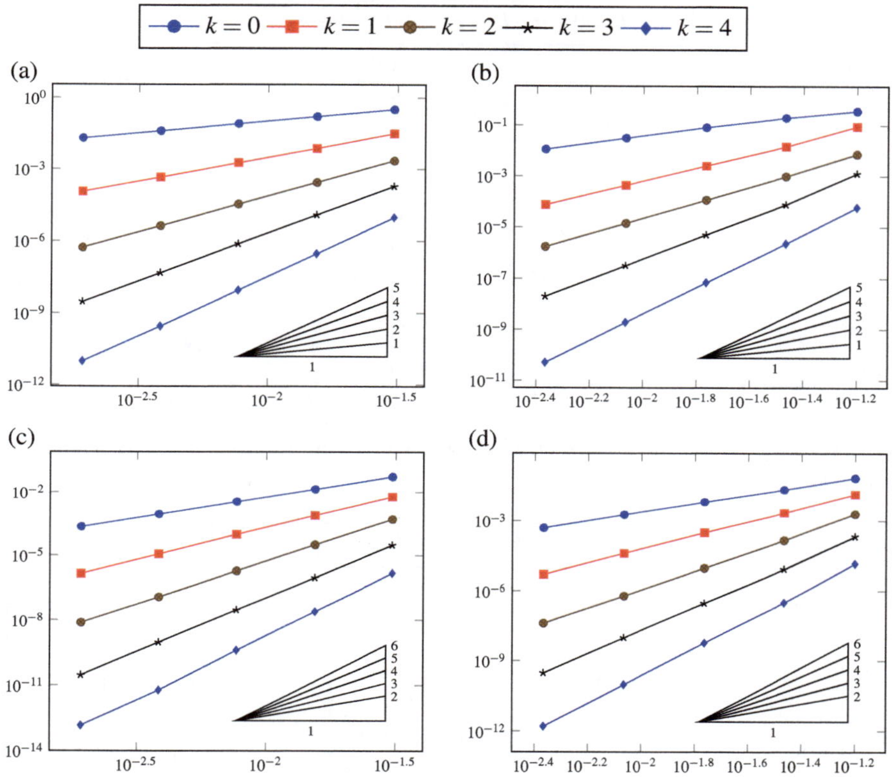

Fig. 4.4 Error vs. h for the test case of Sect. 4.3.5.1. (**a**) $\|\underline{I}_h^k u - \underline{u}_h\|_{a,h}$ vs. h, triangular mesh. (**b**) $\|\underline{I}_h^k u - \underline{u}_h\|_{a,h}$ vs. h, polygonal mesh. (**c**) $\|\pi_h^{0,k} u - u_h\|$ vs. h, triangular mesh. (**d**) $\|\pi_h^{0,k} u - u_h\|$ vs. h, polygonal mesh

can be seen that the energy- and L^2-errors optimally scale as $t_{\text{tot}}^{(k+1)/d}$ and $t_{\text{tot}}^{(k+2)/d}$ (with $d = 3$), respectively.

4.3.5.3 Three-Dimensional Case with Adaptive Mesh Refinement

The third test case, known as Fichera corner benchmark, is taken from [31] and is based on the exact solution of [45] on the etched three-dimensional domain $\Omega = (-1, 1)^3 \setminus [0, 1]^3$:

$$u(\mathbf{x}) = \sqrt[4]{x_1^2 + x_2^2 + x_3^2},$$

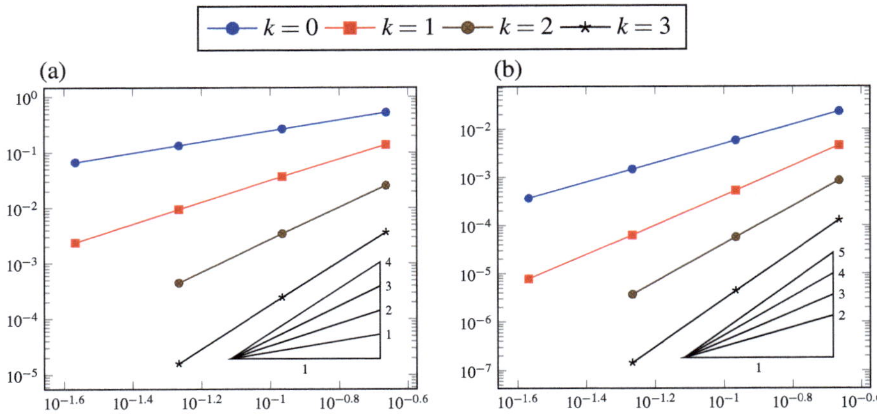

Fig. 4.5 Error vs. h for the test case of Sect. 4.3.5.2. (**a**) $\|\nabla_h(p_h^{k+1}\underline{u}_h - u)\|$ vs. h. (**b**) $\|p_h^{k+1}\underline{u}_h - u\|$ vs. h

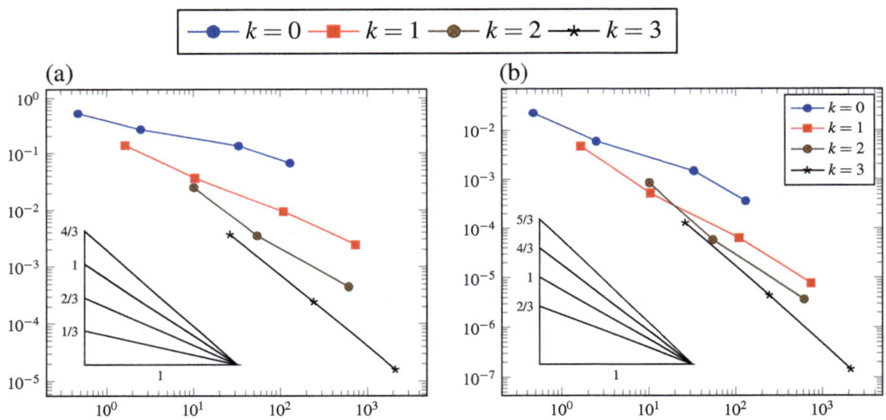

Fig. 4.6 Error vs. total computational time for the test case of Sect. 4.3.5.2. (**a**) $\|\nabla_h(p_h^{k+1}\underline{u}_h - u)\|$ vs. t_{tot}. (**b**) $\|p_h^{k+1}\underline{u}_h - u\|$ vs. t_{tot}

with right-hand side $f(\mathbf{x}) = -3/4(x_1^2 + x_2^2 + x_3^2)^{-3/4}$. In this case, the gradient of the solution has a singularity in the origin which prevents the method from attaining optimal convergence rates even for $k = 0$. In Fig. 4.7 we show a computation comparing the numerical error versus N_{dof} (cf. (4.39c)) for the Fichera problem on uniformly and adaptively refined mesh sequences for polynomial degrees up to 3. Clearly, the order of convergence is limited by the solution regularity when using uniformly refined meshes, while using adaptively refined meshes we recover optimal orders of convergence of $N_{\text{dof}}^{(k+1)/d}$ and $N_{\text{dof}}^{(k+2)/d}$ (with $d = 3$) for the energy- and L^2-errors, respectively.

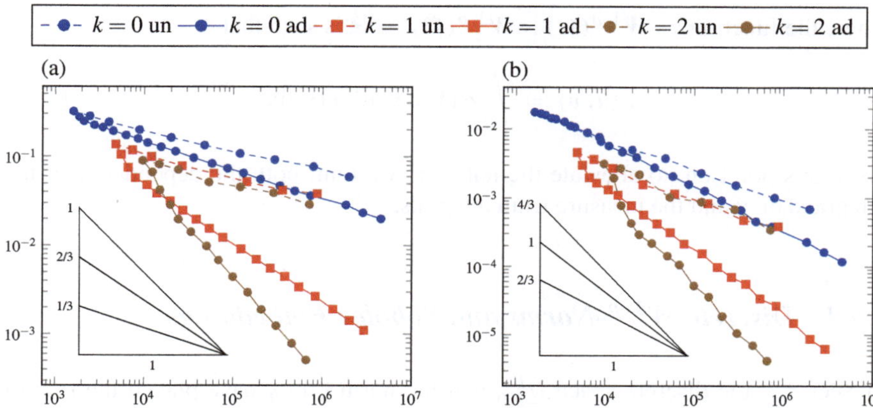

Fig. 4.7 Error vs. N_{dof} for the test case of Sect. 4.3.5.3. (**a**) Energy-error vs. N_{dof}. (**b**) L^2-error vs. N_{dof}

4.4 A Nonlinear Example: The *p*-Laplace Equation

We consider in this section an extension of the HHO method to the p-Laplace equation. This problem will be used to introduce the techniques for the discretization and analysis of nonlinear operators, as well as a set of functional analysis results of independent interest. An additional interesting point is that the p-Laplace problem is naturally posed in a non-Hilbertian setting. This will require to emulate a Sobolev structure at the discrete level.

Let $p \in (1, +\infty)$ be fixed, and set $p' := \frac{p}{p-1}$. The p-Laplace problem reads: Find $u : \Omega \to \mathbb{R}$ such that

$$\begin{aligned} -\boldsymbol{\nabla}\cdot(\boldsymbol{\sigma}(\boldsymbol{\nabla} u)) &= f \qquad \text{in } \Omega, \\ u &= 0 \qquad \text{on } \partial\Omega, \end{aligned} \tag{4.80}$$

where $f \in L^{p'}(\Omega)$ is a volumetric source term and the function $\boldsymbol{\sigma} : \mathbb{R}^d \to \mathbb{R}^d$ is such that

$$\boldsymbol{\sigma}(\boldsymbol{\tau}) := |\boldsymbol{\tau}|^{p-2}\boldsymbol{\tau}. \tag{4.81}$$

The p-Laplace equation is a generalization of the Poisson problem considered in Sect. 4.3, which corresponds to the choice $p = 2$.

Classically, the weak formulation of problem (4.80) reads: Find $u \in W_0^{1,p}(\Omega)$ such that, for all $v \in W_0^{1,p}(\Omega)$,

$$a(u, v) = \int_\Omega f(\mathbf{x})v(\mathbf{x})\mathrm{d}\mathbf{x}, \tag{4.82}$$

where the function $a : W^{1,p}(\Omega) \times W^{1,p}(\Omega) \to \mathbb{R}$ is such that

$$a(u, v) := \int_{\Omega} \boldsymbol{\sigma}(\nabla u(\mathbf{x}))\cdot\nabla v(\mathbf{x})\mathrm{d}\mathbf{x}. \tag{4.83}$$

From this point on, to alleviate the notation, we omit both the dependence of the integrand on $\mathbf{x}$ and the measure from integrals.

4.4.1 Discrete $W^{1,p}$-Norms and Sobolev Embeddings

In Sect. 4.3, the discrete space $\underline{U}_{h,0}^k$ and the norm $\|\cdot\|_{1,h}$ have played the role of the Hilbert space $H_0^1(\Omega)$ and of the seminorm $|\cdot|_{H^1(\Omega)}$, respectively (notice that $|\cdot|_{H^1(\Omega)}$ is a norm on $H_0^1(\Omega)$ by virtue of the continuous Poincaré inequality). For the p-Laplace equation, $\underline{U}_{h,0}^k$ will replace at the discrete level the Sobolev space $W_0^{1,p}(\Omega)$. A good candidate for the role of the corresponding seminorm $|\cdot|_{W^{1,p}(\Omega)}$ is the map $\|\cdot\|_{1,p,h}$ such that, for all $\underline{v}_h \in \underline{U}_h^k$,

$$\|\underline{v}_h\|_{1,p,h}^p := \sum_{T\in\mathcal{T}_h} \|\underline{v}_T\|_{1,p,T}^p, \tag{4.84}$$

where, for all $T \in \mathcal{T}_h$,

$$\|\underline{v}_T\|_{1,p,T}^p := \|\nabla v_T\|_{L^p(T)^d}^p + \sum_{F\in\mathcal{F}_T} h_F^{1-p}\|v_F - v_T\|_{L^p(F)}^p. \tag{4.85}$$

The power of h_F in the second term ensures that both contributions have the same scaling. When $p = 2$, we recover the seminorm $\|\cdot\|_{1,h}$ defined by (4.29).

The following discrete Sobolev embeddings are proved in [22, Proposition 5.4]. The proof hinges on the results of [24, Theorem 6.1] for broken polynomial spaces (based, in turn, on the techniques originally developed in [44] in the context of finite volume methods). Their role in the analysis of HHO methods for problem (4.82) is discussed in Remark 4.9.

Theorem 4.5 (Discrete Sobolev Embeddings) *Let a polynomial degree $k \geq 0$ and an index $p \in (1, +\infty)$ be fixed. Let $(\mathcal{M}_h)_{h\in\mathcal{H}}$ denote a regular sequence of meshes in the sense of Definition 4.3. Let $1 \leq q \leq \frac{dp}{d-p}$ if $1 \leq p < d$ and $1 \leq q < +\infty$ if $p \geq d$. Then, there exists a real number $C > 0$ only depending on Ω, ϱ, l, p, and q such that, for all $\underline{v}_h \in \underline{U}_{h,0}^k$,*

$$\|v_h\|_{L^q(\Omega)} \leq C\|\underline{v}_h\|_{1,p,h}. \tag{4.86}$$

Remark 4.7 (Discrete Poincaré Inequality) The discrete Poincaré inequality (4.30) is a special case of Theorem 4.5 corresponding to $p = q = 2$ (this choice is possible in any space dimension).

4.4.2 Discrete Gradient and Compactness

The analysis of numerical methods for linear problems is usually carried out in the spirit of the Lax–Richtmyer equivalence principle: "For a consistent numerical method, stability is equivalent to convergence"; see for instance [20] for a rigorous proof in the case of linear Cauchy problems. When dealing with nonlinear problems, however, some form of compactness is also required; cf. Remark 4.10 for further insight into this point. In order to achieve it for problem (4.82), we need to introduce a local gradient reconstruction slightly richer than ∇p_T^{k+1}; see (4.16).

Let a mesh element $T \in \mathcal{T}_h$ be fixed. By the principles illustrated in Sect. 4.3.1.1, we define the local gradient reconstruction $\mathbf{G}_T^k : \underline{U}_T^k \to \mathbb{P}^k(T)^d$ such that, for all $\underline{v}_T \in \underline{U}_T^k$,

$$(\mathbf{G}_T^k \underline{v}_T, \boldsymbol{\tau})_T = -(v_T, \nabla\cdot\boldsymbol{\tau})_T + \sum_{F\in\mathcal{F}_T} (v_F, \boldsymbol{\tau}\cdot\mathbf{n}_{TF})_F \quad \forall \boldsymbol{\tau} \in \mathbb{P}^k(T)^d. \tag{4.87}$$

Notice that here we reverted to the L^2-product notation instead of using integrals to emphasize the fact that the definition of $\mathbf{G}_T^k$ is inherently L^2-based.

Remark 4.8 (Relation Between $\mathbf{G}_T^k$ and p_T^{k+1}) Taking $\boldsymbol{\tau} = \nabla w$ with $w \in \mathbb{P}^{k+1}(T)$ in (4.87) and comparing with (4.16a), it is readily inferred that

$$(\mathbf{G}_T^k \underline{v}_T - \nabla p_T^{k+1} \underline{v}_T, \nabla w)_T = 0 \qquad \forall w \in \mathbb{P}^{k+1}(T), \tag{4.88}$$

i.e., $\nabla p_T^{k+1} \underline{v}_T$ is the L^2-orthogonal projection of $\mathbf{G}_T^k \underline{v}_T$ on $\nabla \mathbb{P}^{k+1}(T) \subset \mathbb{P}^k(T)^d$. In passing, we observe that for $k = 0$, using the fact that $\nabla \mathbb{P}^1(T) = \mathbb{P}^0(T)^d$, (4.88) implies that $\mathbf{G}_T^0 \underline{v}_T = \nabla p_T^1 \underline{v}_T$.

Choosing a larger arrival space for $\mathbf{G}_T^k$ has the effect of modifying the commuting property as follows (compare with (4.17)): For all $v \in W^{1,1}(T)$,

$$(\mathbf{G}_T^k \circ \underline{I}_T^k) v = \boldsymbol{\pi}_T^{0,k}(\nabla v). \tag{4.89}$$

At the global level, we define the operator $\mathbf{G}_h^k : \underline{U}_h^k \to \mathbb{P}^k(\mathcal{T}_h)^d$ such that, for all $\underline{v}_h \in \underline{U}_h^k$,

$$(\mathbf{G}_h^k \underline{v}_h)_{|T} := \mathbf{G}_T^k \underline{v}_T \qquad \forall T \in \mathcal{T}_h. \tag{4.90}$$

The commuting property (4.89) is used in conjunction with the properties of the L^2-projector to prove the following lemma, which states the compactness of sequences of HHO functions uniformly bounded in a discrete Sobolev norm.

Lemma 4.5 (Discrete Compactness) *Let a polynomial degree $k \geq 0$ and an index $p \in (1, +\infty)$ be fixed. Let $(\mathcal{M}_h)_{h\in\mathcal{H}}$ denote a regular sequence of meshes in the sense of Definition 4.3. Let $(\underline{v}_h)_{h\in\mathcal{H}} \in (\underline{U}_{h,0}^k)_{h\in\mathcal{H}}$ be a sequence for which there exists a real number $C > 0$ independent of h such that*

$$\|\underline{v}_h\|_{1,p,h} \leq C \qquad \forall h \in \mathcal{H}.$$

Then, there exists $v \in W_0^{1,p}(\Omega)$ such that, up to a subsequence, as $h \to 0$,

(i) $v_h \to v$ and $p_h^{k+1}\underline{v}_h \to v$ strongly in $L^q(\Omega)$ for all $1 \leq q < \frac{dp}{d-p}$ if $1 \leq p < d$ and $1 \leq q < +\infty$ if $p \geq d$;
(ii) $\mathbf{G}_h^k\underline{v}_h \to \nabla v$ weakly in $L^p(\Omega)^d$.

4.4.3 Discrete Problem and Well-Posedness

The discrete counterpart of the function a defined by (4.83) is the function $\mathrm{a}_h : \underline{U}_h^k \times \underline{U}_h^k \to \mathbb{R}$ such that, for all $\underline{u}_h, \underline{v}_h \in \underline{U}_h^k$,

$$\mathrm{a}_h(\underline{u}_h, \underline{v}_h) := \int_\Omega \boldsymbol{\sigma}(\mathbf{G}_h^k\underline{u}_h)\cdot\mathbf{G}_h^k\underline{v}_h + \sum_{T\in\mathcal{T}_h} \mathrm{s}_T(\underline{u}_T, \underline{v}_T). \tag{4.91}$$

Here, for all $T \in \mathcal{T}_h$, $\mathrm{s}_T : \underline{U}_T^k \times \underline{U}_T^k \to \mathbb{R}$ is a local stabilization function which can be obtained, e.g., by generalizing (4.23) to the non-Hilbertian setting:

$$\mathrm{s}_T(\underline{u}_T, \underline{v}_T):= \sum_{F\in\mathcal{F}_T} h_F^{1-p} \int_F |\delta_{TF}^k\underline{u}_T - \delta_T^k\underline{u}_T|^{p-2}(\delta_{TF}^k\underline{u}_T - \delta_T^k\underline{u}_T)(\delta_{TF}^k\underline{v}_T - \delta_T^k\underline{v}_T). \tag{4.92}$$

The discrete problem reads: Find $\underline{u}_h \in \underline{U}_{h,0}^k$ such that

$$\mathrm{a}_h(\underline{u}_h, \underline{v}_h) = \int_\Omega f v_h \qquad \forall \underline{v}_h \in \underline{U}_{h,0}^k. \tag{4.93}$$

The following result summarizes [22, Theorem 4.5, Remark 4.7, and Proposition 6.1].

Lemma 4.6 (Well-Posedness) *Problem* (4.93) *admits a unique solution, and there exists a real number $C > 0$ independent of h, but possibly depending on Ω, d, ϱ, and k, such that, denoting by $p' := \frac{p}{p-1}$ the dual exponent of p, it holds that*

$$\|\underline{u}_h\|_{1,p,h} \le C\|f\|_{L^{p'}(\Omega)}{}^{\frac{1}{p-1}}. \tag{4.94}$$

Remark 4.9 (Role of the Discrete Sobolev Embeddings) The discrete Sobolev embedding (4.86) with $q = p$ is used in the proof of the a priori bound (4.94) to estimate the right-hand side of the discrete problem (4.93) after selecting $\underline{v}_h = \underline{u}_h$ and using Hölder's inequality:

$$\int_\Omega f u_h \le \|f\|_{L^{p'}(\Omega)}\|u_h\|_{L^p(\Omega)} \le \|f\|_{L^{p'}(\Omega)}\|\underline{u}_h\|_{1,p,h}.$$

4.4.4 Convergence and Error Analysis

The following theorem states the convergence of the sequence of solutions to problem (4.93) on a regular mesh sequence. Notice that convergence is proved for exact solutions that display only the minimal regularity $u \in W_0^{1,p}(\Omega)$ required by the weak formulation (4.82). This is an important point when dealing with nonlinear problems, for which further regularity can be hard to prove, and possibly requires assumptions on the data too strong to be matched in practical situations.

Theorem 4.6 (Convergence) *Let a polynomial degree $k \ge 0$ and an index $p \in (1, +\infty)$ be fixed. Let $(\mathcal{M}_h)_{h\in\mathcal{H}}$ denote a regular sequence of meshes in the sense of Definition 4.3. Let $u \in W_0^{1,p}(\Omega)$ denote the unique solution to* (4.82)*, and denote by $(\underline{u}_h)_{h\in\mathcal{H}} \in (\underline{U}_{h,0}^k)_{h\in\mathcal{H}}$ the sequence of solutions to* (4.93) *on $(\mathcal{T}_h)_{h\in\mathcal{H}}$. Then, as $h \to 0$, it holds*

(i) $u_h \to u$ *and* $p_h^{k+1}\underline{u}_h \to u$ *strongly in* $L^q(\Omega)$ *for all* $1 \le q < \frac{dp}{d-p}$ *if* $1 \le p < d$ *and* $1 \le q < +\infty$ *if* $p \ge d$;
(ii) $\mathbf{G}_h^k\underline{u}_h \to \nabla u$ *strongly in* $L^p(\Omega)^d$.

Remark 4.10 (Convergence by Compactness) Convergence proofs by compactness such as that of Theorem 4.6 proceed in three steps: (i) an energy estimate on the discrete solution is established; (ii) compactness of the sequence of discrete solutions is inferred from the energy estimate; (iii) the limit is identified as being a solution to the continuous problem. In our context, the first point corresponds to the a priori bound (4.94), while the second point relies on the compactness result of Lemma 4.5. The third step is carried out adapting the techniques of [50, 51].

When dealing with high-order methods, it is also important to determine the convergence rates attained when the solution is regular enough (or when adaptive mesh refinement is used, cf. Sect. 4.3.5.3). This makes the object of the following result, proved in [23, Theorem 7 and Corollary 10].

Theorem 4.7 (Energy Error Estimate) *Under the assumptions and notations of Theorem 4.6, and further assuming the regularity* $u \in W^{k+2,p}(\Omega)$ *and* $\boldsymbol{\sigma}(\nabla u) \in W^{k+1,p'}(\Omega)^d$ *with* $p' := \frac{p}{p-1}$, *there exists a real number* $C > 0$ *independent of* h *such that the following holds: If* $p \geq 2$,

$$\begin{aligned} &\|\nabla_h(p_h^{k+1}\underline{u}_h - u)\|_{L^p(\Omega)^d} + |\underline{u}_h|_{s,h} \\ &\quad\leq C\left[h^{k+1}|u|_{W^{k+2,p}(\Omega)} + h^{\frac{k+1}{p-1}}\left(|u|_{W^{k+2,p}(\Omega)}^{\frac{1}{p-1}} + |\boldsymbol{\sigma}(\nabla u)|_{W^{k+1,p'}(\Omega)^d}^{\frac{1}{p-1}}\right)\right], \end{aligned} \tag{4.95a}$$

while, if $p < 2$,

$$\begin{aligned} &\|\nabla_h(p_h^{k+1}\underline{u}_h - u)\|_{L^p(\Omega)^d} + |\underline{u}_h|_{s,h} \\ &\quad\leq C\left(h^{(k+1)(p-1)}|u|_{W^{k+2,p}(\Omega)}^{p-1} + h^{k+1}|\boldsymbol{\sigma}(\nabla u)|_{W^{k+1,p'}(\Omega)^d}\right), \end{aligned} \tag{4.95b}$$

where, recalling the definition (4.92) *of the local stabilization function, we have introduced the seminorm on* $\underline{U}_h^k$ *such that, for all* $\underline{v}_h \in \underline{U}_h^k$, $|\underline{v}_h|_{s,h}{}^p := \sum_{T\in\mathcal{T}_h} s_T(\underline{v}_T, \underline{v}_T)$.

Remark 4.11 (Order of Convergence) The asymptotic scaling for the approximation error in the left-hand side of (4.95) is determined by the leading terms in the right-hand side. Using the Bachmann–Landau notation,

$$\|\nabla_h(p_h^{k+1}\underline{u}_h - u)\|_{L^p(\Omega)^d} + |\underline{u}_h|_{s,h} = \begin{cases} \mathcal{O}(h^{\frac{k+1}{p-1}}) & \text{if } p \geq 2, \\ \mathcal{O}(h^{(k+1)(p-1)}) & \text{if } p < 2. \end{cases} \tag{4.96}$$

For a discussion of these orders of convergence and a comparison with other methods studied in the literature, we refer the reader to [23, Remark 3.3].

4.4.5 Numerical Example

To illustrate the performance of the HHO method, we solve the p-Laplace problem corresponding to the exact solution

$$u(\mathbf{x}) = \exp(x_1 + \pi x_2)$$

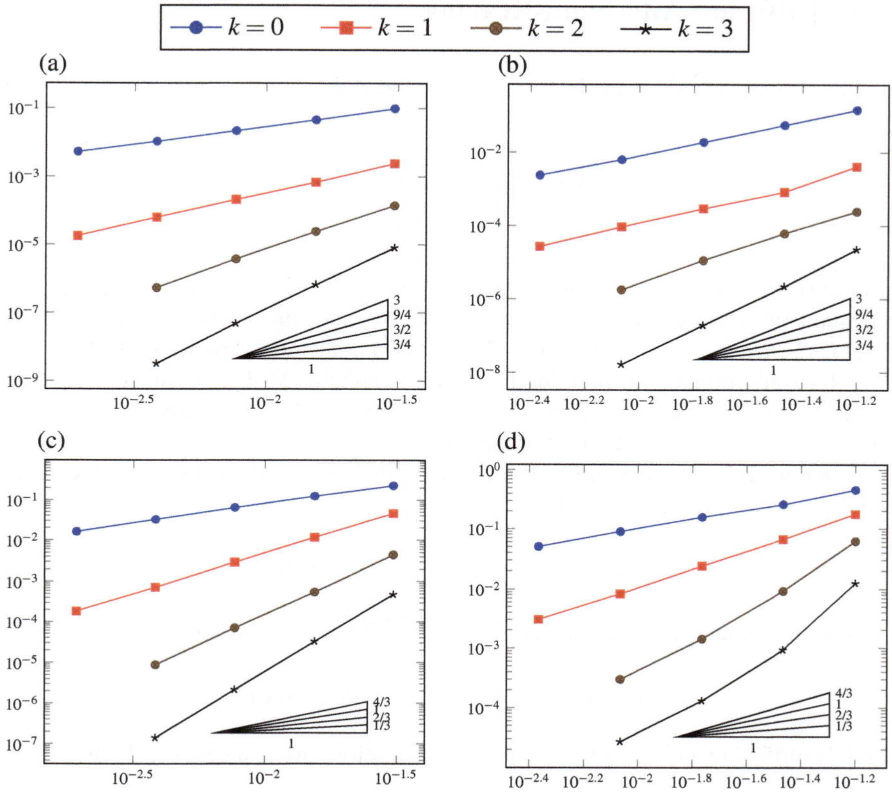

Fig. 4.8 $\|\underline{I}_h^k u - \underline{u}_h\|_{1,p,h}$ vs. h for the test case of Sect. 4.4.5. (**a**) Triangular, $p = 7/4$. (**b**) Hexagonal, $p = 7/4$. (**c**) Triangular, $p = 4$. (**d**) Hexagonal, $p = 4$

for $p \in \{7/4, 4\}$. This test is taken from [22, Section 4.4] and [23, Section 3.5]. The domain is again the unit square $\Omega = (0, 1)^2$, and the volumetric source term f is inferred from (4.80). The convergence results for the same triangular and polygonal mesh families of Sect. 4.3.5.1 (see Fig. 4.1a, c) are displayed in Fig. 4.8. Here, the error is measured by the quantity $\|\underline{I}_h^k u - \underline{u}_h\|_{1,p,h}$, for which analogous estimates as those in Theorem 4.7 hold. The error estimate seem sharp for $p = 7/4$, and the asymptotic orders of convergence match the one predicted by the theory. For $p = 4$, better orders of convergence than the asymptotic ones in (4.96) are observed. One possible explanation is that the lowest-order terms in the right-hand side of (4.95) are not yet dominant for the specific problem data and mesh. Another possibility is that compensations occur among terms that are separately estimated in the proof.

4.5 Diffusion-Advection-Reaction

In this section we extend the HHO method to the scalar diffusion-advection-reaction problem: Find $u : \Omega \to \mathbb{R}$ such that

$$\begin{aligned} \nabla\cdot(-\kappa\nabla u + \boldsymbol{\beta} u) + \mu u &= f \qquad \text{in}\,\Omega, \\ u &= 0 \qquad \text{on } \partial\Omega, \end{aligned}$$

where (i) $\kappa : \Omega \to \mathbb{R}_+^*$ is the diffusion coefficient, which we assume piecewise constant on a fixed partition of the domain P_Ω and uniformly elliptic; (ii) $\boldsymbol{\beta} \in \mathrm{Lip}(\Omega)^d$ (hence, in particular, $\boldsymbol{\beta} \in W^{1,\infty}(\Omega)^d$) is the advective velocity field, for which we additionally assume, for the sake of simplicity, $\nabla\cdot\boldsymbol{\beta} \equiv 0$; (iii) $\mu \in L^\infty(\Omega)$ is the reaction coefficient such that $\mu \geq \mu_0 > 0$ a.e. in Ω for some real number μ_0; (iv) $f \in L^2(\Omega)$ is the volumetric source term.

Having assumed κ uniformly elliptic, the following weak formulation classically holds: Find $u \in H_0^1(\Omega)$ such that

$$a_{\kappa,\boldsymbol{\beta},\mu}(u, v) = (f, v) \qquad \forall v \in H_0^1(\Omega), \tag{4.98}$$

where the bilinear form $a_{\kappa,\boldsymbol{\beta},\mu} : H^1(\Omega) \times H^1(\Omega) \to \mathbb{R}$ is such that

$$a_{\kappa,\boldsymbol{\beta},\mu}(u, v) := a_\kappa(u, v) + a_{\boldsymbol{\beta},\mu}(u, v),$$

and the diffusive and advective-reactive contributions are respectively defined by

$$a_\kappa(u, v) := (\kappa\nabla u, \nabla v), \qquad a_{\boldsymbol{\beta},\mu}(u, v) := \tfrac{1}{2}(\boldsymbol{\beta}\cdot\nabla u, v) - \tfrac{1}{2}(u, \boldsymbol{\beta}\cdot\nabla v) + (\mu u, v).$$

The first novel ingredient introduced in this section is the robust HHO discretization of first-order terms. Problem (4.98) is characterized by the presence of spatially varying coefficients, which can give rise to different regimes in different regions of the domain. In practice, one is typically interested in numerical methods that handle in a robust way locally dominant advection, corresponding to large values of a local Péclet number. As pointed out in [32], this requires that the discrete counterpart of the bilinear form $a_{\boldsymbol{\beta},\mu}$ satisfies a stability condition that guarantees well-posedness even in the absence of diffusion. This is realized here combining a reconstruction of the advective derivative obtained in the HHO spirit with an upwind stabilization that penalizes the differences between face- and element-based DOFs.

The second novelty introduced in this section is a formulation of diffusive terms with weakly enforced boundary conditions. A relevant feature of problem (4.98) is that boundary layers can appear in the vicinity of the outflow portion of $\partial\Omega$ when the diffusion coefficient takes small values. To improve the numerical approximation in this situation, one can resort to weakly enforced boundary conditions, which do not constrain the numerical solution to a fixed boundary value.

The following material is closely inspired by [34], where locally vanishing diffusion is treated (see Remark 4.15), and more general formulations for the advective stabilization term are considered.

4.5.1 Discretization of Diffusive Terms with Weakly Enforced Boundary Conditions

To avoid dealing with jumps of the diffusion coefficient inside the elements when writing the HHO discretization of problem (4.98) on a mesh $\mathcal{M}_h = (\mathcal{T}_h, \mathcal{F}_h)$, we make the following

Assumption 4.2 (Compatible Mesh) *The mesh $\mathcal{M}_h = (\mathcal{T}_h, \mathcal{F}_h)$ is compatible with the diffusion coefficient, i.e., for all $T \in \mathcal{T}_h$, there exists a unique subdomain $\omega \in P_\Omega$ such that $T \subset \omega$. For all $T \in \mathcal{T}_h$ we set, for the sake of brevity, $\kappa_T := \kappa_{|T}$.*

Letting $\varsigma > 0$ denote a user-dependent boundary penalty parameter, we define the discrete diffusive bilinear form $\mathrm{a}_{\kappa,h} : \underline{U}_h^k \times \underline{U}_h^k \to \mathbb{R}$ such that

$$\begin{aligned} \mathrm{a}_{\kappa,h}(\underline{u}_h, \underline{v}_h) &:= \sum_{T\in\mathcal{T}_h} \kappa_T \mathrm{a}_T(\underline{u}_T, \underline{v}_T) \\ &+ \sum_{F\in\mathcal{F}_h^{\mathrm{b}}} \left\{ -(\kappa_{T_F} \nabla p_{T_F}^{k+1} \underline{u}_{T_F}, v_F)_F + (u_F, \kappa_{T_F} \nabla p_{T_F}^{k+1} \underline{v}_{T_F})_F + \frac{\varsigma \kappa_{T_F}}{h_F}(u_F, v_F)_F \right\}, \end{aligned} \tag{4.99}$$

where, for all mesh elements $T \in \mathcal{T}_h$, a_T is the local diffusive bilinear form defined by (4.18) and, for all boundary faces $F \in \mathcal{F}_h^{\mathrm{b}}$, T_F denotes the unique mesh element such that $F \subset \partial T_F$. The terms in the second line of (4.99) are responsible for the weak enforcement of boundary conditions à la Nitsche.

Define the diffusion-weighted norm on $\underline{U}_h^k$ such that, for all $\underline{v}_h \in \underline{U}_h^k$, letting $\|\underline{v}_T\|_{\mathrm{a},T}^2 := \mathrm{a}_T(\underline{v}_T, \underline{v}_T)$,

$$\|\underline{v}_h\|_{\kappa,h}^2 := \sum_{T\in\mathcal{T}_h} \kappa_T \|\underline{v}_T\|_{\mathrm{a},T}^2 + \sum_{F\in\mathcal{F}_h^{\mathrm{b}}} \frac{\kappa_{T_F}}{h_F} \|v_F\|_F^2.$$

It is a simple matter to check that, for all $\varsigma \geq 1$, we have the following coercivity property for $\mathrm{a}_{\kappa,h}$: For all $\underline{v}_h \in \underline{U}_h^k$,

$$\|\underline{v}_h\|_{\kappa,h}^2 \leq \mathrm{a}_{\kappa,h}(\underline{v}_h, \underline{v}_h). \tag{4.100}$$

4.5.2 Discretization of Advective Terms with Upwind Stabilization

We introduce the ingredients for the discretization of first-order terms: a local advective derivative reconstruction and an upwind stabilization term penalizing the differences between face- and element-based DOFs.

4.5.2.1 Local Contribution

Let a mesh element $T \in \mathcal{T}_h$ be fixed. By the principles illustrated in Sect. 4.3.1.1, we define the local discrete advective derivative reconstruction $\mathbf{G}^k_{\boldsymbol{\beta},T} : \underline{U}^k_T \to \mathbb{P}^k(T)$ such that, for all $\underline{v}_T \in \underline{U}^k_T$,

$$(\mathbf{G}^k_{\boldsymbol{\beta},T}\underline{v}_T, w)_T = -(v_T, \boldsymbol{\beta}\cdot\nabla w)_T + \sum_{F\in\mathcal{F}_T} ((\boldsymbol{\beta}\cdot\mathbf{n}_{TF})v_F, w)_F \quad \forall w \in \mathbb{P}^k(T).$$

The local advective-reactive bilinear form $\mathrm{a}_{\boldsymbol{\beta},\mu,T} : \underline{U}^k_T \times \underline{U}^k_T \to \mathbb{R}$ is defined as follows:

$$\mathrm{a}_{\boldsymbol{\beta},\mu,T}(\underline{u}_T, \underline{v}_T) := \frac{1}{2}(\mathbf{G}^k_{\boldsymbol{\beta},T}\underline{u}_T, v_T)_T - \frac{1}{2}(u_T, \mathbf{G}^k_{\boldsymbol{\beta},T}\underline{v}_T)_T + \mathrm{s}_{\boldsymbol{\beta},T}(\underline{u}_T, \underline{v}_T) + (\mu u_T, v_T)_T, \tag{4.101}$$

where the bilinear form

$$\mathrm{s}_{\boldsymbol{\beta},T}(\underline{u}_T, \underline{v}_T) := \frac{1}{2}\sum_{F\in\mathcal{F}_T} (|\boldsymbol{\beta}\cdot\mathbf{n}_{TF}|(u_F - u_T), v_F - v_T)_F, \tag{4.102}$$

can be interpreted as an upwind stabilization term.

Remark 4.12 (Element-Face Upwind Stabilization) Upwinding is realized here by penalizing the difference between face- and element-based DOFs. This is a relevant difference with respect to classical (cell-based) finite volume and discontinuous Galerkin methods, where jumps of element-based DOFs are considered instead. With the choice (4.102) for the stabilization term, the stencil remains the same as for a pure diffusion problem, and static condensation of element-based DOFs in the spirit of Sect. 4.3.2.4 remains possible. In the context of the lowest-order Hybrid Mixed Mimetic methods, face-element upwind terms have been considered in [5].

To express the stability properties of $\mathrm{a}_{\boldsymbol{\beta},\mu,T}$, we define the local seminorm such that, for all $\underline{v}_T \in \underline{U}^k_T$,

$$\|\underline{v}_T\|^2_{\boldsymbol{\beta},\mu,T} := \frac{1}{2}\sum_{F\in\mathcal{F}_T} \||\boldsymbol{\beta}\cdot\mathbf{n}_{TF}|^{1/2}(v_F - v_T)\|^2_F + \hat{\tau}_T^{-1}\|v_T\|^2_T,$$

where, letting $\mathrm{L}_{\beta,T} := \max_{1\le i\le d} \|\nabla\beta_i\|_{L^\infty(T)^d}$, we have introduced the reference time

$$\hat{\tau}_T := \{\max(\|\mu\|_{L^\infty(T)}, \mathrm{L}_{\beta,T})\}^{-1}.$$

Notice that the map $\|{\cdot}\|_{\boldsymbol{\beta},\mu,T}$ is actually a norm on $\underline{U}_T^k$ provided that $\boldsymbol{\beta}_{|F}{\cdot}\mathbf{n}_{TF}$ is nonzero a.e. on each $F \in \mathfrak{F}_T$. For all $\underline{v}_T \in \underline{U}_T^k$, letting $\underline{u}_T = \underline{v}_T$ in (4.101), it can be easily checked that the following coercivity property holds:

$$\min(1, \hat{\tau}_T\mu_0)\|\underline{v}_T\|_{\boldsymbol{\beta},\mu,T}^2 \le \mathrm{a}_{\boldsymbol{\beta},\mu,T}(\underline{v}_T, \underline{v}_T). \tag{4.103}$$

4.5.2.2 Global Advective-Reactive Bilinear Form

The global advective-reactive bilinear form is given by

$$\mathrm{a}_{\boldsymbol{\beta},\mu,h}(\underline{u}_h, \underline{v}_h) := \sum_{T\in\mathfrak{T}_h} \mathrm{a}_{\boldsymbol{\beta},\mu,T}(\underline{u}_T, \underline{v}_T) + \frac{1}{2}\sum_{F\in\mathfrak{F}_h^{\mathrm{b}}} (|\boldsymbol{\beta}{\cdot}\mathbf{n}|u_F, v_F)_F, \tag{4.104}$$

where the first term results from the assembly of elementary contributions, while the second term is responsible for the enforcement of the boundary condition on the inflow portion of $\partial\Omega$.

Remark 4.13 (Link with the Advective-Reactive Bilinear Form of [34]) The bilinear form $\mathrm{a}_{\boldsymbol{\beta},\mu,h}$ defined by (4.104) admits the following equivalent reformulation, which corresponds to [34, Eq. (16)] when the upwind stabilization discussed in Section 4.2 therein is used:

$$\begin{aligned}\mathrm{a}_{\boldsymbol{\beta},\mu,h}(\underline{u}_h, \underline{v}_h) = \sum_{T\in\mathfrak{T}_h}\Big(-(u_T, \mathbf{G}_{\boldsymbol{\beta},T}^k\underline{v}_T)_T + \sum_{F\in\mathfrak{F}_T}((\boldsymbol{\beta}{\cdot}\mathbf{n}_{TF})^-(u_F-u_T), v_F-v_T)_F\Big)\\ + \sum_{T\in\mathfrak{T}_h}(\mu u_T, v_T)_T + \sum_{F\in\mathfrak{F}_h^{\mathrm{b}}}((\boldsymbol{\beta}{\cdot}\mathbf{n})^+u_F, v_F)_F,\end{aligned} \tag{4.105}$$

where, for any real number α, we have set $\alpha^\pm := \frac{1}{2}(|\alpha| \pm \alpha)$. As a matter of fact, recalling the discrete integration by parts formula [34, Eq. (35)],

$$\begin{aligned}\sum_{T\in\mathfrak{T}_h}(u_T, \mathbf{G}_{\boldsymbol{\beta},T}^k\underline{v}_T)_T = &-\sum_{T\in\mathfrak{T}_h}(\mathbf{G}_{\boldsymbol{\beta},T}^k\underline{u}_T, v_T)_T\\ &-\sum_{T\in\mathfrak{T}_h}\sum_{F\in\mathfrak{F}_T}((\boldsymbol{\beta}{\cdot}\mathbf{n}_{TF})(u_F-u_T), v_F-v_T)_F\\ &+\sum_{F\in\mathfrak{F}_h^{\mathrm{b}}}((\boldsymbol{\beta}{\cdot}\mathbf{n}_{TF})u_F, v_F)_F,\end{aligned}$$

we can reformulate the first term in the right-hand side of (4.105) as follows:

$$\begin{aligned}\sum_{T\in\mathcal{T}_h} -(u_T, \mathbf{G}^k_{\boldsymbol{\beta},T}\underline{v}_T)_T &= \sum_{T\in\mathcal{T}_h}\left(-\frac{1}{2}(u_T, \mathbf{G}^k_{\boldsymbol{\beta},T}\underline{v}_T)_T - \frac{1}{2}(u_T, \mathbf{G}^k_{\boldsymbol{\beta},T}\underline{v}_T)_T\right)\\ &= \sum_{T\in\mathcal{T}_h}\left(-\frac{1}{2}(u_T, \mathbf{G}^k_{\boldsymbol{\beta},T}\underline{v}_T)_T + \frac{1}{2}(\mathbf{G}^k_{\boldsymbol{\beta},T}\underline{u}_T, v_T)_T\right)\\ &\quad+\frac{1}{2}\sum_{T\in\mathcal{T}_h}\sum_{F\in\mathcal{F}_T}((\boldsymbol{\beta}\cdot\mathbf{n}_{TF})(u_F-u_T), v_F-v_T)_F\\ &\quad-\frac{1}{2}\sum_{F\in\mathcal{F}_h^{\mathrm{b}}}((\boldsymbol{\beta}\cdot\mathbf{n}_{TF})u_F, v_F)_F.\end{aligned}$$

Inserting this equality into (4.105) and rearranging the terms we recover (4.104). The formulation (4.104) highlights two key properties of the bilinear form $\mathrm{a}_{\boldsymbol{\beta},\mu,h}$: its positivity and the skew-symmetric nature of the consistent term. The reformulation (4.105), on the other hand, has a more familiar look for the reader accustomed to upwind stabilization terms.

Define the global advective-reactive norm such that, for all $\underline{v}_h \in \underline{U}_h^k$,

$$\|\underline{v}_h\|^2_{\boldsymbol{\beta},\mu,h} := \sum_{T\in\mathcal{T}_h}\|\underline{v}_T\|^2_{\boldsymbol{\beta},\mu,T} + \frac{1}{2}\sum_{F\in\mathcal{F}_h^{\mathrm{b}}}\||\boldsymbol{\beta}\cdot\mathbf{n}|^{1/2}v_F\|_F^2.$$

The following coercivity result for $\mathrm{a}_{\boldsymbol{\beta},\mu,h}$ follows from (4.103): For all $\underline{v}_h \in \underline{U}_h^k$

$$\min_{T\in\mathcal{T}_h}(1, \hat{\tau}_T\mu_0)\|\underline{v}_h\|^2_{\boldsymbol{\beta},\mu,h} \le \mathrm{a}_{\boldsymbol{\beta},\mu,h}(\underline{v}_h, \underline{v}_h). \tag{4.106}$$

4.5.3 Global Problem and Inf-Sup Stability

We can now define the global bilinear form $\mathrm{a}_{\kappa,\boldsymbol{\beta},\mu,h} : \underline{U}_h^k \times \underline{U}_h^k \to \mathbb{R}$ combining the diffusive and advective-reactive contributions defined above:

$$\mathrm{a}_{\kappa,\boldsymbol{\beta},\mu,h}(\underline{u}_h, \underline{v}_h) := \mathrm{a}_{\kappa,h}(\underline{u}_h, \underline{v}_h) + \mathrm{a}_{\boldsymbol{\beta},\mu,h}(\underline{u}_h, \underline{v}_h).$$

The HHO approximation of (4.98) then reads: Find $\underline{u}_h \in \underline{U}_h^k$ such that, for all $\underline{v}_h \in \underline{U}_h^k$,

$$\mathrm{a}_{\kappa,\boldsymbol{\beta},\mu,h}(\underline{u}_h, \underline{v}_h) = (f, v_h). \tag{4.107}$$

Let us examine stability. In view of (4.100) and (4.106), the bilinear form $\mathrm{a}_{\kappa,\boldsymbol{\beta},\mu,h}$ is clearly coercive with respect to the norm

$$\|\underline{v}_h\|_{\flat,h}^2 := \|\underline{v}_h\|_{\kappa,h}^2 + \|\underline{v}_h\|_{\boldsymbol{\beta},\mu,h}^2,$$

which guarantees that problem (4.107) has a unique solution. This norm, however, does not convey any information on the discrete advective derivative. A stronger stability result is stated in the following lemma, where we consider the augmented norm

$$\|\underline{v}_h\|_{\sharp,h}^2 := \|\underline{v}_h\|_{\flat,h}^2 + \sum_{T\in\mathcal{T}_h,\hat{\boldsymbol{\beta}}_T\neq 0} h_T\hat{\boldsymbol{\beta}}_T^{-1}\|\mathbf{G}_{\boldsymbol{\beta},\underline{v}_T}^k\|_T^2,$$

with $\hat{\boldsymbol{\beta}}_T := \|\boldsymbol{\beta}\|_{L^\infty(T)^d}$ denoting the reference velocity on T and the summand is taken only if $\hat{\boldsymbol{\beta}}_T \neq 0$.

Lemma 4.7 (Inf-Sup Stability of $\mathrm{a}_{\kappa,\boldsymbol{\beta},\mu,h}$) *Assume that $\zeta \geq 1$ and that, for all $T \in \mathcal{T}_h$,*

$$h_T \max(\mathrm{L}_{\beta,T}, \mu_0) \leq \hat{\boldsymbol{\beta}}_T. \tag{4.108}$$

Then, there exists a real number $C > 0$, independent of h, κ, $\boldsymbol{\beta}$ and μ, but possibly depending on d, ϱ, and k such that, for all $\underline{w}_h \in \underline{U}_h^k$,

$$C \min_{T\in\mathcal{T}_h}(1, \hat{\tau}_T\mu_0)\|\underline{w}_h\|_{\sharp,h} \leq \sup_{\underline{v}_h\in\underline{U}_h^k\setminus\{\underline{0}_h\}} \frac{\mathrm{a}_{\kappa,\boldsymbol{\beta},\mu,h}(\underline{w}_h,\underline{v}_h)}{\|\underline{v}_h\|_{\sharp,h}}.$$

Remark 4.14 (Condition (4.108)*)* Condition (4.108) means (i) that the advective field is well-resolved by the mesh and (ii) that reaction is not dominant.

4.5.4 Convergence

For each mesh element $T \in \mathcal{T}_h$, we introduce the local Péclet number such that

$$\mathrm{Pe}_T := \max_{F\in\mathcal{F}_T} \frac{h_F\|\boldsymbol{\beta}_{|F}\cdot\mathbf{n}_{TF}\|_{L^\infty(F)}}{\kappa_F},$$

where $\kappa_F := \min_{T\in\mathcal{T}_F}\kappa_T$. For the mesh elements where diffusion dominates we have $\mathrm{Pe}_T \leq h_T$, for those where advection dominates we have $\mathrm{Pe}_T \geq 1$, while intermediate regimes correspond to $\mathrm{Pe}_T \in (h_T, 1)$.

The following error estimate accounts for the variation of the convergence rate according to the value of the local Péclet number, showing that diffusion-dominated elements contribute with a term in $\mathcal{O}(h_T^{k+1})$ (as for a pure diffusion problem), whereas convection-dominated elements contribute with a term in $\mathcal{O}(h_T^{k+1/2})$ (as for a pure advection problem).

Theorem 4.8 (Energy Error Estimate) *Let u solve* (4.98) *and $\underline{u}_h$ solve* (4.107). *Under the assumptions of Lemma 4.7, and further assuming the regularity $u_{|T} \in H^{k+2}(T)$ for all $T \in \mathcal{T}_h$, there exists a real number $C > 0$ independent of h, κ, $\boldsymbol{\beta}$, and μ, but possibly depending on ρ, d, and k, such that*

$$C \min_{T\in\mathcal{T}_h} (1, \hat{\tau}_T \mu_0) \|\hat{\underline{u}}_h - \underline{u}_h\|_{\sharp,h} \le \Bigg\{ \sum_{T\in\mathcal{T}_h} \Big[\left(\kappa_T \|u\|^2_{H^{k+2}(T)} + \hat{\tau}_T^{-1} \|u\|^2_{H^{k+1}(T)} \right) h_T^{2(k+1)} + \hat{\boldsymbol{\beta}}_T \min(1, \mathrm{Pe}_T) \|u\|^2_{H^{k+1}(T)} h_T^{2k+1} \Big] \Bigg\}^{1/2}.$$

Remark 4.15 (Extension to Locally Vanishing Diffusion) It has been showed in [34] that the error estimate of Theorem 4.8 extends to locally vanishing diffusion provided that we conventionally set $\mathrm{Pe}_T = +\infty$ for any element $T \in \mathcal{T}_h$ such that $\kappa_F = 0$ for some $F \in \mathcal{F}_T$.

4.5.5 Numerical Example

To illustrate the performance of the HHO method, we solve in the unit square $\Omega = (0, 1)^2$ the Dirichlet problem corresponding to the solution (4.79) with $\boldsymbol{\beta}(\mathbf{x}) = (1/2 - x_2, x_1 - 1/2)$, $\mu \equiv 1$, and a uniform diffusion coefficient κ taking values in $\{1, 1 \cdot 10^{-3}, 0\}$. We take triangular and predominantly hexagonal meshes, as depicted in Fig. 4.1a and c respectively. The convergence results are depicted in Fig. 4.9. We observe that the convergence rate decreases with κ, with a loss slightly less than the half order predicted by the error estimate of Theorem 4.8.

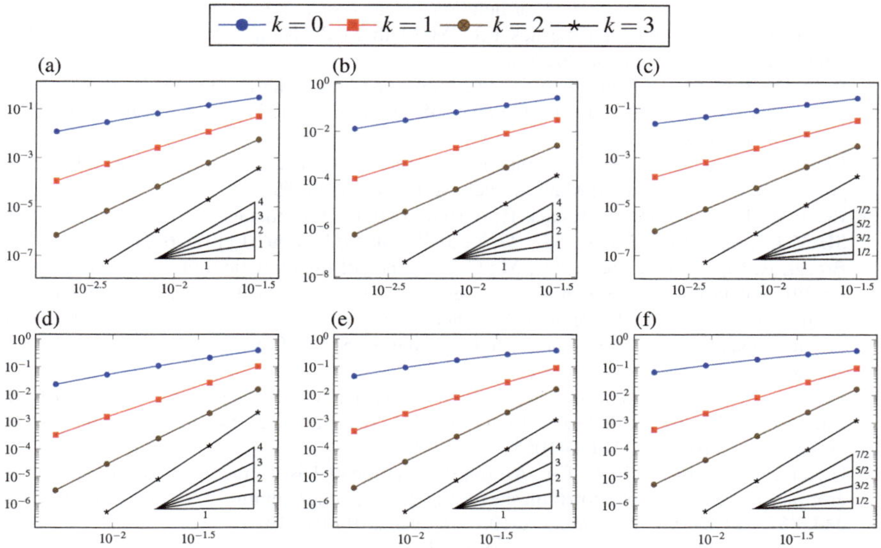

Fig. 4.9 $\|\underline{I}_h^k u - \underline{u}_h\|_{\sharp,h}$ vs. h for the test case of Sect. 4.5.5. (**a**) $\kappa = 1$, triangular. (**b**) $\kappa = 1 \times 10^{-3}$, triangular. (**c**) $\kappa = 0$, triangular. (**d**) $\kappa = 1$, polygonal. (**e**) $\kappa = 1 \times 10^{-3}$, polygonal. (**f**) $\kappa = 0$, polygonal

Acknowledgements This work was funded by Agence Nationale de la Recherche grant HHOMM (ref. ANR-15-CE40-0005-01).

References

1. Aghili, J., Boyaval, S., Di Pietro, D.A.: Hybridization of mixed high-order methods on general meshes and application to the Stokes equations. Comput. Methods Appl. Math. **15**(2), 111–134 (2015). https://doi.org/10.1515/cmam-2015-0004
2. Arnold, D.N., Brezzi, F., Cockburn, B., Marini, L.D.: Unified analysis of discontinuous Galerkin methods for elliptic problems. SIAM J. Numer. Anal. **39**(5), 1749–1779 (2002)
3. Bassi, F., Botti, L., Colombo, A., Di Pietro, D.A., Tesini, P.: On the flexibility of agglomeration based physical space discontinuous Galerkin discretizations. J. Comput. Phys. **231**(1), 45–65 (2012). https://doi.org/10.1016/j.jcp.2011.08.018
4. Bebendorf, M.: A note on the Poincaré inequality for convex domains. Z. Anal. Anwend. **22**(4), 751–756 (2003)
5. Beirão da Veiga, L., Droniou, J., Manzini, G.: A unified approach to handle convection terms in Finite Volumes and Mimetic Discretization Methods for elliptic problems. IMA J. Numer. Anal. **31**(4), 1357–1401 (2011)
6. Beirão da Veiga, L., Brezzi, F., Cangiani, A., Manzini, G., Marini, L.D., Russo, A.: Basic principles of virtual element methods. Math. Models Methods Appl. Sci. **199**(23), 199–214 (2013)
7. Beirão da Veiga, L., Brezzi, F., Marini, L.D.: Virtual elements for linear elasticity problems. SIAM J. Numer. Anal. **2**(51), 794–812 (2013)

8. Boffi, D., Di Pietro, D.A.: Unified formulation and analysis of mixed and primal discontinuous skeletal methods on polytopal meshes. ESAIM: Math. Model Numer. Anal. **52**(1), 1–28 (2018). https://doi.org/10.1051/m2an/2017036
9. Boffi, D., Botti, M., Di Pietro, D.A.: A nonconforming high-order method for the Biot problem on general meshes. SIAM J. Sci. Comput. **38**(3), A1508–A1537 (2016). https://doi.org/10.1137/15M1025505
10. Bonelle, J., Ern, A.: Analysis of compatible discrete operator schemes for elliptic problems on polyhedral meshes. ESAIM: Math. Model. Numer. Anal. **48**, 553–581 (2014)
11. Botti, M., Di Pietro, D.A., Sochala, P.: A Hybrid High-Order method for nonlinear elasticity. SIAM J. Numer. Anal. **55**(6), 2687–2717 (2017). https://doi.org/10.1137/16M1105943
12. Brezzi, F., Lipnikov, K., Shashkov, M.: Convergence of the mimetic finite difference method for diffusion problems on polyhedral meshes. SIAM J. Numer. Anal. **43**(5), 1872–1896 (2005)
13. Castillo, P., Cockburn, B., Perugia, I., Schötzau, D.: An a priori error analysis of the local discontinuous Galerkin method for elliptic problems. SIAM J. Numer. Anal. **38**, 1676–1706 (2000)
14. Chave, F., Di Pietro, D.A., Marche, F., Pigeonneau, F.: A hybrid high-order method for the Cahn–Hilliard problem in mixed form. SIAM J. Numer. Anal. **54**(3), 1873–1898 (2016). https://doi.org/10.1137/15M1041055
15. Cicuttin, M., Di Pietro, D.A., Ern, A.: Implementation of Discontinuous Skeletal methods on arbitrary-dimensional, polytopal meshes using generic programming. J. Comput. Appl. Math. **344**, 852–874 (2008). https://doi.org/10.1016/j.cam.2017.09.017.
16. Cockburn, B., Fu, G.: Superconvergence by M-decompositions. Part III: construction of three-dimensional finite elements. ESAIM Math. Model. Numer. Anal. **51**(1), 365–398 (2017). https://doi.org/10.1051/m2an/2016023
17. Cockburn, B., Gopalakrishnan, J., Lazarov, R.: Unified hybridization of discontinuous Galerkin, mixed, and continuous Galerkin methods for second order elliptic problems. SIAM J. Numer. Anal. **47**(2), 1319–1365 (2009). http://dx.doi.org/10.1137/070706616
18. Cockburn, B., Di Pietro, D.A., Ern, A.: Bridging the Hybrid High-Order and Hybridizable Discontinuous Galerkin methods. ESAIM: Math. Model. Numer. Anal. **50**(3), 635–650 (2016). https://doi.org/10.1051/m2an/2015051
19. Codecasa, L., Specogna, R., Trevisan, F.: A new set of basis functions for the discrete geometric approach. J. Comput. Phys. **19**(299), 7401–7410 (2010)
20. Dahlquist, G.: Convergence and stability in the numerical integration of ordinary differential equations. Math. Scand. **4**, 33–53 (1956)
21. Di Pietro, D.A.: Cell centered Galerkin methods for diffusive problems. ESAIM: Math. Model. Numer. Anal. **46**(1), 111–144 (2012). https://doi.org/10.1051/m2an/2011016
22. Di Pietro, D.A., Droniou, J.: A Hybrid High-Order method for Leray–Lions elliptic equations on general meshes. Math. Comput. **86**(307), 2159–2191 (2017). https://doi.org/10.1090/mcom/3180
23. Di Pietro, D.A., Droniou, J.: $W^{s,p}$-approximation properties of elliptic projectors on polynomial spaces, with application to the error analysis of a Hybrid High-Order discretisation of Leray–Lions problems. Math. Models Methods Appl. Sci. **27**(5), 879–908 (2017). https://doi.org/10.1142/S0218202517500191
24. Di Pietro, D.A., Ern, A.: Discrete functional analysis tools for discontinuous Galerkin methods with application to the incompressible Navier-Stokes equations. Math. Comput. **79**(271), 1303–1330 (2010). https://doi.org/10.1090/S0025-5718-10-02333-1
25. Di Pietro, D.A., Ern, A.: Mathematical aspects of discontinuous Galerkin methods. In: Mathématiques & Applications, vol. 69. Springer, Berlin (2012). https://doi.org/10.1007/978-3-642-22980-0
26. Di Pietro, D.A., Ern, A.: Equilibrated tractions for the Hybrid High-Order method. C. R. Acad. Sci. Paris Ser. I **353**, 279–282 (2015). https://doi.org/10.1016/j.crma.2014.12.009
27. Di Pietro, D.A., Ern, A.: A hybrid high-order locking-free method for linear elasticity on general meshes. Comput. Methods Appl. Mech. Eng. **283**, 1–21 (2015). https://doi.org/10.1016/j.cma.2014.09.009

28. Di Pietro, D.A., Ern, A.: Arbitrary-order mixed methods for heterogeneous anisotropic diffusion on general meshes. IMA J. Numer. Anal. **37**(1), 40–63 (2016). https://doi.org/10.1093/imanum/drw003
29. Di Pietro, D.A., Krell, S.: A Hybrid High-Order method for the steady incompressible Navier–Stokes problem. J. Sci. Comput. **74**(3), 1677–1705 (2018). https://doi.org/10.1007/s10915-017-0512-x
30. Di Pietro, D.A., Lemaire, S.: An extension of the Crouzeix–Raviart space to general meshes with application to quasi-incompressible linear elasticity and Stokes flow. Math. Comput. **84**(291), 1–31 (2015). https://doi.org/10.1090/S0025-5718-2014-02861-5
31. Di Pietro, D.A., Specogna, R.: An a posteriori-driven adaptive Mixed High-Order method with application to electrostatics. J. Comput. Phys. **326**(1), 35–55 (2016). https://doi.org/10.1016/j.jcp.2016.08.041
32. Di Pietro, D.A., Ern, A., Guermond, J.L.: Discontinuous Galerkin methods for anisotropic semidefinite diffusion with advection. SIAM J. Numer. Anal. **46**(2), 805–831 (2008). https://doi.org/10.1137/060676106
33. Di Pietro, D.A., Ern, A., Lemaire, S.: An arbitrary-order and compact-stencil discretization of diffusion on general meshes based on local reconstruction operators. Comput. Methods Appl. Math. **14**(4), 461–472 (2014). https://doi.org/10.1515/cmam-2014-0018
34. Di Pietro, D.A., Droniou, J., Ern, A.: A discontinuous-skeletal method for advection-diffusion-reaction on general meshes. SIAM J. Numer. Anal. **53**(5), 2135–2157 (2015). https://doi.org/10.1137/140993971
35. Di Pietro, D.A., Ern, A., Lemaire, S.: A review of Hybrid High-Order methods: formulations, computational aspects, comparison with other methods. In: Building Bridges: Connections and Challenges in Modern Approaches to Numerical Partial Differential Equations, pp. 205–236. Springer, Cham (2016)
36. Di Pietro, D.A., Ern, A., Linke, A., Schieweck, F.: A discontinuous skeletal method for the viscosity-dependent Stokes problem. Comput. Methods Appl. Mech. Eng. **306**, 175–195 (2016). https://doi.org/10.1016/j.cma.2016.03.033
37. Di Pietro, D.A., Droniou, J., Manzini, G.: Discontinuous Skeletal Gradient Discretisation methods on polytopal meshes. J. Comput. Phys. **355**, 397–425 (2018). https://doi.org/10.1016/j.jcp.2017.11.018
38. Droniou, J.: Finite volume schemes for diffusion equations: introduction to and review of modern methods. Math. Models Methods Appl. Sci. **24**(8), 1575–1619 (2014). https://doi.org/10.1142/S0218202514400041
39. Droniou, J., Eymard, R.: A mixed finite volume scheme for anisotropic diffusion problems on any grid. Numer. Math. **105**, 35–71 (2006)
40. Droniou, J., Eymard, R., Gallouët, T., Herbin, R.: A unified approach to mimetic finite difference, hybrid finite volume and mixed finite volume methods. Math. Models Methods Appl. Sci. **20**(2), 1–31 (2010)
41. Droniou, J., Eymard, R., Gallouet, T., Herbin, R.: Gradient schemes: a generic framework for the discretisation of linear, nonlinear and nonlocal elliptic and parabolic equations. Math. Models Methods Appl. Sci. **23**(13), 2395–2432 (2013). https://doi.org/10.1142/S0218202513500358
42. Droniou, J., Eymard, R., Gallouët, T., Guichard, C., Herbin, R.: The Gradient Discretisation Method: A Framework for the Discretisation and Numerical Analysis of Linear and Nonlinear Elliptic and Parabolic Problems. Mathématiques et Applications. Springer, Berlin (2017). http://hal.archives-ouvertes.fr/hal-01382358.
43. Eymard, R., Gallouët, T., Herbin, R.: Finite Volume Methods, pp. 713–1020. North-Holland, Amsterdam (2000)
44. Eymard, R., Gallouët, T., Herbin, R.: Discretization of heterogeneous and anisotropic diffusion problems on general nonconforming meshes. SUSHI: a scheme using stabilization and hybrid interfaces. IMA J. Numer. Anal. **30**(4), 1009–1043 (2010)

45. Fichera, G.: Asymptotic behaviour of the electric field and density of the electric charge in the neighbourhood of singular points of a conducting surface. Russ. Math. Surv. **30**(3), 107 (1975). http://stacks.iop.org/0036-0279/30/i=3/a=R03
46. Grisvard, P.: Singularities in Boundary Value Problems. Masson, Paris (1992)
47. Herbin, R., Hubert, F.: Benchmark on discretization schemes for anisotropic diffusion problems on general grids. In: Eymard, R., Hérard, J.M. (eds.) Finite Volumes for Complex Applications V, pp. 659–692. Wiley, Hoboken (2008)
48. Karakashian, O.A., Pascal, F.: A posteriori error estimates for a discontinuous Galerkin approximation of second-order elliptic problems. SIAM J. Numer. Anal. **41**(6), 2374–2399 (2003)
49. Lax, P.D., Milgram, A.N.: Parabolic equations. In: Contributions to the Theory of Partial Differential Equations. Annals of Mathematics Studies, vol. 33, pp. 167–190. Princeton University Press, Princeton (1954)
50. Leray, J., Lions, J.L.: Quelques résultats de Višik sur les problèmes elliptiques non linéaires par les méthodes de Minty-Browder. Bull. Soc. Math. France **93**, 97–107 (1965)
51. Minty, G.J.: On a "monotonicity" method for the solution of non-linear equations in Banach spaces. Proc. Natl. Acad. Sci. U.S.A. **50**, 1038–1041 (1963)
52. Payne, L.E., Weinberger, H.F.: An optimal Poincaré inequality for convex domains. Arch. Ration. Mech. Anal. **5**, 286–292 (1960)
53. Verfürth, R.: A Review of a Posteriori Error Estimation and Adaptive Mesh-Refinement Techniques. Wiley-Teubner, Stuttgart (1996)
54. Vohralík, M., Wohlmuth, B.I.: Mixed finite element methods: implementation with one unknown per element, local flux expressions, positivity, polygonal meshes, and relations to other methods. Math. Models Methods Appl. Sci. **23**(5), 803–838 (2013). https://doi.org/10.1142/S0218202512500613
55. Whitney, H.: Geometric Integration Theory. Princeton University Press, Princeton (1957)

Chapter 5
Distributed Lagrange Multiplier for Fluid-Structure Interactions

Daniele Boffi, Frédéric Hecht, and Olivier Pironneau

Abstract In this paper we make preliminary numerical tests to assess the performance of the scheme introduced in Boffi et al. (SIAM J Numer Anal 53(6):2584–2604, 2015) and analyzed in Boffi and Gastaldi (Numer Math 135(3):711–732, 2017) for the approximation of fluid-structure interaction problems. We show how to implement the scheme within the `FreeFem++` framework (Hecht, J Numer Math 20(3–4):251–265, 2012) and validate our code with respect to some known problems and benchmarks. The main conclusion is that a simple implementation can provide quite accurate results for non trivial applications.

5.1 Introduction

The use of a distributed Lagrange multiplier for the modeling and approximation of interface problems has a long history within approaches based on fictitious domain techniques [6].

The applications of this methodology for fluid-structure interaction problems has been rediscovered and discussed in recent research [1, 4] originating from the immersed boundary method [2, 9]. The theoretical properties of this approach are quite good, showing unconditional stability for a semi-implicit time discretization and inf-sup stability for the global saddle point problem under suitable conditions on the underlying meshes. Our formulation and some of the main results about it will be summarized in Sect. 5.2.

In this paper we present a series of numerical tests performed with the help of `FreeFem++` [7]. All results are collected in Sect. 5.3. In all tests, the agreement

D. Boffi (✉)
Dipartimento di Matematica "F. Casorati", University of Pavia, Pavia, Italy
e-mail: daniele.boffi@unipv.it; http://www-dimat.unipv.it/boffi/

F. Hecht · O. Pironneau
Sorbonne Universités, UPMC (Paris VI), Laboratoire Jacques-Louis Lions, Paris, France
e-mail: frederic.hecht@upmc.fr; http://www.ljll.math.jussieu.fr/hecht;
Olivier.Pironneau@upmc.fr; http://www.ljll.math.jussieu.fr/pironneau

D. A. Di Pietro et al. (eds.), *Numerical Methods for PDEs*,
SEMA SIMAI Springer Series 15, https://doi.org/10.1007/978-3-319-94676-4_5

with the analytical solution (if known) or with solutions present in the literature is quite good.

One of the main difficulties in the implementation of any fluid-structure interaction model, consists in the appropriate treatment of the exchange of information between fluid and solid. In our formulation the fluid mesh is fixed, while the solid mesh is defined on a reference configuration and mapped to the actual solid domain via the (unknown) transformation which defines the position of the body. It turns out that some terms in our variational formulation need to combine quantities defined on the fluid and solid meshes. Actually, `FreeFem++` has a built in function that allows the computation of such terms. Our codes are listed in Appendix 1 and some comments are provided in Appendix 2.

5.2 Problem Setting

The model introduced in [4] can deal with co-dimension zero (thick) or co-dimension one (thin) bodies immersed in a fluid of two or three space dimensions. Our numerical tests involve thick bodies in two space dimensions; we recall the related formulation.

Let Ω be a bounded domain in $\mathbb{R}^2$ with Lipschitz continuous boundary. We assume that the domain is partitioned into a fluid part Ω_f and a solid part Ω_s (both subdomains are time dependent). The solid domain Ω_s is the image of a reference domain $\mathcal{B} \subset \mathbb{R}^2$. More precisely, the mapping $\mathbf{X} : \mathcal{B} \to \mathbb{R}^2$ associates to each point $\mathbf{s} \in \mathcal{B}$ its image $\mathbf{x} = \mathbf{X}(\mathbf{s}, t) \in \Omega_s$ at time t. We denote by ρ^f and ρ^s the fluid and solid densities, respectively, by ν the fluid viscosity, and by λ and μ the Lamé constants.

The problem considered in [4] is the following one: given an initial velocity $\mathbf{u}_0 \in (H_0^1(\Omega))^2$, an initial body position $\mathbf{X}_0 \in (W^{1,\infty}(\mathcal{B}))^2$, find velocity and pressure $(\mathbf{u}(t), p(t)) \in (H_0^1(\Omega))^2 \times L_0^2(\Omega)$, body position $\mathbf{X}(t) \in (H^1(\mathcal{B}))^2$, and a Lagrange multiplier $\boldsymbol{\lambda}(t) \in \Lambda$ such that for almost every $t \in]0, T[$ is holds

$$
\begin{aligned}
&\int_\Omega \left(\rho^f \mathbb{D}_t \mathbf{u}(t) \cdot \hat{\mathbf{u}} - \hat{p}\nabla \cdot \mathbf{u}(t) - p(t)\nabla \cdot \hat{\mathbf{u}} + \frac{\nu}{2}\mathrm{D}\mathbf{u}(t) : \mathrm{D}\hat{\mathbf{u}}\right) \\
&\qquad + \int_{\mathcal{B}} \left(c_1 \mathrm{D}\boldsymbol{\lambda}(t) : \mathrm{D}(\hat{\mathbf{u}}(\mathbf{X}(t))) + c_2 \boldsymbol{\lambda}(t) \cdot \hat{\mathbf{u}}(\mathbf{X}(t))\right) = 0 \\
&\qquad \forall (\hat{\mathbf{u}}, \hat{p}) \in (H_0^1(\Omega))^2 \times L_0^2(\Omega) \\
&\int_{\mathcal{B}} \Big((\rho^s - \rho^f)\partial_{tt}\mathbf{X}(t) \cdot \hat{\mathbf{X}} + \frac{\mu}{2}\mathrm{D}\mathbf{X}(t) : \mathrm{D}\hat{\mathbf{X}} + \lambda \nabla \cdot \mathbf{X}(t) \nabla \cdot \hat{\mathbf{X}} \\
&\qquad - c_1 \mathrm{D}\boldsymbol{\lambda}(t) : \mathrm{D}\hat{\mathbf{X}} - c_2 \boldsymbol{\lambda}(t) \cdot \hat{\mathbf{X}}
\end{aligned}
$$

$$+c_1\mathrm{D}\hat{\boldsymbol{\lambda}} : \mathrm{D}(\mathbf{u}(t)(\mathbf{X}(t)) - \partial_t\mathbf{X}(t)) + c_2\hat{\boldsymbol{\lambda}} \cdot (\mathbf{u}(t)(\mathbf{X}(t)) - \partial_t\mathbf{X}(t))\Big) = 0$$

$$\forall(\hat{\mathbf{X}}, \hat{\boldsymbol{\lambda}}) \in (H^1(\mathcal{B}))^2 \times \Lambda, \tag{5.1}$$

where $\mathbb{D}_t$ is the total derivative and D is the symmetric gradient.

The constants c_1, c_2 and the space Λ are crucial for the definition of the model and the subsequent numerical scheme: in our computations we consider both c_1 and c_2 positive and different from zero (H^1-based Lagrange multiplier), so that the space Λ is $(H^1(\mathcal{B}))^2$.

5.2.1 Numerical Approximation

The time semi-discretization of Problem (5.1) is constructed as follows: in the first equation the total derivative is approximated by the Galerkin-characteristic method (see [10]); the second derivative $\partial_{tt}\mathbf{X}(t)$ in the second equation is approximated by $(\mathbf{X}^{n+1} - 2\mathbf{X}^n + \mathbf{X}^{n-1})/\delta t^2$; $\partial_t\mathbf{X}$ is approximated by $(\mathbf{X}^{n+1} - \mathbf{X}^n)/\delta t$; all other quantities are evaluated implicitly at time $n + 1$ with the following exception. Clearly, there is a problem when a term involving $\hat{\mathbf{u}}(\mathbf{X})$ has to be integrated on $\mathcal{B}$. Treating this term fully implicitly would imply the use of the mapping $\mathbf{X}^{n+1}$ which is not yet available; for this reason we use a semi-implicit scheme where $\hat{\mathbf{u}}(\mathbf{X}^n)$ is used, instead.

In [4, Prop. 3] it has been shown that the resulting semi-discrete scheme is unconditionally stable with respect to the time step δt. The proof is based on a discrete energy estimate which is analogous to the stability estimate for the continuous problem:

$$\begin{aligned}
&\frac{\rho_f}{2\delta t}\left(\|\mathbf{u}^{n+1}\|_0^2 - \|\mathbf{u}^n\|_0^2\right) + \nu\|\mathrm{D}\mathbf{u}^{n+1}\|_0^2 \\
&\quad + \frac{\rho_s - \rho_f}{2\delta t}\left(\left\|\frac{\mathbf{X}^{n+1} - \mathbf{X}^n}{\delta t}\right\|_{0,\mathcal{B}}^2 - \left\|\frac{\mathbf{X}^n - \mathbf{X}^{n-1}}{\delta t}\right\|_{0,\mathcal{B}}^2\right) \\
&\quad + \frac{\mathbf{E}(\mathbf{X}^{n+1}) - \mathbf{E}(\mathbf{X}^n)}{\delta t} \le 0
\end{aligned}$$

where the energy $\mathbf{E}$ is defined in terms of the energy density $W(\mathbb{F})$ ($\mathbb{F}$ being the deformation gradient)

$$\mathbf{E}(\mathbf{X}(t)) = \int_{\mathcal{B}} W(\mathbb{F}(\mathbf{s}, t))\, d\mathbf{s}$$

The numerical approximation of Problem (5.1) is based on a set of four finite element spaces: $V_h \subset (H_0^1(\Omega))^2$ and $Q_h \subset L_0^2(\Omega)$ are inf-sup stable finite element spaces, while $S_h \subset (H^1(\mathcal{B}))^2$ and $\Lambda_h \subset \Lambda$ are finite elements defined in the solid domain. More precisely, V_h and Q_h are finite elements defined according to a triangulation $\mathcal{T}_{h_f}$ of Ω, while S_h and Λ_h are finite elements defined on a mesh $\mathcal{T}_{h_s}$ of the reference solid configuration $\mathcal{B}$.

5.3 Numerical Tests

In all tests the solid is hyper-elastic with Young modulus E, Poisson ratio κ and density ρ^s. The shear modulus is then given by $\mu = E/(1+\kappa)/2$.

The fluid is Newtonian incompressible with density ρ^f and viscosity ν.

5.3.1 Disk Falling in a Liquid

A disk of diameter d is at rest initially centred at $x_c = W/2,\ y_c = H - h$ in a rectangular channel of width W and height H. Only the disk is subject to gravity g, not the fluid. No slip conditions are applied on the walls of the channel.

This test was proposed by Zhao et al. in [12] and more recently by Wang et al. in [11]. Here we chose Wang's values for the parameters:

$$\begin{aligned} &W = 2,\ d = 0.125,\ h = 0.5,\ H = 4, \\ &\rho^s = 1.2,\ \kappa = 0.3,\ \mu = 10^8,\ \rho^f = 1,\ \nu = 1,\ g = 981 \end{aligned} \tag{5.2}$$

The asymptotic vertical velocity is known to be -0.3567. Figure 5.1 shows the evolution of the vertical velocity versus time for four meshes: a coarse mesh with

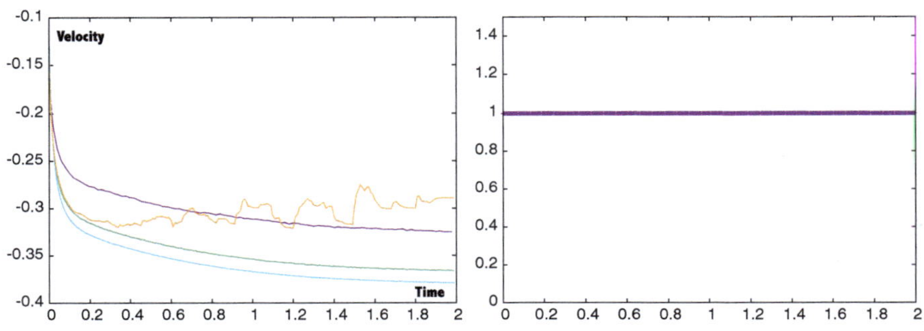

Fig. 5.1 Left: vertical velocity of the solid versus time computed with three body fitted mesh and one non-body fitted mesh. Convergence seems monotone towards a limit curve for the first three meshes; the coarse mesh is the highest, the middle mesh is in the middle and the finest mesh is below. Right: area of the solid divided by $\pi d^2/4$, as a function of time

1182 vertices, a middle mesh with 4693 vertices and a fine mesh with 18,661 vertices. The corresponding time steps are 0.02, 0.01, 0.005.

For these three cases the fluid mesh is modified at each time step to include the fluid-structure interface as an interline curve made of edges of the triangulation. Computation is also made on a fourth mesh with 4693 and the initial fluid-solid interface at time zero but not changed with time. On this fourth mesh precision degrades with time, probably due to mesh interpolations. On the three previous body fitted mesh there is convergence to a limit curve, but the asymptotic value seems to be -0.3788 rather than -0.3567. It could be due to the fact that the fluid model is extended in the solid leading to an error proportional to $\rho^s - \rho^f$. But it could also be the effect of interpolation needed to computed variables earlier defined on the mesh before motion. We are currently intersecting meshes to reduce this interpolation error and preliminary tests (to be published later) point to the direction of a more accurate falling velocity. On the other hand, mass is remarkably preserved as shown on Fig. 5.1-right.

A pressure map at $t = 0.7$ is given on Fig. 5.2-left. Next the same simulation is done with a very soft material having $\mu = 10$. The shape of the solid at $t = 0.7$ is given with a color map of the yy component of the stress on Fig. 5.2-right. The computing time for this last test is 434" on a Core i7-2.5GHz on a single core.

For these two simulations the influence of the coefficients c_1 and c_2 are small, as long as c_2 is not zero. Here both are set at 1. The influence of the degree of the finite element spaces is also surprisingly small. Both the P2/P1 element for velocity pressure or the P1-bubble/P1 element gave the same results. Changing P1 into P2 for the Lagrangian coordinates also didn't make a difference. It seems that the precision of the method is entirely driven by the quadrature formula used for the mixed integrals involving a function on the fluid mesh times a function on the transformed solid mesh.

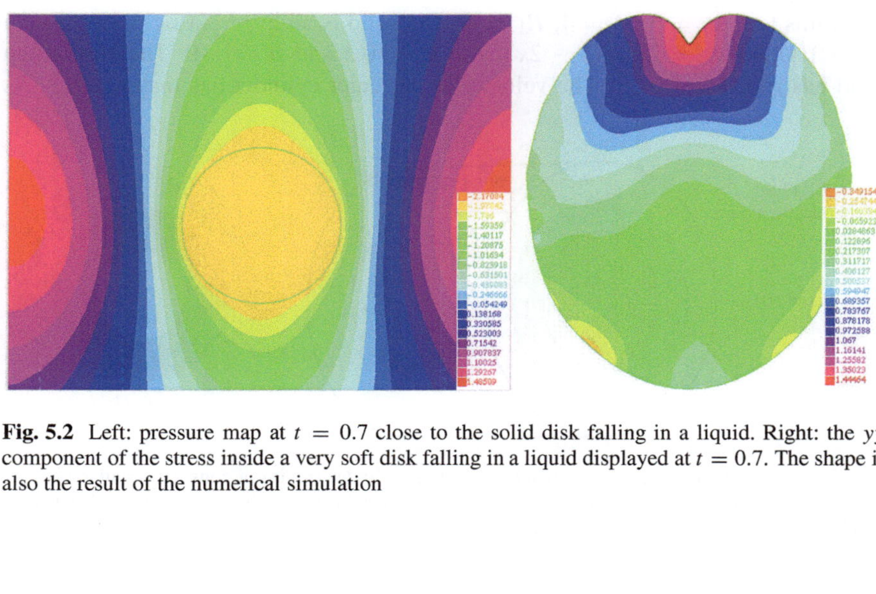

Fig. 5.2 Left: pressure map at $t = 0.7$ close to the solid disk falling in a liquid. Right: the yy component of the stress inside a very soft disk falling in a liquid displayed at $t = 0.7$. The shape is also the result of the numerical simulation

Some integrals in the variational formulation involve piecewise polynomial functions defined on different meshes. We have developed a special quadrature formula in `FreeFem++` (see [7]) to handle them. For instance let u be defined on mesh T_h^u and v be defined on mesh T_h^v obtained from T_h^u by convecting the vertices $q^i \in T_h^u$ with X, namely $X(q^i)$ is a vertex of T_h^v. Then for a triangle T of T_h^v the integral on T of $u \circ X \cdot v$ is approximated by $\sum_{j=1}^J u(X(\xi_j))v(\xi_j)\omega_j$ where ξ_j, ω_j are a valid set of quadrature points and coefficients for a quadrature on T (shown in the `FreeFem++` code by a parameter in the integral like `int2d(Ths,qft=qf9pT,mapu=[Xo,Yo])`.

5.3.2 Validation with a Rotating Disk

The purpose of this test is to compare the numerical solution with a semi-analytical solution which can be computed to any desired accuracy.

A cylinder contains a fixed rigid cylindrical rod in its center, a cylindrical layer of hyperelastic material around the rod and the rest is filled with a fluid (see Fig. 5.3). First the system is at rest and then a constant rotation is given to the outer cylinder. This cause the fluid to rotate with an angular velocity which depends on the distance r to the main axis; in turn, because the friction of the fluid at the interface the hyperelastic material will be dragged into a angular velocity ω which is also only a function of r and time . Due to elasticity ω will oscillate with time until numerical dissipation and fluid viscosity damps it.

In a two dimensional cut perpendicular to the main axis, the velocities and displacements are two dimensional as well. Hence the geometry is a ring of inner and outer radii, R_0 and R_1, with hyperelastic material between R_0 and R and fluid between R and R_1. Because of axial symmetry, R is constant, so the geometry does not change.

In this test $R_0 = 3$, $R = 4$, $R_1 = 5$. The solid is an hyperelastic material with $\mu = 100$ and $\lambda = 2\kappa\mu/(1 - 2\kappa)$ with $\kappa = 0.3$ and $\rho^s = 10$. The Newtonian fluid has $\nu = 1$, $\rho^f = 1$. The velocity of the outer cylinder has magnitude 3. As

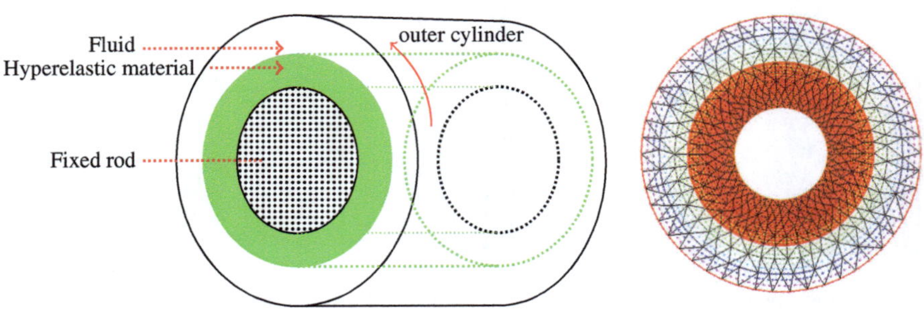

Fig. 5.3 A fluid-structure system inside a rotating cylinder (giving a constant angular velocity to the fluid outer boundary) with a fixed rod in its center. Left: sketch of the system. Right: a 2d calculation showing the velocity vectors at time 0.85 for the coarser mesh

everything is axisymmetric the computation can be done in polar coordinates r, θ, and the fluid-solid system reduces to

$$\rho\partial_t v - \frac{1}{r}\partial_r[\xi^f r\partial_r v + \xi^s r\partial_r d] = 0,$$
$$\partial_t d = v, \ r \in (R_0, R_1), \quad v_{|R_0} = 0, \ v_{|R_1} = 3, \tag{5.3}$$

with $\rho = \rho^s \mathbf{1}_{r\le R} + \rho^f \mathbf{1}_{r>R}$, $\xi^s = \mu\mathbf{1}_{r\le R}$, $\xi^f = \nu\mathbf{1}_{r>R}$, and with $d(r, 0) = 0$.

In all 2d computations $c_1 = c_2 = 10$ and $\delta t = 0.005$. We have verified that a smaller time step does not improve the precision.

Comparison between this one dimensional approach and the numerical solution of system (5.1) is given on Fig. 5.3—right, at $T = 0.5$ and a coarse mesh with 505 vertices. Then the same is computed on a finer mesh having 1986 vertices and finally with a mesh with 7433 vertices. Results are displayed on Fig. 5.4.

This test has two qualities: (a) the exact solution is easy to compute to any precision; (b) the geometry does not change and quadrature errors are due only to quadrature for integrals involving functions on the same domain but with two different triangulations.

5.3.2.1 Flow Past a Cylinder with a Flagella Attached

This test is known as FLUSTRUK-FSI-3 in [5]. The geometry is shown on Fig. 5.5. The inflow velocity is $\bar{U} = 2$, $\mu = 210^6$ and $\rho^s = \rho^f$. After some time a Karman-Vortex alley develops and the flagella beats accordingly. Results are shown on Figs. 5.5 and 5.6 with a mesh of 9692 vertices and a time step size of 0.0015; the first one displays a snapshot of the velocity vector norms and the second the y-coordinate versus time of the top right corner of the flagella.

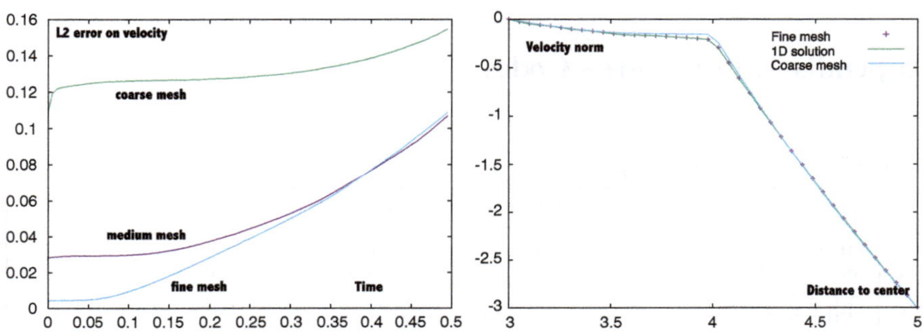

Fig. 5.4 Rotating cylinder. Left: Evolution of the L^2 error versus time for the three meshes. Right: velocities normal to the ray at $\theta = \pi/4$ versus $r - 3$, computed on the coarsest meshes shown in green with continuous line and crosses. The "exact" solution of the one dimensional equation is shown in blue

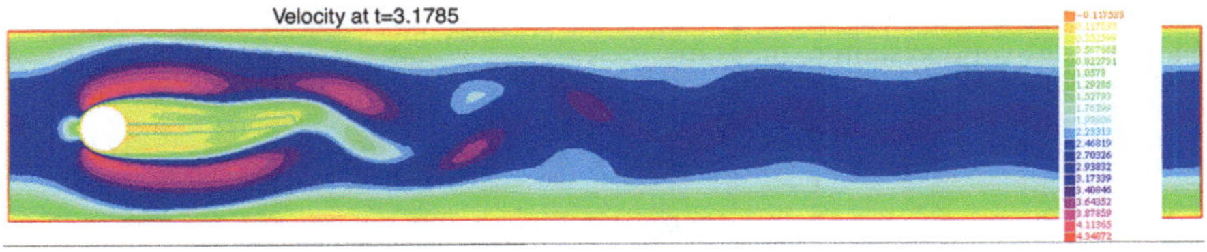

Fig. 5.5 FLUSTRUK-FSI-3 Test. Color map based on the norm of the fluid and solid velocity vectors

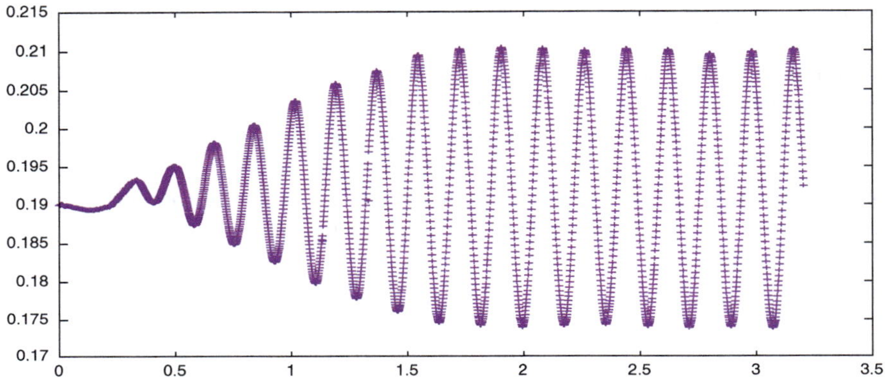

Fig. 5.6 FLUSTRUK-FSI-3 Test. Vertical position of the upper right tip of the flagella versus time shown up to $t = 5$

These numerical results compare reasonably well with those of [5]. The frequency is 5.6 compared to 5.04 and the maximum amplitude 0.018 compared with 0.032. Amplitude is very sensitive to the method (see [8]).

Appendix 1: FreeFem++ Codes

Listing 5.1 Code for the falling disk

```
int n=20, m=8*n;  // higher for fine mesh

int H=4, W=2; // vertical length of fluid box
real h=0.5, R1=0.125*2, R2=R1, xc=W/2, yc=H-h;  // elliptic radii and
    center of disk
real rhof=1, rhos=1.2, nu=1, penal=1e-9;
// rho, mu, rescaled : divided by 1e6
real  kappa=0.3, /*E=1e4, mu=E/(1+kappa)/2 */ mu=1e1, lambda=2*kappa*mu
    /(1-2*kappa);
real gravity=981;
real T=0.7, dt=0.1/n, dt2=dt*dt;
real c1=1, c2=10; // Lagrange multiplier constants: H1 -> 1,1, L2 -> 0,1

// mesh Thf=square(10*n,H*n,[x,H*y]); // fluid + solid domain
```

```

border a1(t=0,1){x=W*t; y=0;}
border a2(t=0,1){x=W; y=H*t;}
border a3(t=1,0){x=W*t; y=H;}
border a4(t=1,0){x=0;y=H*t;}
border C(t=0,2*pi){x=xc+R1*cos(t); y=yc+R2*sin(t);}
mesh Thsi = buildmesh(C(m)); // Initial solid domain
fespace Whi(Thsi,P1);
Whi Xoi=x,Yoi=y;
real[int] xx(m), yy(m);
int[int] nn(m);
for(int i=0;i<m;i++){xx[i]=xc+R1*cos(2*i*pi/(m)); yy[i]=yc+R2*sin(2*i*pi/(m)); nn[i]=2;}
border D(t=0,1;i){
        int ii = (i+1)%m; real t1 = 1-t; real x1 = xx[i]*t1 + xx[ii]*t;
        real y2 = yy[i]*t1 + yy[ii]*t; x= Xoi(x1,y2); y=Yoi(x1,y2);
}
plot(D(nn));
mesh Ths = buildmesh(D(nn));
mesh Thso=Ths;
mesh Thf= buildmesh(a1(W*n/2)+a2(H*n)+a3(W*n)+a4(H*n)+D(nn));
//plot(a1(10*n)+a2(H*n)+a3(10*n)+a4(H*n)+D(nn));
plot(Thf,Thso, cmm="Initial configuration");


fespace Vh(Thf,P1b);  // velocity space
fespace Qh(Thf,P1);  // pressure space
fespace Wh(Ths,P1b);  // Lagrangian coordinates X,Y space
fespace Lh(Ths,P1);  // Lagrangian multiplier space
fespace Zh(Thf,[P1b,P1b,P1]); // fluid space
fespace Rh(Ths,[P1,P1,P1b,P1b]);// solid space

Vh u,v,uh,vh;
Qh p,ph;
Wh  Xoo=x-0.*(x-xc),Yoo= y-0.*(y-yc); // the X,Y are now the displacements
Rh [lamx,lamy,X,Y],[lamxo,lamyo,Xo,Yo]=[0,0,x,y];//x-(x-xc)/2,y+(y-yc)/2];
Zh [uo,vo,po]=[0,0,0];


macro div(u,v) ( dx(u)+dy(v) ) // EOM
macro Grad(u,v)[[dx(u),dy(u)],[dx(v),dy(v)]] // EOM
macro DD(u,v)  [[2*dx(u),div(v,u)],[div(v,u),2*dy(v)]] // EOM

varf aa([u,v,p],[uh,vh,ph]) =
          int2d(Thf)(rhof*[u,v]'*[uh,vh]/dt- div(uh,vh)*p -div(u,v)*ph
                    + penal*p*ph + nu*trace(DD(uh,vh)'*DD(u,v))/2)
          + on(1,3,u=0,v=0) + on(2,4,u=0) ;

varf bb([lamx,lamy,X,Y],[lamxh,lamyh,Xh,Yh]) =
          int2d(Ths)((rhos-rhof)*(X*Xh+Y*Yh)/dt2
                        - c1*trace(Grad(lamx,lamy)'*Grad(Xh,Yh)) - c2*(lamx*Xh+lamy*Yh)
                        - c1*trace(Grad(lamxh,lamyh)'*Grad(X,Y))/dt - c2*(lamxh*X+lamyh*Y)/dt
                        + mu*trace(DD(X,Y)'*DD(Xh,Yh))/2 + lambda*div(X,Y)*div(Xh,Yh)
                                                + penal*(lamx*lamxh+lamy*lamyh));

varf ab([u,v,p],[lamxh,lamyh,Xh,Yh]) =
          int2d(Ths,qft=qf9pT,mapu=[Xo,Yo])(
                        c1*trace(Grad(lamxh,lamyh)'*(Grad(Xo,Yo)*Grad(u,v)))
                        + c2*(lamxh*u+lamyh*v));

```

```
varf ba([lamx,lamy,X,Y],[uh,vh,ph]) =
        int2d(Ths,qft=qf9pT,mapt=[Xo,Yo])(
                  c1*trace(Grad(lamx,lamy)'*(Grad(Xo,Yo)*Grad(uh,vh)))
                + c2*(lamx*uh+lamy*vh));

varf rhs1([u,v,p],[uh,vh,ph]) =
        int2d(Thf)(-rhof*gravity*vh * 0 // 0=no gravity in the fluid
                + rhof*convect([uo,vo],-dt,uo)*uh/dt
                + rhof*convect([uo,vo],-dt,vo)*vh/dt)
         + on(1,3,u=0,v=0) + on(2,4,u=0,v=0) ;

varf rhs2([lamx,lamy,X,Y],[lamxh,lamyh,Xh,Yh]) =
        int2d(Ths)(-(rhos-rhof)*gravity*Yh + 2*(mu+lambda)*div(Xh,Yh)
                - c1*trace(Grad(lamxh,lamyh)'*Grad(Xo,Yo))/dt
                - c2*(lamxh*Xo+lamyh*Yo)/dt
                - (rhos-rhof)*((Xoo-2*Xo)*Xh+(Yoo-2*Yo)*Yh)/dt2);

for(int i=0;i<T/dt;i++){
//        cout<<"time= "<<i*dt<<endl;
            real[int] RHS1 = rhs1(0,Zh);
            real[int] RHS2 = rhs2(0,Rh);
            matrix AA = aa(Zh,Zh);
            matrix AB = ab(Zh,Rh);
            matrix BA = ba(Rh,Zh);
            matrix BB = bb(Rh,Rh);

            matrix AABB = [ [AA,BA], [AB,BB] ];
            set(AABB,solver=sparsesolver,master=-1);
            real[int] rhs = [RHS1, RHS2];
            real[int] w=AABB^-1*rhs;
            Xoo=Xo; Yoo=Yo; Xoi=Xo; Yoi=Yo;
                    [uo[],lamxo[]] = w;
            Thso = buildmesh(D(nn));
                       Thf= buildmesh(a1(W*n/2)+a2(H*n)+a3(W*n)+a4(H*n)+D(nn));
                       [uo,vo,po]=[uo,vo,po]; [lamxo,lamyo,Xo,Yo]=[lamxo,lamyo,Xo,Yo];
                       cout<<i*dt<<" "<<int2d(Thso)(Yo-Yoo)/R1/R1/pi/dt<<" "<<int2d(Thso)(vo)/R1/R1/pi<<" area= "<<int2d(Thso)(1.)/R1/R1/pi<<" velocity "<<endl;
            uh=sqrt(uo*uo+vo*vo);
            plot(Thf, Thso, uh, value=1, fill=0, coef=100, cmm="t="+i*dt+"      Velocity");
}
Thf = buildmesh(a1(W*n/2)+a2(H*n)+a3(W*n)+a4(H*n)+D(nn));
[uo,vo,po]=[uo,vo,po];
plot(vo, value=1, fill=0, coef=100, cmm="t="+T+" Pressure ");
fespace Bh(Thso,P1);
Bh s,sh;
solve bbc(s,sh)= int2d(Thso)(s*sh + 0.01*(dx(s)*dx(sh)+dy(s)*dy(sh)))
-int2d(Thso)(sh*(mu*dy(vo) + lambda*(dx(uo)+dy(vo))));
plot(s);
```

Listing 5.2 Code for the rotating disk test

```
load "MUMPS"
load "pipe"
verbosity=0;

int n=20;  // higher for fine mesh

real R0=1, R1=2, R2=3,  gravity=0, ringvelocity=-3;
real  kappa=0.3, /*E=1e4, mu=E/(1+kappa)/2 */ mu=1e2, lambda=2*kappa*mu/(1-2*kappa);
```

```
real rhof=1, rhos=10, nu=1, penal=1e-6;
real c1=10,c2=10,T=0.5, dt=0.005, dt2=dt*dt;

real delta=0.05;
border CF(t=0,2*pi){x=R2*cos(t); y=R2*sin(t);} // fluid
border CS(t=0,2*pi){x=R1*cos(t); y=R1*sin(t);}  // solid
border CC(t=0,2*pi){x=R0*cos(t); y=R0*sin(t);}  // clamped
border C1(t=0,2*pi){x=(R1-delta)*cos(t); y=(R1-delta)*sin(t);}  // solid
border C2(t=0,2*pi){x=(R1-delta/2)*cos(t); y=(R1-delta/2)*sin(t);}  // solid
border C3(t=0,2*pi){x=(R1+delta/2)*cos(t); y=(R1+delta/2)*sin(t);}  // solid
border C4(t=0,2*pi){x=(R1+delta)*cos(t); y=(R1+delta)*sin(t);}  // solid
mesh Thf = buildmesh(CC(-10*n) /*+C1(10*n)+C2(10*n)+C3(10*n) +C4(10*n)*/ +CS(10*n)+CF(10*n));
mesh Ths = buildmesh(CC(-10*n)+CS(10*n));
mesh Thso=Ths;
// plot(Thf,Ths);
fespace Vh(Thf,P2);
fespace Qh(Thf,P1);
fespace Wh(Ths,P1);
fespace Lh(Ths,P1);
fespace Zh(Thf,[P2,P2,P1]);
fespace Rh(Ths,[P1,P1,P1,P1]);

Vh u,v,uh,vh;
Qh p,ph;
Wh  Xoo=x,Yoo=y;
Rh [lamx,lamy,X,Y],[lamxo,lamyo,Xo,Yo]=[0,0,x,y];//x-(x-xc)/2,y+(y-yc)/2];
Zh [uo,vo,po]=[0,0,0];

macro div(u,v) ( dx(u)+dy(v) ) // EOM
macro Grad(u,v)[[dx(u),dy(u)],[dx(v),dy(v)]] // EOM
macro DD(u,v)  [[2*dx(u),div(v,u)],[div(v,u),2*dy(v)]] // EOM

varf aa([u,v,p],[uh,vh,ph]) =
          int2d(Thf)(rhof*[u,v]'*[uh,vh]/dt- div(uh,vh)*p -div(u,v)*ph
                     + penal*p*ph + nu*trace(DD(uh,vh)'*DD(u,v))/2)
          + on(CC,u=0,v=0) + on(CF,u=-ringvelocity*y/R2,v=ringvelocity*x/R2) ;

varf bb([lamx,lamy,X,Y],[lamxh,lamyh,Xh,Yh]) =
          int2d(Ths)((rhos-rhof)*(X*Xh+Y*Yh)/dt2
                     - c1*trace(Grad(lamx,lamy)'*Grad(Xh,Yh)) - c2*(lamx*Xh+lamy*Yh)
                     - c1*trace(Grad(lamxh,lamyh)'*Grad(X,Y))/dt - c2*(lamxh*X+lamyh*Y)/dt
                     + mu*trace(DD(X,Y)'*DD(Xh,Yh))/2 + lambda*div(X,Y)*div(Xh,Yh)
                                         + penal*(lamx*lamxh+lamy*lamyh));

varf ab([u,v,p],[lamxh,lamyh,Xh,Yh]) =
          int2d(Ths,qft=qf9pT,mapu=[Xo,Yo])(
                       c1*trace(Grad(lamxh,lamyh)'*(Grad(Xo,Yo)*Grad(u,v)))
                     + c2*(lamxh*u+lamyh*v));

varf ba([lamx,lamy,X,Y],[uh,vh,ph]) =
          int2d(Ths,qft=qf9pT,mapt=[Xo,Yo])(
                       c1*trace(Grad(lamx,lamy)'*(Grad(Xo,Yo)*Grad(uh,vh)))
                     + c2*(lamx*uh+lamy*vh));

varf rhs1([u,v,p],[uh,vh,ph]) =
          int2d(Thf)( rhof*convect([uo,vo],-dt,uo)*uh/dt
```

```
                        + rhof*convect([uo,vo],-dt,vo)*vh/dt)
            + on(CC,u=0,v=0) + on(CF,u=-ringvelocity*y/R2,v=ringvelocity*x/R2) ;

varf rhs2([lamx,lamy,X,Y],[lamxh,lamyh,Xh,Yh]) =
            int2d(Ths)( 2*(mu+lambda)*div(Xh,Yh)
                        - c1*trace(Grad(lamxh,lamyh)'*Grad(Xo,Yo))/dt
                        - c2*(lamxh*Xo+lamyh*Yo)/dt
                        - (rhos-rhof)*((Xoo-2*Xo)*Xh+(Yoo-2*Yo)*Yh)/dt2);

/////////////////////////////////////////////////////////////////
//// semi-analytic solution by solving a 1d problem /////
mesh Th=square(100,1,[R0+(R2-R0)*x,0.1*y]);
fespace WWh(Th,P2,periodic=[[1,x],[3,x]]);
fespace W0(Th,P1dc);
WWh d=0,wo,wh,wold=0;
W0 nnu=nu*(x>R1)+mu*dt*(x<=R1), Rho=rhof*(x>R1)+(rhos-rhof)*(x<=R1);
problem AA1d(wo,wh) = int2d(Th)(Rho*x*wo*wh/dt+x*nnu*dx(wo)*dx(wh) )
                        + int2d(Th)( -Rho*x*wold*wh/dt + mu*(x<=R1)*x*dx(d)*dx(wh) )
                        +on(2,wo=ringvelocity)+on(4,wo=0);// this is the one-d axisymmetric problem
/////////////////////////////////////////////

pstream pgnuplot("gnuplot" ); // prepare gnuplot //////////////
int NT=T/dt,J=40;
real l2error=0, dr = (R2-R0)/(J-1);

//////////////////////////////// time loop ////////////////////////
for(int i=0;i<NT;i++){
//      cout<<"time= "<<i*dt<<endl;
                real[int] RHS1 = rhs1(0,Zh);
                real[int] RHS2 = rhs2(0,Rh);
                matrix AA = aa(Zh,Zh);
                matrix AB = ab(Zh,Rh);
                matrix BA = ba(Rh,Zh);
                matrix BB = bb(Rh,Rh);

                matrix AABB = [ [AA,BA], [AB,BB] ];
                set(AABB,solver=sparsesolver,master=-1);
                real[int] rhs = [RHS1, RHS2];
                real[int] w=AABB^-1*rhs;
                Xoo=Xo; Yoo=Yo;
                        [uo[],lamxo[]] = w;
//                              cout<<i*dt;<<" "<<int2d(Thso)(Yo-Yoo)/R1/R1/pi/dt<<" "<<int2d(Thso)(vo)/R1/R1/pi<<" area= "<<int2d(Thso)(1.)/R1/R1/pi<<" velocity "<<endl;
//                uh=sqrt(uo*uo+vo*vo);
                ///////////////// for error plot //////////////////////
                AA1d;
                d=d+wo*dt;
                wold=wo; // this is for the one-d axisymmetric problem
                ofstream f("aux.gp");
                for(int j=0;j<J;j++) {
                        f << j*dr <<"   " << vo(R0+j*dr,0)<<"   "<< wo(R0+j*dr,0.05)<< endl;
                        l2error += (vo(R0+j*dr,0)-wo(R0+j*dr,0.05))^2*dt;
                }
                pgnuplot << " plot [0:2][-3:0.51]'aux.gp' u 1:2 w l,'aux.gp' u 1:3 w l"<< endl;
                cout<<i*dt<<"  "<<sqrt(l2error)/T<<endl;
                flush(pgnuplot);
}
plot(Ths,Thf,[uo,vo],fill=1, coef=0.1, cmm="t="+T, wait=1)      ;

for(int j=0;j<J;j++)
```

```
cout<< j*dr<<"  "<< vo(R0+j*dr,0)<<"  "<<wo(R0+j*dr,0.05)<<"  "<<vo(R0+j*dr,0)-wo(R0+j*dr,0.05) <<endl;
// copy past the numbers in a file "results.txt", call gnuplot and do in the gnuplot terminal window:
// plot"results.txt"using 1:2,"results.txt"using 1:3 w l
```

Listing 5.3 Code for FSI-3 test

```
verbosity=0;

int n=2, m=2*n, NN=500*m;  // higher for fine mesh

real rhof=1, rhos=1, nu=0.001, penal=1e-9;
// rho, mu, rescaled : divided by 1e6
real  kappa=0.4, E=1e6, mu=2e3, lambda=2*kappa*mu/(1-2*kappa);
real Ubar=2, gravity=0*981;
real T=6, dt=T/NN, dt2=dt*dt;
real c1=1, c2=1; // Lagrange multiplier constants: H1 -> 1,1, L2 -> 0,1

// mesh Thf=square(10*n,H*n,[x,H*y]); // fluid + solid domain

int la1=10, la2=11;
real cx0 = 0.2, cy0 = 0.2; // center of cyl.
real r=0.05, H=0.41, L=2.5; // radius of cylinder, size of domain
real l1=0.35, h2=0.01;//flagella length and half thickness
real la=asin(h2/r), x0=sqrt(r*r-h2*h2);

border fr1(t=0,L){x=t; y=0; label=1;} // outer box begins
border fr2(t=0,H){x=L; y=t; label=2;}
border fr3(t=L,0){x=t; y=H; label=1;}
border fr4(t=H,0){x=0; y=t; label=3;} // outer box ends
border fr5(t=la,2*pi-la){x=cx0+r*cos(-t); y=cy0+r*sin(-t); label=4;}
border br1(t=-la,la){x=cx0+r*cos(-t); y=cy0+r*sin(-t); label=4;} // flag begins
border br2(t=0,l1){x=cx0+x0+t; y=cy0-h2;label=la1;}
border br3(t=-h2,h2){x=x0+cx0+l1; y=cy0+t;label=la1;}
border br4(t=l1,0){x=cx0+x0+t; y=cy0+h2;label=la1;}                    // flag ends

mesh Thf=buildmesh(fr1(20*m)+fr2(3.25*m)+fr3(20*m)+fr4(4*m)+fr5(12*m) + br1(m)+br2(12*m)+br3(2*m)+br4(12*m));

mesh Thsi = buildmesh(br1(m)+br2(12*m)+br3(m)+br4(12*m)); // Initial solid domain

fespace Whi(Thsi,P1);
Whi Xoi=x,Yoi=y;
real[int] xx(25*m+1), yy(25*m+1);
int[int] nn(25*m);
for(int i=0;i<=25*m;i++){
        if(i<=12*m){ real t=l1*i/(12.0*m); xx[i]=cx0+x0+t; yy[i]=cy0-h2; }
                else if(i<=13*m){real t=2*h2*(i-12.0*m)/m-h2; xx[i]=x0+cx0+l1; yy[i]=cy0+t;}
                else { real t=l1-l1*(i-13*m)/(12.0*m); xx[i]=cx0+x0+t; yy[i]=cy0+h2; }
                if(i<25*m) nn[i]=2;
        }
border D(t=0,1;i){
        int ii = (i+1)%(25*m+1); real t1 = 1-t; real x1 = xx[i]*t1 + xx[ii]*t;
        real y2 = yy[i]*t1 + yy[ii]*t; x= Xoi(x1,y2); y=Yoi(x1,y2);
        label=la1;
}
plot(br1(m) + D(nn));
mesh Ths = buildmesh(br1(m) + D(nn));
```

```
int nbA;
real xnear=x0+cx0+l1+3*h2, ynear=cy0-3*h2, distmin=10;
plot(Ths);

mesh Thso=Ths;
//mesh Thf= buildmesh(a1(W*n/2)+a2(H*n)+a3(W*n)+a4(H*n)+D(nn));
//plot(a1(10*n)+a2(H*n)+a3(10*n)+a4(H*n)+D(nn));
plot(Thf,Thso, cmm="Initial configuration");


fespace Vh(Thf,P1b);  // velocity space
fespace Qh(Thf,P1);  // pressure space
fespace Wh(Ths,P1b);  // Lagrangian coordinates X,Y space
fespace Lh(Ths,P1);  // Lagrangian multiplier space
fespace Zh(Thf,[P1b,P1b,P1]); // fluid space
fespace Rh(Ths,[P1,P1,P1b,P1b]);// solid space

Vh u,v,uh,vh;
Qh p,ph;
Wh  Xoo=x,Yoo= y; // the X,Y are now the displacements
Rh [lamx,lamy,X,Y],[lamxo,lamyo,Xo,Yo]=[0,0,x,y];//x-(x-xc)/2,y+(y-yc)
    /2];
Zh [uo,vo,po]=[0,0,0];


macro div(u,v) ( dx(u)+dy(v) ) // EOM
macro Grad(u,v)[[dx(u),dy(u)],[dx(v),dy(v)]] // EOM
macro DD(u,v)  [[2*dx(u),div(v,u)],[div(v,u),2*dy(v)]] // EOM

varf aa([u,v,p],[uh,vh,ph]) =
          int2d(Thf)(rhof*[u,v]'*[uh,vh]/dt- div(uh,vh)*p -div(u,v)*ph
                    + penal*p*ph + nu*trace(DD(uh,vh)'*DD(u,v))/2)
          + on(1,4, u=0,v=0) +  on(3,u=Ubar*y*(H-y)*6/H/H,v=0) ;

varf bb([lamx,lamy,X,Y],[lamxh,lamyh,Xh,Yh]) =
          int2d(Ths)((rhos-rhof)*(X*Xh+Y*Yh)/dt2
                    - c1*trace(Grad(lamx,lamy)'*Grad(Xh,Yh)) - c2*(lamx
                        *Xh+lamy*Yh)
                    - c1*trace(Grad(lamxh,lamyh)'*Grad(X,Y))/dt - c2*(
                        lamxh*X+lamyh*Y)/dt
                    + mu*trace(DD(X,Y)'*DD(Xh,Yh))/2 + lambda*div(X,Y)*
                        div(Xh,Yh)
                                   + penal*(lamx*lamxh+lamy*lamyh)
                                       )
                                   + on(4,X=x,Y=y);

varf ab([u,v,p],[lamxh,lamyh,Xh,Yh]) =
          int2d(Ths,qft=qf9pT,mapu=[Xo,Yo])(
                    c1*trace(Grad(lamxh,lamyh)'*(Grad(Xo,Yo)*Grad(u,v
                        )))
                    + c2*(lamxh*u+lamyh*v));

varf ba([lamx,lamy,X,Y],[uh,vh,ph]) =
          int2d(Ths,qft=qf9pT,mapt=[Xo,Yo])(
                    c1*trace(Grad(lamx,lamy)'*(Grad(Xo,Yo)*Grad(uh,vh
                        )))
                    + c2*(lamx*uh+lamy*vh));

varf rhs1([u,v,p],[uh,vh,ph]) =
          int2d(Thf)(-rhof*gravity*vh * 0 // 0=no gravity in the fluid
                    + rhof*convect([uo,vo],-dt,uo)*uh/dt
                    + rhof*convect([uo,vo],-dt,vo)*vh/dt)
           + on(1,4, u=0,v=0) +  on(3,u=Ubar*y*(H-y)*6/H/H,v=0) ;

varf rhs2([lamx,lamy,X,Y],[lamxh,lamyh,Xh,Yh]) =
          int2d(Ths)(-(rhos-rhof)*gravity*Yh + 2*(mu+lambda)*div(Xh,Yh)
                    - c1*trace(Grad(lamxh,lamyh)'*Grad(Xo,Yo))/dt
```

```
                    - c2*(lamxh*Xo+lamyh*Yo)/dt
                    - (rhos-rhof)*((Xoo-2*Xo)*Xh+(Yoo-2*Yo)*Yh)/dt2)
                                + on(4,X=x,Y=y);

real t0=0, MMCL=-100, MCL=-100,maxCL=-100,minCL=100;
for(int i=0;i<T/dt;i++){
//        cout<<"time= "<<i*dt<<endl;
                real[int] RHS1 = rhs1(0,Zh);
                real[int] RHS2 = rhs2(0,Rh);
                matrix AA = aa(Zh,Zh);
                matrix AB = ab(Zh,Rh);
                matrix BA = ba(Rh,Zh);
                matrix BB = bb(Rh,Rh);

                matrix AABB = [ [AA,BA], [AB,BB] ];
                set(AABB,solver=sparsesolver,master=-1);
                real[int] rhs = [RHS1, RHS2];
                real[int] w=AABB^-1*rhs;
                Xoo=Xo; Yoo=Yo;
                          Xoi=Xo; Yoi=Yo;
                      [uo[],lamxo[]] = w;
                Thso = buildmesh(br1(m) + D(nn));
                          Thf= buildmesh(fr1(20*m)+fr2(3.25*m)+fr3
                             (20*m)+fr4(4*m)+fr5(12*m) + br1(m) +
                              D(nn));
//                        plot(br1(m) + D(nn));
                          [uo,vo,po]=[uo,vo,po]; [lamxo,lamyo,Xo,
                             Yo]=[lamxo,lamyo,Xo,Yo];
                uh=sqrt(uo*uo+vo*vo);
                plot(Thf, Thso, uh, value=1, fill=1, coef=100, cmm="t="+
                   i*dt+"      Velocity");
                          int nbA;
                          real xnear=x0+cx0+l1+3*h2, ynear=cy0-3*
                             h2, distmin=10;
                          for(int k=0;k<Ths.nv;k++)
                                  if((Thso(k).x-xnear)^2 +
                                     (Thso(k).y-ynear)^2
                                     < distmin)
                                      {distmin=(Thso(k
                                          ).x-xnear)^2
                                           + (Thso(k).
                                          y-ynear)^2;
                                          nbA=k;}
                      real CL=Thso(nbA).y;
                      if(minCL>CL) minCL=CL;
                      if(MMCL<MCL && CL<MCL && MCL>0.2 && MMCL
                         >0.2) {
                              cout<<"ft= "<<1./(i*dt-t0)<<"
                                 minCL= "<<minCL
                              <<" maxCL= "<<MCL<<" (max-min)
                                 /2= "<< (MCL-minCL)/2<< endl
                                 ;
                              MCL=-100; MMCL=-100;
                              t0=i*dt; minCL=10;
                          }
                       MMCL=MCL; MCL=CL;
                          cout<<i*dt<<"  "<<Thso(nbA).x<<" "<<Thso
                             (nbA).y<<" " << int2d(Thso)(1.)<<
                             endl;
}
```

Appendix 2: Some Comments on the Codes

We assume that the reader has a basic experience with `FreeFem++`. The codes are then pretty straightforward to be understood. Given the fluid mesh `Thf`, for instance, the definition of the finite element spaces for the approximation of the Navier–Stokes equation is performed with the following lines

```
fespace Vh(Thf,P1b);
fespace Qh(Thf,P1);
```

in the case of the MINI element [3], or with the following lines

```
fespace Vh(Thf,P2);
fespace Qh(Thf,P1);
```

if the user prefers Taylor–Hood element. Analogous definition can be used for the two finite element spaces based on the solid mesh `Ths`, for instance,

```
fespace Wh(Ths,P1);
fespace Lh(Ths,P1);
```

Like in all fluid-structure interaction problems, one of the crucial parts of the code consists in the evaluation of the terms involving an interaction between the fluid and solid meshes. This occurs in two places of our code: in the assembly of the bilinear form `ab` and of `ba`. Let us look in more detail at the assembly of `ab`, for instance. The bilinear form to be approximated is

$$ab((\mathbf{u}, p); (\hat{\boldsymbol{\lambda}}, \hat{\mathbf{X}})) = \int_{\mathcal{B}} \left(c_1 \mathrm{D}\hat{\boldsymbol{\lambda}} : \mathrm{D}\mathbf{u}(\mathbf{X}) + c_2 \hat{\boldsymbol{\lambda}} \cdot \mathbf{u}(\mathbf{X})\right) ds$$

and the crucial terms involve $\mathbf{u}(\mathbf{X})$ where the finite element function $\mathbf{u}$, defined on the mesh `Thf` has to be evaluated on $\mathbf{X}$ which is defined on the mesh `Ths`. This interpolation problem is naturally solved by using the option `mapu` in the evaluation of the integral as follows:

```
varf ab([u,v,p],[lamxh,lamyh,Xh,Yh]) =
     int2d(Ths,qft=qf9pT,mapu=[Xo,Yo])(
          c1*trace(Grad(lamxh,lamyh)'
              *(Grad(Xo,Yo)*Grad(u,v)))
          + c2*(lamxh*u+lamyh*v));
```

Analogously, when the mapping between the meshes involves the test function, as in the case of the definition of `ba`, the option `mapt` comes into play. In this particular case, the bilinear form

$$ab((\boldsymbol{\lambda}, \mathbf{X}); (\hat{\mathbf{u}}, \hat{p}) = \int_{\mathcal{B}} (c_1 \mathrm{D}\boldsymbol{\lambda} : \mathrm{D}\hat{\mathbf{u}}(\mathbf{X}) + c_2 \boldsymbol{\lambda} \cdot \hat{\mathbf{u}}(\mathbf{X}))\, ds$$

can be described with the following code instruction

```
varf ba([lamx,lamy,X,Y],[uh,vh,ph]) =
    int2d(Ths,qft=qf9pT,mapt=[Xo ,Yo])(
        c1*trace(Grad(lamx,lamy)'
           *(Grad(Xo,Yo)*Grad(uh,vh)))
       + c2*(lamx*uh+lamy*vh));
```

References

1. Boffi, D., Gastaldi, L.: A fictitious domain approach with Lagrange multiplier for fluid-structure interactions. Numer. Math. **135**(3), 711–732 (2017)
2. Boffi, D., Gastaldi, L., Heltai, L., Peskin, C.S.: On the hyper-elastic formulation of the immersed boundary method. Comput. Methods Appl. Mech. Eng. **197**(25–28), 2210–2231 (2008)
3. Boffi, D., Brezzi, F., Fortin, M.: Mixed Finite Element Methods and Applications. Springer Series in Computational Mathematics, vol. 44. Springer, New York (2013)
4. Boffi, D., Cavallini, N., Gastaldi, L.: The finite element immersed boundary method with distributed Lagrange multiplier. SIAM J. Numer. Anal. **53**(6), 2584–2604 (2015)
5. Dunne, Th., Rannacher, R., Richter, Th.: Numerical simulation of fluid-structure interaction based on monolithic variational formulations. In: Fundamental Trends in Fluid-Structure Interaction. Contemporary Challenges in Mathematical Fluid Dynamics and its Applications, vol. 1, pp. 1–75. World Scientific Publishing, Hackensack (2010)
6. Girault, V., Glowinski, R.: Error analysis of a fictitious domain method applied to a Dirichlet problem. Jpn. J. Ind. Appl. Math. **12**(3), 487–514 (1995)
7. Hecht, F.: New development in FreeFem++. J. Numer. Math. **20**(3–4), 251–265 (2012)
8. Hecht, F., Pironneau, O.: An energy stable monolithic Eulerian fluid-structure finite element method. Int. J. Numer. Methods Fluids **85**(7), 430–446 (2017)
9. Peskin, C.S.: The immersed boundary method. Acta Numer. **11**, 479–517 (2002)
10. Pironneau, O.: On the transport-diffusion algorithm and its applications to the Navier-Stokes equations. Numer. Math. **38**(3), 309–332 (1981/82)
11. Wang, Y., Jimack, P.K., Walkley, M.A.: A one-field monolithic fictitious domain method for fluid–structure interactions. Comput. Methods Appl. Mech. Eng. **317**, 1146–1168 (2017)
12. Zhao, H., Freund, J.B., Moser, R.D.: A fixed-mesh method for incompressible flow-structure systems with finite solid deformations. J. Comput. Phys. **227**(6), 3114–3140 (2008)

References

Chapter 6
Generalization of the Pythagorean Eigenvalue Error Theorem and Its Application to Isogeometric Analysis

Michael Bartoň, Victor Calo, Quanling Deng, and Vladimir Puzyrev

Abstract This chapter studies the effect of the quadrature on the isogeometric analysis of the wave propagation and structural vibration problems. The dispersion error of the isogeometric elements is minimized by optimally blending two standard Gauss-type quadrature rules. These blending rules approximate the inner products and increase the convergence rate by two extra orders when compared to those with fully-integrated inner products. To quantify the approximation errors, we generalize the Pythagorean eigenvalue error theorem of Strang and Fix. To reduce the computational cost, we further propose a two-point rule for C^1 quadratic isogeometric elements which produces equivalent inner products on uniform meshes and yet requires fewer quadrature points than the optimally-blended rules.

6.1 Introduction

Partial differential eigenvalue problems arise in a wide variety of applications, for example the vibration of elastic bodies (structural vibration) or multi-group diffusion in nuclear reactors [58]. Finite element analysis of these differential eigenvalue problems leads to the matrix eigenvalue problem with the entries of the matrices which are usually approximated by numerical integration. The effect of

M. Bartoň
Basque Center for Applied Mathematics, Bilbao, Basque Country, Spain
e-mail: Michael.Barton@centrum.cz

V. Calo
Department of Applied Geology, Western Australian School of Mines, Curtin University, Bentley, Perth, WA, Australia

Mineral Resources, Commonwealth Scientific and Industrial Research Organisation (CSIRO), Kensington, Perth, WA, Australia

Q. Deng (✉) · V. Puzyrev
Department of Applied Geology, Western Australian School of Mines, Curtin University, Bentley, Perth, WA, Australia
e-mail: Quanling.Deng@curtn.edu.au

D. A. Di Pietro et al. (eds.), *Numerical Methods for PDEs*,
SEMA SIMAI Springer Series 15, https://doi.org/10.1007/978-3-319-94676-4_6

these numerical integration methods on the eigenvalue and eigenfunction errors has been investigated in the literature; see for example Fix [29], Strang and Fix [58], and others [8–10]. Sharp and optimal estimates of the numerical eigenfunctions and eigenvalues of finite element analysis are established in [8, 9].

Hughes et al. [41] unified the analysis of the spectrum properties of the eigenvalue problem with the dispersion analysis for wave propagation problems. They established a duality principle between them: any numerical scheme that reduces the dispersion error of the wave propagation problems reduces the eigenvalue errors of the different eigenvalues problems and vice versa. Moreover, they share the same convergence property in the sense of convergence rates [15, 43, 54]. In this work, we focus on developing quadrature rules to optimize the dispersion errors and then apply these rules to the approximation of differential eigenvalue problems.

The dispersion analysis of the finite element method and spectral element method has been studied extensively; see for example Thomson and Pinsky[59, 60], Ihlenburg and Babuska [44], Ainsworth [1–3], and many others [23, 28, 35–38, 45, 46, 63]. Thomson and Pinsky studied the dispersive effects of using the Legendre, spectral, and Fourier local approximation basis functions for finite element methods when applied to the Helmholtz equation [59]. The choice of the basis functions has a negligible effect on the dispersion errors. Nevertheless, the continuity of the basis functions has a significant impact. Hughes et al. [41] showed that high continuities (up to C^{p-1} for p-th order isogeometric elements) on the basis functions result in dramatically smaller dispersion errors than that of finite elements.

Ainsworth [1] and [2] established that the optimal convergence rate, which is of order $2p$, of the dispersion error for the p-th order standard finite elements and spectral elements, respectively. The work was complete as they established the analysis for arbitrary polynomial order. The dispersive properties of these methods have been studied in detail and the most effective scheme was conjectured to be a mixed one of these two [3, 49, 56]. Ainsworth and Wajid beautifully established the optimal blending of these two methods for arbitrary polynomial order in 2010 in [3]. The blending was shown to provide two orders of extra accuracy (superconvergence) in the dispersion error, which includes the fourth order superconvergence result obtained by a modified integration rule for linear finite elements in [35]. Also, this blending scheme is equivalent to the use of nonstandard quadrature rules and therefore it can be efficiently implemented by replacing the standard Gaussian quadrature by a nonstandard rule [3].

This blending idea can be extended to isogeometric analysis (IGA), a numerical method that bridges the gap between computer aided design (CAD) and finite element analysis (FEA). We refer to [13, 19, 21, 40] for its initial development and to [20, 26, 33, 34, 41–43, 47, 48, 50] for its applications. The feature that distinguishes isogeometric elements from finite and spectral elements is the fact that the basis functions have up to $p-1$ continuous derivatives across element boundaries, where p is the order of the underlying polynomial. The publications [4, 19, 20, 41–43, 55] show that highly continuous isogeometric analysis delivers more robustness and better accuracy per degree of freedom than standard finite elements. Nevertheless, a detailed analysis of the solution cost reveals that IGA is more expensive to solve on a per degree of freedom basis than the lower continuous counterparts, such as finite element analysis [16–18, 52]. To exploit the reduction in cost, a set of solution strategies which control the continuity of the basis functions to deliver optimal solution costs were proposed [31, 32].

The dispersion analysis of isogeometric elements is studied in [41, 43, 54], presenting significant advantages over finite elements. Hughes et al. [41] showed that the dispersion error of the isogeometric analysis with high continuity (up to C^{p-1} for p-th order basis function) on the basis functions is smaller than that of the lower continuity finite element counterparts. Dedè et al. [24] study the dispersion analysis of the isogeometric elements for the two-dimensional harmonic plane waves in an isotropic and homogeneous elastic medium. The anisotropic curves are represented using NURBS-based IGA and the errors associated with the compressional and shear wave velocities for different directions of the wave vector are modeled. Recently, the dispersion error minimization for isogeometric analysis has been performed numerically in Puzyrev et al. [54] and analytically in Calo et al. [15].

In this work, we seek blending quadrature rules for isogeometric element to minimize the dispersion error of the scheme and hence increase its accuracy and robustness. We focus on the dispersion analysis of isogeometric elements and apply the blending ideas introduced by [3] for finite and spectral elements to isogeometric elements by using a modified inner product. The new blending schemes reduce the errors in the approximation of the eigenvalues (and, in some cases, the eigenfunctions). Using the optimal blending, convergence rates of the dispersion error is increased by two additional orders. To analyze the errors, we characterize the errors in the eigenvalues and the eigenfunctions for all the modes. The total "error budget" of the numerical method consists of the errors arising from the approximation of eigenvalues and eigenfunctions. When the stiffness and mass terms are fully integrated, for each eigenvalue, the sum of the eigenvalue error and the square of the eigenfunction error in the L^2-norm scaled by the exact eigenvalue equals the square of the error in the energy norm. Once one of these terms are not fully integrated, this is not true any more. To account for the error of the approximated/modified inner product, we generalize Strang's Pythagorean eigenvalue theorem to include the effect of inexact integration.

The outline of the remainder of this chapter is as follows. We first describe the model problem in Sect. 6.2. In Sect. 6.3, we present a generalization of the Pythagorean eigenvalue error theorem that accounts for the error of the modified inner products. In Sect. 6.4, we describe the optimal blending of finite and spectral elements and present an optimal blending scheme for isogeometric analysis. In Sect. 6.5, we develop a two-point quadrature rule for periodic boundaries. Numerical examples for one-dimensional and two-dimensional problems are given in Sect. 6.6. Finally, Sect. 6.7 summarizes our findings and describes future research directions.

6.2 Problem Setting

We begin with the differential eigenvalue problem

$$\begin{aligned} -\Delta u &= \lambda u \quad \text{in} \quad \Omega, \\ u &= 0 \quad \text{on} \quad \partial\Omega, \end{aligned} \tag{6.1}$$

where $\Delta = \nabla^2$ is the Laplacian and $\Omega \subset \mathbb{R}^d, d = 1, 2, 3$ is a bounded open domain with Lipschitz boundary. This eigenvalue problem has a countable infinite set of eigenvalues $\lambda_j \in \mathbb{R}$

$$0 < \lambda_1 \leq \lambda_2 \leq \cdots \leq \lambda_j \leq \cdots \tag{6.2}$$

and an associated set of orthonormal eigenfunctions u_j

$$(u_j, u_k) = \int_\Omega u_j(x) u_k(x)\, \mathrm{d}x = \delta_{jk}, \tag{6.3}$$

where δ_{jk} is the Kronecker delta which is equal to 1 when $i = j$ and 0 otherwise (see for example [58]). The normalized eigenfunctions form an L^2-orthonormal basis. Moreover, using integration by parts and (6.1), they are orthogonal also in the energy inner product

$$(\nabla u_j, \nabla u_k) = (-\Delta u_j, u_k) = (\lambda_j u_j, u_k) = \lambda_j (u_j, u_k) = \lambda_j \delta_{jk}. \tag{6.4}$$

Let V be the solution space, a subspace of the Hilbert space $H_0^1(\Omega)$. The standard weak form for the eigenvalue problem: Find all eigenvalues $\lambda_j \in \mathbb{R}$ and eigenfunctions $u_j \in V$ such that,

$$a(u_j, w) = \lambda_j (u_j, w), \quad \forall\, w \in V \tag{6.5}$$

where

$$a(w, v) = \int_\Omega \nabla w \cdot \nabla v\, \mathrm{d}x, \tag{6.6}$$

and $(\cdot, \cdot)$ is the L^2 inner product. These two inner products are associated with the following energy and L^2 norms

$$\|w\|_E = \sqrt{a(w, w)}, \quad \|w\| = \sqrt{(w, w)}. \tag{6.7}$$

The Galerkin-type formulation of the eigenvalue problem (6.1) is the discrete form of (6.5): Seek $\lambda_j^h \in \mathbb{R}$ and $u_j^h \in V^h \subset V$ such that

$$a(u_j^h, w^h) = \lambda_j^h (u_j^h, w^h), \quad \forall\, w^h \in V^h, \tag{6.8}$$

which results in the generalized matrix eigenvalue problem

$$\mathbf{K}\,\mathbf{u}^h = \lambda^h \mathbf{M}\,\mathbf{u}^h, \tag{6.9}$$

where $\mathbf{K}$ is referred as the stiffness matrix, $\mathbf{M}$ is referred as the mass matrix, and $(\lambda^h, \mathbf{u}^h)$ are the unknown eigenpairs.

We described the differential eigenvalue problem and its Galerkin discretization above. For dispersion analysis, we study the classical wave propagation equation

$$-\Delta u + \frac{1}{c^2}\frac{\partial^2 u}{\partial^2 t} = 0, \tag{6.10}$$

where c is the wave propagation speed. We abuse the notation of unknown u here. Assuming time-harmonic solutions of the form $u(\mathbf{x}, t) = e^{-i\omega t} u(\mathbf{x})$ for a given temporal frequency ω, the wave equation reduces to the well-known Helmholtz equation

$$-\Delta u - k^2 u = 0, \tag{6.11}$$

where the wavenumber $k = \omega/c$ represents the ratio of the angular frequency ω to the wave propagation speed c. The wavelength is equal to $2\pi/k$. The discretization of (6.11) leads to the following linear equation system

$$\left(\mathbf{K} - k^2\mathbf{M}\right)\mathbf{u}^h = 0. \tag{6.12}$$

The equivalence between (6.1) and (6.11) or (6.9) and (6.12) is established by setting λ or $\lambda^h = k^2$. Based on this equivalence, a duality principle between the spectrum analysis of the differential eigenvalue problem and the dispersion analysis of the wave propagation is established in [41]. In practice, the wavenumber is approximated and we denote it as k^h. In general, $k^h \neq k$. Then the solution of (6.12) is a linear combination of plane waves with numerical wavenumbers k^h. Hence the discrete and exact waves have different wavelengths. The goal of the dispersion analysis is to quantify this difference and define this difference as the dispersion error of a specific numerical method. That is, dispersion analysis seeks to quantify how well the discrete wavenumber k^h approximates the continuous/exact k. Finally, in the view of unified analysis in [41], this dispersion error describes the errors of the approximated eigenvalues to the exact ones for (6.8) or (6.9).

6.3 Pythagorean Eigenvalue Error Theorem and Its Generalization

The theorem was first described in Strang and Fix [58] and was referred as the Pythagorean eigenvalue error theorem in Hughes [43]. In this section, we revisit this theorem in detail and generalize it.

6.3.1 The Theorem

Following Strang and Fix [58], the Rayleigh-Ritz idea for the steady-state equation $\mathcal{L}u = f$ ($\mathcal{L}$ is a differential operator) was extended to the differential eigenvalue problem. The idea leads to the finite element approximation of the eigenvalue problem. Equation (6.5) resembles the variational formulation for the steady-state equation. Hence, one expects the approximated eigenfunction errors are of the same convergence rates as those in steady-state problems. Definitely, the a priori error estimation of the eigenfunction will depend on the index j (as in j-th eigenvalue) and the accuracy will deteriorate as j increases. In fact, the errors of the approximated eigenvalues also increase and hence deteriorate the accuracy as j increases [7, 41, 58].

The a priori error analysis for the approximation of eigenfunctions and eigenvalues has a prominent connection. The motivation to derive the Pythagorean eigenvalue error theorem as stated below (see also Lemma 6.3 in [58]) is to elucidate the relation the between the eigenvalue and eigenvector errors to the total approximation error.

Theorem 6.1 *For each discrete mode, with the normalization* $\|u_j\| = 1$ *and* $\|u_j^h\| = 1$*, we have*

$$\|u_j - u_j^h\|_E^2 = \lambda_j \|u_j - u_j^h\|^2 + \lambda_j^h - \lambda_j. \tag{6.13}$$

By the Minmax Principle (discovered by Poincaré, Courant, and Fischer; referred by Strang and Fix), all finite element approximated eigenvalues bound the exact ones from above, that is

$$\lambda_j^h \geq \lambda_j \quad \forall\, j. \tag{6.14}$$

This allows us to write (6.13) in the conventional Pythagorean theorem formulation

$$\|u_j - u_j^h\|_E^2 = \left(\sqrt{\lambda_j}\|u_j - u_j^h\|\right)^2 + \left(\sqrt{\lambda_j^h - \lambda_j}\right)^2. \tag{6.15}$$

This theorem was established with a simple proof in [58]. Alternatively, we present here

$$\begin{aligned}
\|u_j - u_j^h\|_E^2 &= a(u_j - u_j^h, u_j - u_j^h) \\
&= a(u_j, u_j) - 2a(u_j, u_j^h) + a(u_j^h, u_j^h) \\
&= \lambda_j(u_j, u_j) - 2\lambda_j(u_j, u_j^h) + \lambda_j^h(u_j^h, u_j^h) \\
&= \lambda_j\big((u_j, u_j) - 2(u_j, u_j^h) + (u_j^h, u_j^h)\big) + \lambda_j^h - \lambda_j \\
&= \lambda_j \|u_j - u_j^h\|^2 + \lambda_j^h - \lambda_j.
\end{aligned} \tag{6.16}$$

This theorem tells that for each discrete mode, the square of the error in the energy norm consists of the eigenvalue error and the product of the eigenvalue and the square of the eigenfunction error in the L^2-norm. We can rewrite (6.13) as

$$\frac{\lambda_j^h - \lambda_j}{\lambda_j} + \|u_j^h - u_j\|^2 = \frac{\|u_j^h - u_j\|_E^2}{\lambda_j}, \tag{6.17}$$

which implies

$$\lambda_j^h - \lambda_j \leq \|u_j^h - u_j\|_E^2, \tag{6.18}$$

$$\|u_j^h - u_j\|^2 \leq \frac{\|u_j^h - u_j\|_E^2}{\lambda_j}. \tag{6.19}$$

This tells further the relation among the eigenvalue errors, eigenfunction error in L^2 norm, and eigenfunction error in energy norm. Once error estimation for eigenfunction error in energy norm is established, the other two are obvious. Also, the inequality (6.19) does not hold for methods that do not approximate all eigenvalues from above (that is violating (6.14)), for example, the spectral element method [2]. In general, the spectral element method is realized by using the Gauss-Legendre-Lobatto nodes to define the interpolation nodes for Lagrange basis functions in each element. This quadrature rule induces an error in the approximation of the inner products, but preserves the optimal order of convergence of the scheme. In fact, these errors in the inner product allow the numerical scheme to approximate eigenvalues from below. If the discrete method does not fully reproduce the inner products associated with the stiffness and mass matrices or these inner products are approximated using numerical integration, this theorem needs to be extended to account for the errors introduced by the approximations of the inner products.

6.3.2 The Quadrature

Now to derive the generalized Pythagorean eigenvalue error theorem, we first introduce the numerical integration with quadratures. The entries of the stiffness and mass matrices **K** and **M** in (6.9) are given by the inner products

$$\mathbf{M}_{ij} = \int_{\Omega} \phi_i(x)\phi_j(x)\,\mathrm{d}x, \tag{6.20}$$

$$\mathbf{K}_{ij} = \int_{\Omega} \nabla\phi_i(x)\cdot\nabla\phi_j(x)\,\mathrm{d}x, \tag{6.21}$$

where $\phi_i(x)$ are the piecewise polynomial basis functions. Here, we consider basis functions for finite elements, spectral elements, and isogeometric analysis. **M** and **K** are symmetric positive definite matrices. Moreover, in the 1D matrices have $2p+1$ diagonal entries.

In practice, the integrals in (6.20) and (6.21) are evaluated numerically, that is, approximated by quadrature rules. Now we give a brief description of the quadrature rules for approximating the inner products (6.20) and (6.21). On a reference element $\hat{K}$, an $(n+1)$-point quadrature rule for a function $f(x)$ is of the form

$$\int_{\hat{K}} \hat{f}(\hat{\boldsymbol{x}})\,\mathrm{d}\hat{\boldsymbol{x}} = \sum_{l=0}^{n} \hat{\varpi}_l \hat{f}(\hat{n}_l) + \hat{E}_{n+1}, \tag{6.22}$$

where $\hat{\varpi}_l$ are the weights, $\hat{n}_l$ are the nodes, and $\hat{E}_{n+1}$ is the error of the quadrature rule. For each element K, there is an invertible affine map σ such that $K = \sigma(\hat{K})$, which leads to the correspondence between the functions on K and $\hat{K}$. Let J_K be the corresponding Jacobian of the mapping. Then (6.22) induces a quadrature rule over the element K given by

$$\int_{K} f(\boldsymbol{x})\,\mathrm{d}\boldsymbol{x} \approx \sum_{l=0}^{n} \varpi_{l,K} f(n_{l,K}) + E_{n+1}, \tag{6.23}$$

where $\varpi_{l,K} = \det(J_K)\hat{\varpi}_l$ and $n_{l,K} = \sigma(\hat{n}_l)$.

The quadrature rule is exact for a given function $f(x)$ when the remainder E_{n+1} is exactly zero. For example, the standard $(n+1)$-point Gauss-Legendre (GL or Gauss) quadrature is exact for the linear space of polynomials of degree at most $2n+1$ (see, for example, [12, 57]).

The classical Galerkin finite element analysis typically employs the Gauss quadrature with $p+1$ (where p is the polynomial order) quadrature points per parametric direction that fully integrates every term in the bilinear forms defined by the weak form. A quadrature rule is optimal if the function is evaluated with the minimal number of nodes (for example, Gauss quadrature with $n + 1$ evaluations is optimal for polynomials of order $2n + 1$ in one dimension).

Element-level integrals may be approximated using other quadrature rules, for example the Gauss-Lobatto-Legendre (GLL or Lobatto) quadrature rule that is used in the spectral element method (SEM). The Lobatto quadrature evaluated at $n + 1$ nodes is accurate for polynomials up to degree $2n+1$. However, selecting a rule with $p + 1$ evaluations for a polynomial of order p and collocating the Lagrange nodes with the quadrature positions renders the mass matrix diagonal in 1D, 2D and 3D for arbitrary geometrical mappings. This resulting diagonal mass matrix is a more relevant result than the reduction in the accuracy of the calculation. Particularly, given that this property preserves the optimal convergence order for these higher-order schemes. Lastly, the spectral elements possess a superior phase accuracy when compared with the standard finite elements of the same polynomial order [2].

Isogeometric analysis based on NURBS (Non-Uniform Rational B-Splines) has been described in a number of papers (e.g. [13, 19, 20, 41]). Isogeometric analysis employs piecewise polynomial curves composed of linear combinations of B-spline basis functions. B-spline curves of polynomial order p may have up to $p - 1$ continuous derivatives across element boundaries. Three different refinement mechanisms are commonly used in isogeometric analysis, namely the h-, p- and k-refinement, as detailed in [20]. We refer the reader to [53] for the definition of common concepts of isogeometric analysis such as knot vectors, B-spline functions, and NURBS.

The derivation of optimal quadrature rules for NURBS-based isogeometric analysis with spaces of high polynomial degree and high continuity has attracted significant attention in recent years [5, 6, 11, 12, 14, 39, 42]. The efficiency of Galerkin-type numerical methods for partial differential equations depends on the formation and assembly procedures, which, in turn, largely depend on the efficiency of the quadrature rule employed. Integral evaluations based on full Gauss quadrature are known to be efficient for standard C^0 finite element methods, but inefficient for isogeometric analysis that uses higher-order continuous spline basis functions [51].

Hughes et al. [42] studied the effect of reduced Gauss integration on the finite element and isogeometric analysis eigenvalue problems. By using p Gauss points (i.e., underintegrating using one point less), one modifies the mass matrix only (in 1D). By using less than p Gauss points (i.e., underintegrating using several points less), both mass and stiffness matrices are underintegrated. Large underintegration errors may lead to the loss of stability since the stiffness matrix becomes singular. As shown in [42], this kind of underintegration led to the results that were worse than the fully integrated ones and the highest frequency errors diverged as the mesh was refined. However, as we show in the next sections, using properly designed alternative quadratures may lead to more accurate results.

The assembly of the elemental matrices into the global stiffness and mass matrices is done in a similar way for all Galerkin methods we analyze in this chapter. Similarly, the convergence rate for all Galerkin schemes we analyze is the same. However, the heterogeneity of the high-order finite element (C^0 elements, i.e., SEM and FEA) basis functions leads to a branching of the discrete spectrum and a fast degradation of the accuracy for higher frequencies. In fact, the degraded frequencies in 1D are about half of all frequencies, while in 3D this proportion reduces to about seven eighths. On uniform meshes, B-spline basis functions of the highest $p-1$ continuity, on the contrary, are homogeneous and do not exhibit such branching patterns other than the outliers that correspond to the basis functions with support on the boundaries of the domain.

6.3.3 The Generalization

Now we consider the generalization. Applying quadrature rules to (6.8), we have the approximated form

$$a_h(\tilde{u}_j^h, w^h) = \tilde{\lambda}_j^h(\tilde{u}_j^h, w^h)_h \quad \forall\, w_h \in V^h, \tag{6.24}$$

where

$$a_h(w, v) = \sum_{K\in\mathcal{T}_h} \sum_{l=1}^{N_q} \varpi_{l,K}^{(1)} \nabla w(n_{l,K}^{(1)}) \cdot \nabla v(n_{l,K}^{(1)}), \tag{6.25}$$

and

$$(w, v)_h = \sum_{K\in\mathcal{T}_h} \sum_{l=1}^{N_q} \varpi_{l,K}^{(2)} w(n_{l,K}^{(2)}) v(n_{l,K}^{(2)}), \tag{6.26}$$

where $\{\varpi_{l,K}^{(1)}, n_{l,K}^{(1)}\}$ and $\{\varpi_{l,K}^{(2)}, n_{l,K}^{(2)}\}$ specify two (possibly different) quadrature rules. This leads to the matrix eigenvalue problem

$$\mathbf{K}^h\tilde{\mathbf{u}}^h = \tilde{\lambda}^h\mathbf{M}^h\tilde{\mathbf{u}}^h, \tag{6.27}$$

where the superscripts on **K** and **M** and the tildes specify the effect of the quadratures.

Remark 6.1 For multidimensional problems on tensor product grids, the stiffness and mass matrices can be expressed as Kronecker products of 1D matrices [30]. For example, in the 2D case, the components of **K** and **M** can be represented as fourth-order tensors using the definitions of the matrices and the basis functions for the 1D case [22, 30]

$$\mathbf{M}_{ijkl} = \mathbf{M}_{ik}^{1D}\mathbf{M}_{jl}^{1D}, \tag{6.28}$$

$$\mathbf{K}_{ijkl} = \mathbf{K}_{ik}^{1D}\mathbf{M}_{jl}^{1D} + \mathbf{K}_{jl}^{1D}\mathbf{M}_{ik}^{1D}, \tag{6.29}$$

where $\mathbf{M}_{ij}^{1D}$ and $\mathbf{K}_{ij}^{1D}$ are the mass and stiffness matrices of the 1D problem as given by (6.20) and (6.21). We refer the reader to [22] for the description of the summation rules.

To understand the errors of the approximations of eigenvalues and eigenfunctions when quadratures are applied, we measure the errors they induce in the inner products. The following theorem generalizes the Pythagorean eigenvalue error theorem to account for these modified inner products [54].

Theorem 6.2 *For each discrete mode, with the normalization* $\|u_j\| = 1$ *and* $(\widetilde{u}_j^h, \widetilde{u}_j^h)_h = 1$, *we have*

$$\|u_j - \widetilde{u}_j^h\|_E^2 = \widetilde{\lambda}_j^h - \lambda_j + \lambda_j \|u_j - \widetilde{u}_j^h\|^2 + \|\widetilde{u}_j^h\|_E^2 - \|\widetilde{u}_j^h\|_{E,h}^2 + \lambda_j \Big(1 - \|\widetilde{u}_j^h\|^2\Big), \tag{6.30}$$

where $\|\cdot\|_{E,h}$ *is the energy norm evaluated by a quadrature rule.*

Proof By definition and linearity of the bilinear forms, we have

$$\|u_j - \widetilde{u}_j^h\|_E^2 = a(u_j - \widetilde{u}_j^h, u_j - \widetilde{u}_j^h) = a(u_j, u_j) - 2a(u_j, \widetilde{u}_j^h) + a(\widetilde{u}_j^h, \widetilde{u}_j^h). \tag{6.31}$$

From (6.5), we have

$$\begin{aligned} a(u_j, u_j) &= \lambda_j (u_j, u_j), \\ a(u_j, \widetilde{u}_j^h) &= \lambda_j (u_j, \widetilde{u}_j^h). \end{aligned}$$

Thus, adding and subtracting a term $\lambda_j(\tilde{u}_j^h, \tilde{u}_j^h)$, (6.31) is rewritten as

$$\begin{aligned} \|u_j - \widetilde{u}_j^h\|_E^2 &= \lambda_j (u_j, u_j) - 2\lambda_j (u_j, \widetilde{u}_j^h) + \lambda_j (\tilde{u}_j^h, \tilde{u}_j^h) - \lambda_j (\tilde{u}_j^h, \tilde{u}_j^h) + a(\widetilde{u}_j^h, \widetilde{u}_j^h) \\ &= \lambda_j \Big((u_j, u_j) - 2(u_j, \widetilde{u}_j^h) + (\tilde{u}_j^h, \tilde{u}_j^h) \Big) - \lambda_j \|\widetilde{u}_j^h\|^2 + \|\widetilde{u}_j^h\|_E^2 \\ &= \lambda_j \|u_j - \tilde{u}_j^h\|^2 - \lambda_j \|\widetilde{u}_j^h\|^2 + \|\widetilde{u}_j^h\|_E^2. \end{aligned}$$

From (6.24) and the definition of the modified energy norm $\|\cdot\|_{E,h}$, we have

$$\|\widetilde{u}_j^h\|_{E,h}^2 = a_h(\tilde{u}_j^h, \tilde{u}_j^h) = \tilde{\lambda}_j^h (\tilde{u}_j^h, \tilde{u}_j^h)_h.$$

Noting that $(\widetilde{u}_j^h, \widetilde{u}_j^h)_h = 1$, we have

$$\widetilde{\lambda}_j^h - \lambda_j = \Big(\widetilde{\lambda}_j^h - \lambda_j\Big)(\widetilde{u}_j^h, \widetilde{u}_j^h)_h = \|\widetilde{u}_j^h\|_{E,h}^2 - \lambda_j.$$

Thus, adding and subtracting a term $\widetilde{\lambda}_j^h - \lambda_j$ gives

$$\|u_j - \widetilde{u}_j^h\|_E^2 = \lambda_j\|u_j - \widetilde{u}_j^h\|^2 - \lambda_j\|\widetilde{u}_j^h\|^2 + \|\widetilde{u}_j^h\|_E^2 + \left(\widetilde{\lambda}_j^h - \lambda_j\right) - \left(\|\widetilde{u}_j^h\|_{E,h}^2 - \lambda_j\right)$$
$$= \widetilde{\lambda}_j^h - \lambda_j + \lambda_j\|u_j - \widetilde{u}_j^h\|^2 + \|\widetilde{u}_j^h\|_E^2 - \|\widetilde{u}_j^h\|_{E,h}^2 + \lambda_j\left(1 - \|\widetilde{u}_j^h\|^2\right),$$

which completes the proof.

The equation in (6.30) can be rewritten as

$$\frac{\|u_j - \widetilde{u}_j^h\|_E^2}{\lambda_j} = \frac{\widetilde{\lambda}_j^h - \lambda_j}{\lambda_j} + \|u_j - \widetilde{u}_j^h\|^2 + \frac{\|\widetilde{u}_j^h\|_E^2 - \|\widetilde{u}_j^h\|_{E,h}^2}{\lambda_j} + \left(1 - \|\widetilde{u}_j^h\|^2\right),$$

in which the first term on the right-hand side is the relative error of the approximated eigenvalue, the second term represent the error of eigenfunction in L^2 norm, the third term shows the eigenvalue-scaled error due to the modification of the inner product associated with the stiffness, and the last term shows the error due to the modification of the inner product associated with the mass.

The left-hand side and the first two terms on the right-hand side resemble the Pythagorean eigenvalue error theorem, while the extra two terms reveal the effect of numerical integration of the inner products associated with the stiffness and the mass. In the cases when these inner products are integrated exactly, these two extra terms are zeros. Consequently, Theorem 6.2 reduces to the standard Pythagorean eigenvalue error theorem.

6.4 Optimal Blending for Finite Elements and Isogeometric Analysis

Several authors (e.g. [3, 27, 56]) studied the blended spectral-finite element method that uses nonstandard quadrature rules to achieve an improvement of two orders of accuracy compared with the fully integrated schemes. This method is based on blending the full Gauss quadrature, which exactly integrates the bilinear forms to produce the mass and stiffness matrices, with the Lobatto quadrature, which underintegrates them. This methodology exploits the fact that the fully integrated finite elements exhibit phase lead when compared with the exact solutions, while the underintegrated with Lobatto quadrature methods, such as, spectral elements have phase lag.

Ainsworth and Wajid [3] chose the blending parameter to maximize the order of accuracy in the phase error. They showed that the optimal choice for the blending parameter is given by weighting the spectral element and the finite element methods in the ratio $\frac{p}{p+1}$. As mentioned above, this optimally blended scheme improves by two orders the convergence rate of the blended method when compared against the finite or spectral element methods that were the ingredients used in the blending. The blended scheme can be realized in practice without assembling the mass matrices for either of the schemes, but instead by replacing the standard Gaussian quadrature rule by an alternative rule, as Ainsworth and Wajid clearly explained in [3]. Thus, no additional computational cost is required by the blended scheme although the ability to generate a diagonal mass matrix by the underintegrated spectral method is lost.

To show how an improvement in the convergence rate is achieved, consider, for example, the approximate eigenfrequencies written as a series in $\Lambda = \omega h$ for the linear finite and spectral elements, respectively [3]

$$\omega_{FE}^h h = \Lambda - \frac{\Lambda^3}{24} + O(\Lambda^5), \tag{6.32}$$

$$\omega_{SE}^h h = \Lambda + \frac{\Lambda^3}{24} + O(\Lambda^5). \tag{6.33}$$

When these two schemes are blended using a blending parameter τ, the approximate eigenfrequencies become

$$\omega_{BL}^h h = \Lambda + \frac{\Lambda^3}{24}(2\tau - 1) + O(\Lambda^5). \tag{6.34}$$

For $\tau = 0$ and $\tau = 1$, the above expression reduces to the ones obtained by the finite element and spectral element schemes, respectively. The choice of $\tau = 1/2$ allows the middle term of (6.34) to vanish and adds two additional orders of accuracy to the phase approximation when compared with the standard schemes. Similarly, by making the optimal choice of blending parameter $\tau = \frac{p}{(p+1)}$ in high-order schemes, they removed the leading order term from the error expansion.

The numerical examples in Sect. 6.6 show that a similar blending can be applied to the isogeometric mass and stiffness matrices to reduce the eigenvalue error. For C^1 quadratic elements, the approximate eigenfrequencies are

$$\omega_{GL}^h h = \Lambda - \frac{1}{5!}\frac{\Lambda^5}{12} + O(\Lambda^7), \tag{6.35}$$

$$\omega_{GLL}^h h = \Lambda + \frac{1}{5!}\frac{\Lambda^5}{24} + O(\Lambda^7). \tag{6.36}$$

Similarly, blending these two rules utilizing a parameter τ gives

$$\omega_{BL}^h h = \Lambda + \frac{3\tau - 2}{5! \cdot 24}\Lambda^5 + O(\Lambda^7). \tag{6.37}$$

Thus the optimal ratio of the Lobatto and Gauss quadratures is $2:1$ ($\tau = 2/3$) similar to the optimally blended spectral-finite element scheme. For C^2 cubic elements, we determine that a non-convex blending with $\tau = 5/2$ allows us to remove the leading error term and thus achieve two additional orders of accuracy.

Remark 6.2 In general, for C^0 elements such as the finite elements and spectral elements, the optimal blending is [3]: $\tau = \frac{p}{(p+1)}$ for arbitrary p. This is, however, not true for isogeometric C^k elements, where $1 \leq k \leq p-1$ and $p \geq 3$. Finding the optimal blending parameter for $p \geq 3$ with $k > 0$ remains an open question. For $p \leq 7$ with $k = p-1$ and the discussion on its generalization, we refer the reader to [15].

Equations (6.32)–(6.36) show that the *absolute* errors in the eigenfrequencies converge with the rates of $O\left(\Lambda^{2p+1}\right)$ and $O\left(\Lambda^{2p+3}\right)$ for the standard and optimal schemes, respectively. If we consider the *relative* eigenfrequency errors, from Eqs. (6.35) and (6.36), these take the form

$$\frac{\omega^h h}{\Lambda} = 1 \pm \frac{\Lambda^4}{\alpha} + \cdots, \tag{6.38}$$

that is, the convergence rate for frequencies computed using IGA approximations is $O\left(\Lambda^{2p}\right)$ as shown in [19, 55]. The optimal blending in IGA leads to a $O\left(\Lambda^{2p+2}\right)$ convergence rate for the relative eigenfrequencies. This superconvergence result is similar to the one achieved by the optimally-blending of the spectral and finite element methods of [3].

Remark 6.3 Wang et al. [61, 62] constructed super-convergent isogeometric finite elements for dispersion by blending two alternative quadrature methods. They used full Gauss and a method which reduces the bandwidth of the mass and stiffness method. Although the construction is different, algebraically the resulting algebraic system is identical for uniform meshes.

6.5 Two-Point Rules for C^1 Quadratic Isogeometric Analysis

The optimally-blended rules presented above first introduce an auxiliary parameter for combining two different standard quadrature rules. Then the parameter is determined by eliminating the highest order term in the error expansion. We can achieve a similar result by designing a nonstandard quadrature rule here.

For C^1 quadratic isogeometric analysis, the blending requires evaluations of the function at two sets of quadrature nodes on each element, which is not computationally efficient. In this section, we present a two-point rule which eliminates the leading order term in the error expansion hence results in an equivalent but computationally efficient scheme for the C^1 quadratic isogeometric elements.

We consider uniform meshes with periodic boundary conditions for the eigenvalue problem in 1D. In the reference interval $[-1, 1]$, the two point rules are listed in Table 6.1.

Table 6.1 Two-point rules in the reference interval $[-1, 1]$ for C^1 quadratic isogeometric analysis

Rules	n_1	n_2	ϖ_1	ϖ_2
Rule 1	$-\frac{1}{5}\sqrt{11-\frac{2\sqrt{266}}{3}}$	$\frac{1}{5}\sqrt{11+\frac{2\sqrt{266}}{3}}$	$1+\frac{2\sqrt{266}}{133}$	$1-\frac{2\sqrt{266}}{133}$
Rule 2	$\frac{1}{5}\sqrt{11-\frac{2\sqrt{266}}{3}}$	$-\frac{1}{5}\sqrt{11+\frac{2\sqrt{266}}{3}}$	$1+\frac{2\sqrt{266}}{133}$	$1-\frac{2\sqrt{266}}{133}$
Rule 3	$-\frac{1}{5}\sqrt{11+\frac{2\sqrt{266}}{3}}$	$\frac{1}{5}\sqrt{11-\frac{2\sqrt{266}}{3}}$	$1-\frac{2\sqrt{266}}{133}$	$1+\frac{2\sqrt{266}}{133}$
Rule 4	$\frac{1}{5}\sqrt{11+\frac{2\sqrt{266}}{3}}$	$-\frac{1}{5}\sqrt{11-\frac{2\sqrt{266}}{3}}$	$1-\frac{2\sqrt{266}}{133}$	$1+\frac{2\sqrt{266}}{133}$

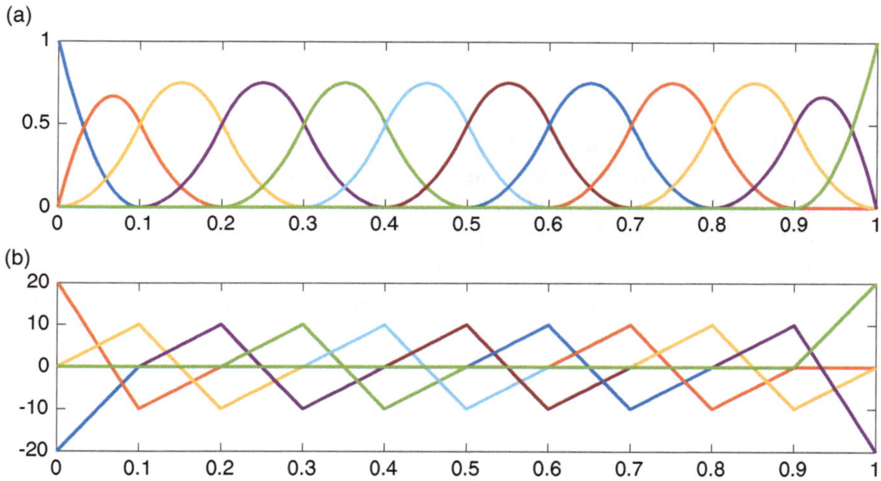

Fig. 6.1 Isogeometric C^1 quadratic B-spline basis functions and their derivatives. (**a**) Basis functions. (**b**) Derivatives of basis functions

These two-point rules share some sense of symmetry and lead to the same matrix eigenvalue problem. On uniform meshes with periodic boundary conditions, all these rules give the same dispersion errors.

In a periodic boundary domain discretized with a uniform mesh, we show numerically that these two-point rules lead to the same set of eigenvalues and eigenfunctions as these obtained by the optimally-blended schemes. In fact, they result in the same stiffness and mass matrices. The two-point rules fail when we use a boundary condition other than periodic, for example, Dirichlet or Neumann conditions. This happens since the two-point rule does not integrate the stiffness terms exactly near the boundary elements where the derivatives of the B-splines basis functions do not vanish; see Fig. 6.1. We will understand and address this shortcoming in future work.

For multidimensional cases, we assume that a tensor product grid is placed on the domain Ω. Then generalize these two-point rules to be 2^d-point rules for d-dimensional problems by simple tensor construction. We conclude that these two-point rules developed above remain valid for higher dimensional problems. More details are referred to [15, 25].

6.6 Numerical Examples

In this section, we present numerical examples of the one- and two-dimensional problems described in Sect. 6.2 to show how the use of optimal quadratures reduce the approximation errors in isogeometric analysis.

The 1D elliptic eigenvalue problem has the following exact eigenvalues and their corresponding eigenfunctions

$$\lambda_j = j^2\pi^2, \quad u_j = \sqrt{2}\sin(j\pi x), \tag{6.39}$$

for $j = 1, 2, \ldots$. The approximate eigenvalues λ_j^h are sorted in ascending order and are compared to the corresponding exact eigenvalues λ_j. The total number of degrees of freedom (discrete modes) is $N = 1000$.

Figure 6.2 compares the approximation errors of C^1 quadratic isogeometric elements using the standard Gaussian quadrature and the optimal rule. We show the relative eigenvalue errors $\frac{\mu_l^h - \lambda_l}{\lambda_l}$, the L_2-norm eigenfunction errors $\|u_l - v_l^h\|_0^2$

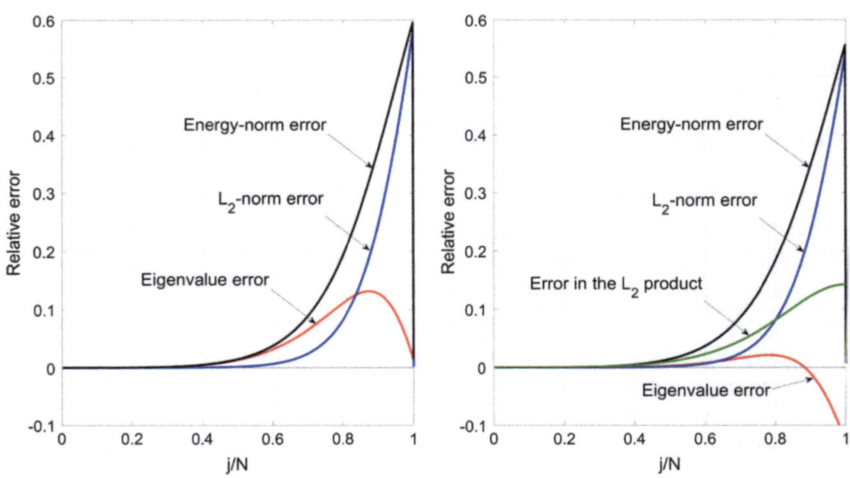

Fig. 6.2 Approximation errors for C^1 quadratic isogeometric elements with standard Gauss quadrature rule (left) and optimal blending (right)

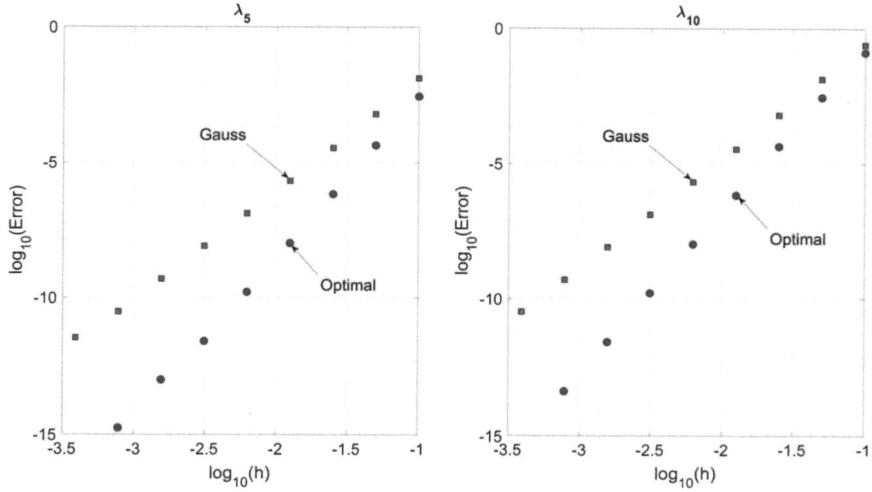

Fig. 6.3 Convergence of the errors in the eigenvalue approximation using C^1 quadratic isogeometric elements with standard and optimal quadratures. The fifth (left) and tenth (right) eigenvalues are shown

and the relative energy-norm errors $\frac{\|u_l - v_l^h\|_E^2}{\lambda_l}$. This format of error representation clearly illustrates the budget of the generalized Pythagorean eigenvalue theorem. The error in the L_2 norm $1 - \left\| v_l^h \right\|_0^2$ is shown only in the case when it is not zero.

In Fig. 6.2, the use of the optimal quadrature leads to more accurate results. Surprisingly, not only the eigenvalues, but also the eigenfunctions of the problem are better approximated in this particular case. The optimal ratio of blending of the Lobatto and Gauss quadrature rules in this case is 2:1, which is the same to the ratio proposed by Ainsworth and Wajid (2010) for the finite element case.

Figure 6.3 shows the dispersion errors in the eigenvalue approximation with C^1 quadratic isogeometric elements. The size of the meshes used in these simulations increases from 10 to 2560 elements. These results confirm two extra orders of convergence in the eigenvalue errors.

To study the behavior of discrete eigenfunctions from different parts of the spectrum, in Fig. 6.4 we compare the discrete and analytical eigenfunctions for C^1 quadratic elements. We show the 200th and the 400th eigenfunctions, where the error is low, and the 600th and the 800th eigenfunctions, for which the approximation is worse. As expected, both the fully- and under-integrated methods provide similar eigenfunctions. There is no loss of accuracy in eigenfunction approximation due to the use of the non-standard optimal quadrature rules.

We also note that for practical applications, one may look for a scheme that reduces errors in the desired intervals of wavenumber (frequency) for a given mesh size. Such blending schemes are also possible and (though not being optimal, i.e. not delivering superconvergence) they are superior in the eigenvalue approximation

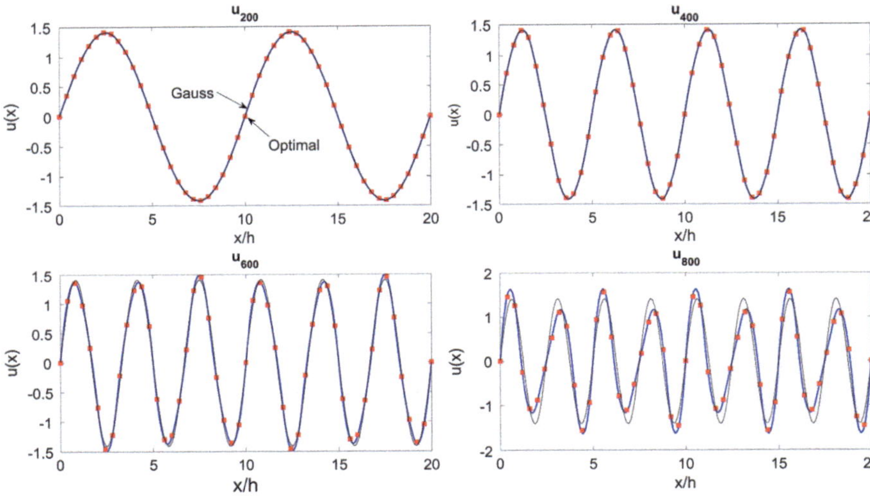

Fig. 6.4 Discrete 200th (top left), 400th (top right), 600th (bottom left) and 800th (bottom right) eigenfunctions for C^1 quadratic elements. The discrete eigenfunctions resulting from the optimal (red squares) and the standard scheme (blue line) are compared with the analytical eigenfunctions (black line). The total number of discrete modes is 1000

compared to the optimal blending in certain ranges of wavenumber that are of practical interest in wave propagation problems. We refer the reader to [54] for further details.

Next, we continue our study with the dispersion properties of the two-dimensional eigenvalue problem on tensor product meshes. Optimal quadratures for multidimensional problems are formed by tensor product of the one dimensional case. The exact eigenvalues and eigenfunctions of the 2D eigenvalue problem are given by

$$\lambda_{kl} = (k^2 + l^2)\pi^2, \quad u_{kl} = 2\sin(k\pi x)\sin(l\pi y), \tag{6.40}$$

for $k, l = 1, 2, \ldots$. Again, the approximate eigenvalues λ^h_{kl} are sorted in ascending order.

Figure 6.5 compares the eigenvalue errors of the standard Gauss using C^1 quadratic elements with the optimal scheme ($\tau = 2/3$). The latter has significantly better approximation properties.

These results demonstrate that the use of optimal quadratures in isogeometric analysis significantly improves the accuracy of the discrete approximations when compared to the fully-integrated Gauss-based method.

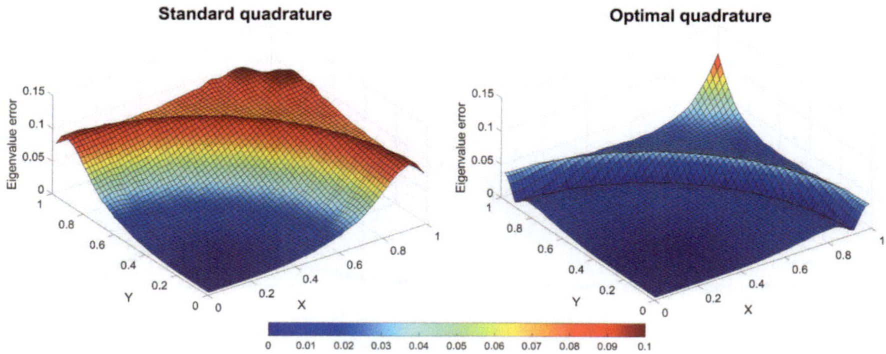

Fig. 6.5 Approximation errors for C^1 quadratic isogeometric elements with standard Gauss (left) and optimal quadrature rule (right). Color represents the absolute value of the relative error

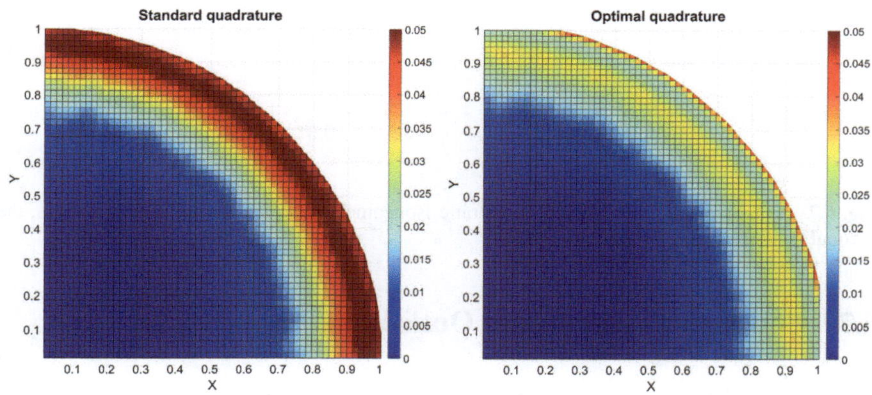

Fig. 6.6 Approximation errors for C^2 cubic isogeometric elements with standard Gauss (left) and optimal quadrature rule (right). Color represents the absolute value of the relative error

Figure 6.6 compares the eigenvalue errors for C^2 cubic isogeometric elements. Again, the optimal scheme has significantly better approximation properties than the standard method. The scale and representation format are different from those of Fig. 6.5.

Figure 6.7 compares the dispersion errors of the standard Gauss fully-integrated method with the optimally-blended scheme and the two-point rule described in the previous section. In this example, we use periodic knots at the boundaries of the domain. As can be seen from Fig. 6.7, the two-point rule leads to the same results as those obtained by the optimally-blended scheme. At the same time, this rule is computationally cheaper than the three-point Gauss rule or any blended scheme.

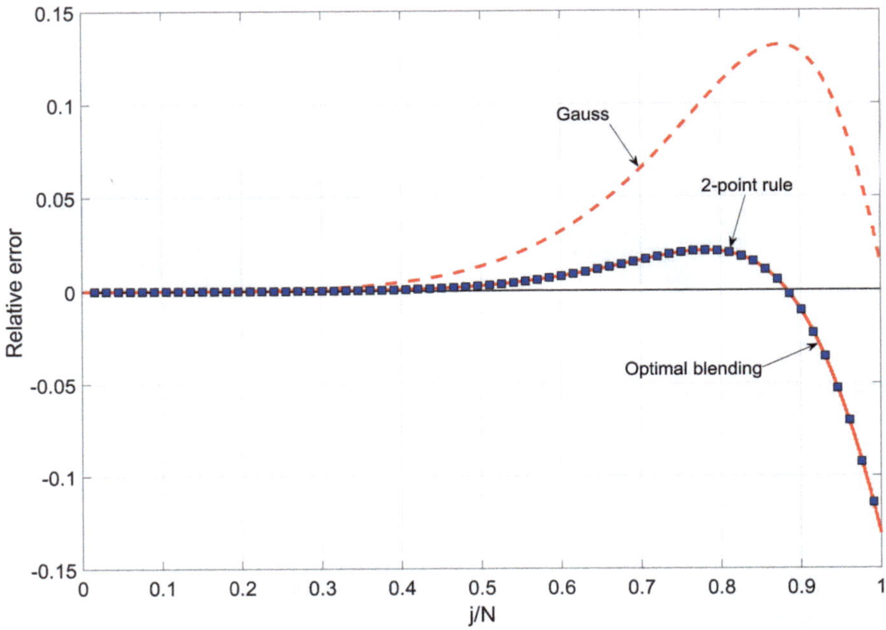

Fig. 6.7 Approximation errors for C^1 quadratic isogeometric elements with standard Gauss, the optimal quadrature rule, and the two-point

6.7 Conclusions and Future Outlook

To understand the dispersion properties of isogeometric analysis and to improve them, we generalize the Pythagorean eigenvalue error theorem to account for the effects of the modified inner products on the resulting weak forms. We show that the blended quadrature rules reduce the phase error of the numerical method for the eigenvalue problems.

The proposed optimally-blended scheme further improves the superior spectral accuracy of isogeometric analysis. We achieve two extra orders of convergence in the eigenvalues by applying these blended rules. We present and test two-point rules which reduce the number of quadrature nodes and the computational cost, and at the same time, produce the same eigenvalues and eigenfunctions. We believe that one can extend the method to arbitrary high-order C^{p-1} isogeometric elements by identifying suitable quadrature rules. Nevertheless, for higher-order polynomial approximations the only known optimal quadratures are the result of blending a Gauss rule and a Lobatto quadrature rule. The search for this class of quadratures that result in super-convergent dispersion properties and use fewer quadrature points will be the subject of our future work.

Another future direction is the study on the non-uniform meshes and non-constant coefficient wave propagation problems. The study with variable continuity is also of interest. We will study the impact of the variable continuities of the basis functions on the dispersion properties of the numerical methods and how the dispersion can be minimized by designing goal-oriented quadrature rules.

Acknowledgements This publication was made possible in part by the CSIRO Professorial Chair in Computational Geoscience at Curtin University and the Deep Earth Imaging Enterprise Future Science Platforms of the Commonwealth Scientific Industrial Research Organisation, CSIRO, of Australia. Additional support was provided by the National Priorities Research Program grant 7-1482-1-278 from the Qatar National Research Fund (a member of The Qatar Foundation), by the European Union's Horizon 2020 Research and Innovation Program of the Marie Skłodowska-Curie grant agreement No. 644202, and by the Projects of the Spanish Ministry of Economy and Competitiveness MTM2016-76329-R (AEI/FEDER, EU). The Spring 2016 Trimester on "Numerical methods for PDEs", organized with the collaboration of the Centre Emile Borel at the Institut Henri Poincare in Paris supported VMC's visit to IHP in October, 2016.

References

1. Ainsworth, M.: Discrete dispersion relation for hp-version finite element approximation at high wave number. SIAM J. Numer. Anal. **42**(2), 553–575 (2004)
2. Ainsworth, M., Wajid, H.A.: Dispersive and dissipative behavior of the spectral element method. SIAM J. Numer. Anal. **47**(5), 3910–3937 (2009)
3. Ainsworth, M., Wajid, H.A.: Optimally blended spectral-finite element scheme for wave propagation and nonstandard reduced integration. SIAM J. Numer. Anal. **48**(1), 346–371 (2010)
4. Akkerman, I., Bazilevs, Y., Calo, V.M., Hughes, T.J.R., Hulshoff, S.: The role of continuity in residual-based variational multiscale modeling of turbulence. Comput. Mech. **41**(3), 371–378 (2008)
5. Antolin, P., Buffa, A., Calabro, F., Martinelli, M., Sangalli, G.: Efficient matrix computation for tensor-product isogeometric analysis: the use of sum factorization. Comput. Methods Appl. Mech. Eng. **285**, 817–828 (2015)
6. Auricchio, F., Calabro, F., Hughes, T.J.R., Reali, A., Sangalli, G.: A simple algorithm for obtaining nearly optimal quadrature rules for NURBS-based isogeometric analysis. Comput. Methods Appl. Mech. Eng. **249**, 15–27 (2012)
7. Babuska, I.M., Sauter, S.A.: Is the pollution effect of the fem avoidable for the Helmholtz equation considering high wave numbers? SIAM J. Numer. Anal. **34**(6), 2392–2423 (1997)
8. Banerjee, U.: A note on the effect of numerical quadrature in finite element eigenvalue approximation. Numer. Math. **61**(1), 145–152 (1992)
9. Banerjee, U., Osborn, J.E.: Estimation of the effect of numerical integration in finite element eigenvalue approximation. Numer. Math. **56**(8), 735–762 (1989)
10. Banerjee, U., Suri, M.: Analysis of numerical integration in p-version finite element eigenvalue approximation. Numer. Methods Partial Differ. Equ. **8**(4), 381–394 (1992)
11. Bartoň, M., Calo, V.M.: Gaussian quadrature for splines via homotopy continuation: rules for C2 cubic splines. J. Comput. Appl. Math. **296**, 709–723 (2016)
12. Bartoň, M., Calo, V.M.: Optimal quadrature rules for odd-degree spline spaces and their application to tensor-product-based isogeometric analysis. Comput. Methods Appl. Mech. Eng. **305**, 217–240 (2016)
13. Bazilevs, Y., Calo, V.M., Cottrell, J., Hughes, T.J.R., Reali, A., Scovazzi, G.: Variational multiscale residual-based turbulence modeling for large eddy simulation of incompressible flows. Comput. Methods Appl. Mech. Eng. **197**(1), 173–201 (2007)

14. Calabrò, F., Sangalli, G., Tani, M.: Fast formation of isogeometric Galerkin matrices by weighted quadrature. Comput. Methods Appl. Mech. Eng. **316**, 606–622 (2017)
15. Calo, V.M., Deng, Q., Puzyrev, V.: Dispersion optimized quadratures for isogeometric analysis. arXiv:1702.04540 (2017, preprint)
16. Collier, N., Pardo, D., Dalcin, L., Paszynski, M., Calo, V.M.: The cost of continuity: a study of the performance of isogeometric finite elements using direct solvers. Comput. Methods Appl. Mech. Eng. **213**, 353–361 (2012)
17. Collier, N., Dalcin, L., Pardo, D., Calo, V.M.: The cost of continuity: performance of iterative solvers on isogeometric finite elements. SIAM J. Sci. Comput. **35**(2), A767–A784 (2013)
18. Collier, N., Dalcin, L., Calo, V.M.: On the computational efficiency of isogeometric methods for smooth elliptic problems using direct solvers. Int. J. Numer. Methods Eng. **100**(8), 620–632 (2014)
19. Cottrell, J.A., Reali, A., Bazilevs, Y., Hughes, T.J.R.: Isogeometric analysis of structural vibrations. Comput. Methods Appl. Mech. Eng. **195**(41), 5257–5296 (2006)
20. Cottrell, J., Hughes, T.J.R., Reali, A.: Studies of refinement and continuity in isogeometric structural analysis. Comput. Methods Appl. Mech. Eng. **196**(41), 4160–4183 (2007)
21. Cottrell, J.A., Hughes, T.J.R., Bazilevs, Y.: Isogeometric Analysis: Toward Integration of CAD and FEA. Wiley, Hoboken (2009)
22. De Basabe, J.D., Sen, M.K.: Grid dispersion and stability criteria of some common finite-element methods for acoustic and elastic wave equations. Geophysics **72**(6), T81–T95 (2007)
23. De Basabe, J.D., Sen, M.K.: Stability of the high-order finite elements for acoustic or elastic wave propagation with high-order time stepping. Geophys. J. Int. **181**(1), 577–590 (2010)
24. Dedè, L., Jäggli, C., Quarteroni, A.: Isogeometric numerical dispersion analysis for two-dimensional elastic wave propagation. Comput. Methods Appl. Mech. Eng. **284**, 320–348 (2015)
25. Deng, Q., Bartoň, M., Puzyrev, V., Calo, V.M.: Dispersion-minimizing optimal quadrature rules for c^1 quadratic isogeometric analysis. Comput. Methods Appl. Mech. Eng. **328**, 554–564 (2018)
26. Elguedj, T., Bazilevs, Y., Calo, V.M., Hughes, T.J.R.: B-bar and F-bar projection methods for nearly incompressible linear and non-linear elasticity and plasticity using higher-order NURBS elements. Comput. Methods Appl. Mech. Eng. **197**(33), 2732–2762 (2008)
27. Esterhazy, S., Melenk, J.: An analysis of discretizations of the Helmholtz equation in L^2 and in negative norms. Comput. Math. Appl. **67**(4), 830–853 (2014). https://doi.org/10.1016/j.camwa.2013.10.005
28. Ewing, R., Heinemann, R., et al.: Incorporation of mixed finite element methods in compositional simulation for reduction of numerical dispersion. In: SPE Reservoir Simulation Symposium. Society of Petroleum Engineers (1983)
29. Fix, G.J.: Effect of quadrature errors in finite element approximation of steady state, eigenvalue and parabolic problems. In: Aziz, A.K. (ed.) The Mathematical Foundation of the Finite Element Method with Applications to Partial Differential Equations, pp. 525–556 (1972)
30. Gao, L., Calo, V.M.: Fast isogeometric solvers for explicit dynamics. Comput. Methods Appl. Mech. Eng. **274**, 19–41 (2014)
31. Garcia, D., Pardo, D., Dalcin, L., Paszyski, M., Collier, N., Calo, V.M.: The value of continuity: refined isogeometric analysis and fast direct solvers. Comput. Methods Appl. Mech. Eng. **316**, 586–605 (2016)
32. Garcia, D., Bartoň, M., Pardo, D.: Optimally refined isogeometric analysis. Proc. Comput. Sci. **108**, 808–817 (2017)
33. Gómez, H., Calo, V.M., Bazilevs, Y., Hughes, T.J.R.: Isogeometric analysis of the Cahn–Hilliard phase-field model. Comput. Methods Appl. Mech. Eng. **197**(49), 4333–4352 (2008)
34. Gomez, H., Hughes, T.J.R., Nogueira, X., Calo, V.M.: Isogeometric analysis of the isothermal Navier–Stokes–Korteweg equations. Comput. Methods Appl. Mech. Eng. **199**(25), 1828–1840 (2010)
35. Guddati, M.N., Yue, B.: Modified integration rules for reducing dispersion error in finite element methods. Comput. Methods Appl. Mech. Eng. **193**(3), 275–287 (2004)

36. Harari, I.: Reducing spurious dispersion, anisotropy and reflection in finite element analysis of time-harmonic acoustics. Comput. Methods Appl. Mech. Eng. **140**(1–2), 39–58 (1997)
37. Harari, I., Slavutin, M., Turkel, E.: Analytical and numerical studies of a finite element PML for the Helmholtz equation. J. Comput Acoust. **8**(1), 121–137 (2000)
38. He, Z., Cheng, A., Zhang, G., Zhong, Z., Liu, G.: Dispersion error reduction for acoustic problems using the edge-based smoothed finite element method (ES-FEM). Int. J. Numer. Methods Eng. **86**(11), 1322–1338 (2011)
39. Hiemstra, R.R., Calabrò, F., Schillinger, D., Hughes, T.J.R.: Optimal and reduced quadrature rules for tensor product and hierarchically refined splines in isogeometric analysis. Comput. Methods Appl. Mech. Eng. **316**, 966–1004 (2016)
40. Hughes, T.J.R., Cottrell, J.A., Bazilevs, Y.: Isogeometric analysis: CAD, finite elements, NURBS, exact geometry and mesh refinement. Comput. Methods Appl. Mech. Eng. **194**(39), 4135–4195 (2005)
41. Hughes, T.J.R., Reali, A., Sangalli, G.: Duality and unified analysis of discrete approximations in structural dynamics and wave propagation: comparison of p-method finite elements with k-method NURBS. Comput. Methods Appl. Mech. Eng. **197**(49), 4104–4124 (2008)
42. Hughes, T.J.R., Reali, A., Sangalli, G.: Efficient quadrature for NURBS-based isogeometric analysis. Comput. Methods Appl. Mech. Eng. **199**(5), 301–313 (2010)
43. Hughes, T.J.R., Evans, J.A., Reali, A.: Finite element and NURBS approximations of eigenvalue, boundary-value, and initial-value problems. Comput. Methods Appl. Mech. Eng. **272**, 290–320 (2014)
44. Ihlenburg, F., Babuška, I.: Dispersion analysis and error estimation of Galerkin finite element methods for the Helmholtz equation. Int. J. Numer. Methods Eng. **38**(22), 3745–3774 (1995)
45. Komatitsch, D., Tromp, J.: Introduction to the spectral element method for three-dimensional seismic wave propagation. Geophys. J. Int. **139**(3), 806–822 (1999)
46. Komatitsch, D., Vilotte, J.P.: The spectral element method: an efficient tool to simulate the seismic response of 2d and 3d geological structures. Bull. Seismol. Soc. Am. **88**(2), 368–392 (1998)
47. Lipton, S., Evans, J.A., Bazilevs, Y., Elguedj, T., Hughes, T.J.R.: Robustness of isogeometric structural discretizations under severe mesh distortion. Comput. Methods Appl. Mech. Eng. **199**(5), 357–373 (2010)
48. Liu, J., Dedè, L., Evans, J.A., Borden, M.J., Hughes, T.J.R.: Isogeometric analysis of the advective Cahn–Hilliard equation: spinodal decomposition under shear flow. J. Comput. Phys. **242**, 321–350 (2013)
49. Marfurt, K.J.: Accuracy of finite-difference and finite-element modeling of the scalar and elastic wave equations. Geophysics **49**(5), 533–549 (1984)
50. Motlagh, Y.G., Ahn, H.T., Hughes, T.J.R., Calo, V.M.: Simulation of laminar and turbulent concentric pipe flows with the isogeometric variational multiscale method. Comput. Fluids **71**, 146–155 (2013)
51. Nguyen, L.H., Schillinger, D.: A collocated isogeometric finite element method based on Gauss–Lobatto Lagrange extraction of splines. Comput. Methods Appl. Mech. Eng. **316**, 720–740 (2016)
52. Pardo, D., Paszynski, M., Collier, N., Alvarez, J., Dalcin, L., Calo, V.M.: A survey on direct solvers for Galerkin methods. SeMA J. **57**(1), 107–134 (2012)
53. Piegl, L., Tiller, W.: The NURBS Book. Springer, New York (1997)
54. Puzyrev, V., Deng, Q., Calo, V.M.: Dispersion-optimized quadrature rules for isogeometric analysis: modified inner products, their dispersion properties, and optimally blended schemes. Comput. Methods Appl. Mech. Eng. **320**, 421–443 (2017). http://dx.doi.org/10.1016/j.cma.2017.03.029. http://www.sciencedirect.com/science/article/pii/S004578251631920X
55. Reali, A.: An isogeometric analysis approach for the study of structural vibrations. Master's Thesis, University of Pavia (2004)
56. Seriani, G., Oliveira, S.P.: Optimal blended spectral-element operators for acoustic wave modeling. Geophysics **72**(5), SM95–SM106 (2007)
57. Stoer, J., Bulirsch, R.: Introduction to Numerical Analysis, vol. 12. Springer, New York (2013)

58. Strang, G., Fix, G.J.: An Analysis of the Finite Element Method, vol. 212. Prentice-Hall, Englewood Cliffs (1973)
59. Thompson, L.L., Pinsky, P.M.: Complex wavenumber Fourier analysis of the p-version finite element method. Comput. Mech. **13**(4), 255–275 (1994)
60. Thompson, L.L., Pinsky, P.M.: A Galerkin least-squares finite element method for the two-dimensional Helmholtz equation. Int. J. Numer. Methods Eng. **38**(3), 371–397 (1995)
61. Wang, D., Liu, W., Zhang, H.: Novel higher order mass matrices for isogeometric structural vibration analysis. Comput. Methods Appl. Mech. Eng. **260**, 92–108 (2013)
62. Wang, D., Liu, W., Zhang, H.: Superconvergent isogeometric free vibration analysis of Euler–Bernoulli beams and Kirchhoff plates with new higher order mass matrices. Comput. Methods Appl. Mech. Eng. **286**, 230–267 (2015)
63. Yue, B., Guddati, M.N.: Dispersion-reducing finite elements for transient acoustics. J. Acoust. Soc. Am. **118**(4), 2132–2141 (2005)

Chapter 7
Weakly Consistent Regularisation Methods for Ill-Posed Problems

Erik Burman and Lauri Oksanen

Abstract This Chapter takes its origin in the lecture notes for a 9 h course at the Institut Henri Poincaré in September 2016. The course was divided in three parts. In the first part, which is not included herein, the aim was to first recall some basic aspects of stabilised finite element methods for convection-diffusion problems. We focus entirely on the second and third parts which were dedicated to ill-posed problems and their approximation using stabilised finite element methods. First we introduce the concept of conditional stability. Then we consider the elliptic Cauchy-problem and a data assimilation problem in a unified setting and show how stabilised finite element methods may be used to derive error estimates that are consistent with the stability properties of the problem and the approximation properties of the finite element space. Finally, we extend the result to a data assimilation problem subject to the heat equation.

7.1 Introduction

In these notes we will give an overview of some recent work on finite element methods for ill-posed problems. For well-posed problems it is known that, in the presence of non-symmetric operators, approximation using Galerkin finite element methods may have poor accuracy, due to the lack of H^1-coercivity. A popular remedy is then to add some stabilising terms that should be balanced in such a way that they cure the stability issue, but vanish quickly enough under mesh-refinement so that optimal error estimates can be obtained. For ill-posed problems on the other hand the state of the art is to add some regularising terms on the continuous level to obtain a well-posed continuous problem that can then typically be discretised using standard finite element methods.

E. Burman (✉) · L. Oksanen
Department of Mathematics, University College London, London, UK
e-mail: e.burman@ucl.ac.uk; l.oksanen@ucl.ac.uk

D. A. Di Pietro et al. (eds.), *Numerical Methods for PDEs*,
SEMA SIMAI Springer Series 15, https://doi.org/10.1007/978-3-319-94676-4_7

Here our aim is to make the ideas from the former class of problems carry over to the ill-posed case, using weakly consistent regularisation that is defined on the discrete level. Indeed prior to discretisation no regularisation is applied, instead the ill-posed problem and associated data are discretised in the form of a minimisation problem, where some suitable distance between the discrete solution and the measured data is minimised under the constraint of the discrete form of the partial differential equation. Regularisation terms may then be devised that are in some sense the minimal choice necessary to achieve a well-posed discrete system. To analyse the resulting approximation we rely on conditional stability estimates for the continuous problem, typically obtained through Carleman estimates.

Compared to the state of the art methods such as the quasi-reversibility method by Lattès and Lions (and the recent improvements on this technique by Bourgeois et al. [7, 8, 21]) or the penalty method by Kohn and Vogelius [4, 30], the present framework has some attractive features. Since no regularised continuous problem is involved the only (nontrivial) regularisation parameter present is the mesh size. This is not the case for more traditional methods where the discretisation parameter and the regularisation parameter must be matched carefully, or as is usually assumed, the mesh size is chosen substantially smaller than the regularisation parameter. Maybe more importantly, in the present framework, the regularisation is independent of the stability of the underlying physical problem while still having a convergence order with respect to the mesh size that is consistent with the stability of the physical problem. On the contrary, balancing regularisation and discretisation errors in the framework of conventional Tikhonov regularisation appears to inevitably lead to a nontrivial relation between the regularisation, the mesh size and the specific form of the stability of the physical problem.

With the recent increased understanding of the stability properties of ill-posed problems, in particular, in the context of inverse and data assimilation problems, we believe that these considerations are important. For instance, data assimilation problems with Hölder, or even Lipschitz, stability will have that precise order reproduced for the convergence order of the approximation error. To the best of our knowledge, apart from the work reviewed here, there exists no results in the literature reporting on such estimates even in Lipschitz stable cases that allow error estimates as good as those for classical well-posed problems. For other work on regularized methods for the Cauchy problem we refer to [2, 3, 6, 29].

The paper consists of two main parts. In the first we consider stationary ill-posed elliptic problems, such as the elliptic Cauchy problem and the so-called data assimilation problem, where measured data is available in some subdomain of the bulk, but not on the boundary. For these problems interior estimates with Hölder stability are known to hold and we show how to make these estimates translate into error estimates for the computational method. In the second chapter we consider the extension of these ideas to a data assimilation problem subject to the heat equation. In this case a Lipschitz-continuous stability estimate holds for the reconstruction of the solution away from the (unknown) initial datum. Also in this case we show, in a space semi-discretised framework, error estimates that reflect the stability of the physical problem. In this second case the estimates obtained are optimal with respect to the approximation order of the finite element space.

7.2 Preliminary Results

In this section we will introduce the geometrical setting of the problems that we will consider, the associated finite element spaces and some technical results, including discrete inequalities and approximation results. We will stay in the simplest of settings, considering only piecewise affine finite element spaces.

Let $\Omega \subset \mathbb{R}^d$, $d = 2, 3$, be a convex polygonal (polyhedral) domain, with boundary $\partial\Omega$ and outward pointing normal n. By $\mathcal{T}$ we denote a quasi-uniform decomposition of Ω in simplices T such that the intersection of two simplices in $\mathcal{T}$ is either the empty set, a shared vertex, a shared face or a shared edge. We also introduce the mesh parameter associated to $\mathcal{T}$, $h_T = \operatorname{diam}(T)$ where the diameter of T is defined as the diameter of the smallest ball circumscribing T. Setting $h = \max_{T\in\mathcal{T}} h_T$ we consider the family of tesselations $\{\mathcal{T}\}_h$ indexed by h. The simplices are shape regular in the sense that the ratio between the smallest circumscribed ball and the largest inscribed ball of any $T \in \mathcal{T}$ is bounded uniformly, with a constant independent of h. The boundary of T will be denoted ∂T with outward pointing normal n_T. We denote the set of element faces by $\mathcal{F}$ and let $\mathcal{F}_i$ and $\mathcal{F}_b$ denote the set of faces in the interior of Ω and on its boundary, respectively. To each interior face we associate a normal n_F that is fixed, but with arbitrary orientation. The normal on faces on the boundary will be chosen pointing outwards.

We define the finite dimensional space

$$\mathbb{V}_h = \{v_h \in H^1(\Omega) : v|_T \in \mathbb{P}_1(T),\ \forall T \in \mathcal{T}\},$$

with $\mathbb{P}_1(T)$ the set of polynomials of degree less than or equal to 1. For a subspace $V \subset H^1(\Omega)$, we denote by V_h the intersection $\mathbb{V}_h \cap V$. In particular, we use the notation $V^0 = H_0^1(\Omega)$ and

$$V_h^0 := \mathbb{V}_h \cap V^0.$$

We will denote the L^2 scalar product over a set Ξ by

$$(v, w)_\Xi := \int_\Xi xy \,\mathrm{d}\Xi, \quad \forall v, w \in L^2(\Omega),$$

and the associated norm by

$$\|x\|_\Xi := (x, x)_\Xi^{\frac{1}{2}}.$$

The subscript will be dropped whenever $\Xi \equiv \Omega$.

7.2.1 Inequalities

We will need a few auxiliary results on how different norms or semi norms are related. In particular we will need the following so-called inverse inequality and trace inequalities (see for instance [22])

$$\|\nabla v_h\|_T \leq C_i h_T^{-1} \|v_h\|_T \quad \forall v_h \in \mathbb{P}_k(T),\ k \geq 0 \tag{7.1}$$

$$\|v\|_{\partial T} \leq C_t h_T^{-1/2} (\|v_h\|_T + h_T \|\nabla v\|_T), \quad \forall v \in H^1(T) \tag{7.2}$$

$$\|v_h\|_{\partial T} \leq C_t h_T^{-1/2} \|v_h\|_T, \quad \forall v_h \in \mathbb{P}_k(T),\ k \geq 0. \tag{7.3}$$

We also define the broken norm

$$\|v\|_h := \left(\sum_{T \in \mathcal{T}} \|v\|_T^2 \right)^{\frac{1}{2}}.$$

7.2.2 Interpolants and Approximation

We will use an interpolant $i_h : H^1(\Omega) \rightarrow \mathbb{V}_h$, that preserves homogeneous boundary conditions and satisfies the following estimates [33]

$$\|u - i_h u\| + h \|\nabla(u - i_h u)\| \leq C h^s \|u\|_{H^s(\Omega)}, \quad s = 1, 2. \tag{7.4}$$

Combining (7.4) and (7.2) allows us to prove the estimates

$$\|h^{-\frac{1}{2}}(u - i_h u_h)\|_{\mathcal{F}} + \|h^{\frac{1}{2}} \nabla(u - u_h)\|_{\mathcal{F}} \leq C h^{s-1} \|u\|_{H^s(\Omega)}, \quad s = 1, 2. \tag{7.5}$$

We will also make use of the H^1-projection $\pi_h : H_0^1(\Omega) \rightarrow V_h^0$ defined by

$$(\nabla \pi_h u, \nabla v_h) = (\nabla u, \nabla v_h), \quad \forall v_h \in V_h^0. \tag{7.6}$$

We note that under the assumption of quasi uniformity and convexity of the domain also this approximation satisfies (7.4) and (7.5).

7.3 Ill-Posed Problems

It is well known that instabilities may cause suboptimality for approximations of convection-diffusion equations when the standard Galerkin method is applied. Examples of how stabilised methods can improve on the situation include the Galerkin Least Squares method [10, 27], subgrid viscosity [26] or the continuous interior penalty method [15]. This is an example of a problem that is well-posed on the continuous level, but where the discrete system may be ill-conditioned and produce poor quality approximations, unless all the scales of the problem have been resolved, something which may be difficult to achieve in practice. The arguments to analyse such methods use the positivity of the bilinear operator $a(\cdot,\cdot)$ defining the problem.

In many practical cases however the problem is indefinite, for instance, this is the case for Helmholtz equation and for non-coercive convection-diffusion. Then the bilinear form does not satisfy such a positivity property, and the inf-sup condition that underpins well-posedness on the continuous level can be difficult to reproduce on the discrete level. This led the first author to develop a method which does not rely on coercivity or inf-sup stability for its analysis [11]. As the method does not rely on the well-posedness structure for its design, it can also be applied to ill-posed problems. This case was then analysed in [12] and applied to a series of different ill-posed problems in [13, 16, 17, 19].

In this section we will discuss how to apply stabilised finite elements to the approximation of ill-posed problems. Of course the class of ill-posed problems is very large and most of these problems are not tractable to the type of high resolution methods that we wish to apply here, so first we will discuss what type of ill-posed problems we are interested in and give some examples. For readers interested in delving deeper into the theory of inverse and ill-posed problems and their regularisation, we refer to [5, 24, 28, 31, 34].

Ill-posed problems are those problems that fail to be well-posed in the sense of the definition due to Hadamard. In order to make this precise we introduce the abstract problem

$$\mathcal{K}u = \mathfrak{f} \tag{7.7}$$

where $\mathcal{K} : V \to \mathcal{X}$ is a linear map between two Hilbert (or Banach) spaces and $\mathfrak{f} \in \mathcal{X}$.

Definition 7.1 (Well-Posed Problem) The problem (7.7) is well-posed if

1. For every $\mathfrak{f} \in \mathcal{X}$ there exists $u \in V$ satisfying (7.7). This means that $\mathcal{X}$ is the range of $\mathcal{L}$.
2. The solution u is unique in V. That is, $\mathcal{L}^{-1}$ exists.
3. The solution u depends continuously on data.

$$\|u\|_V \leq C\|\mathfrak{f}\|_{\mathcal{X}}.$$

Definition 7.2 (Ill-Posed Problem) The problem (7.7) is said to be ill-posed if at least one of the three points in Definition 7.1 fails.

It was recognised by Tikhonov that some ill-posed problems are better behaved than others, and conditionally stable problems are an important class of such problems. We give a definition that is a variation of [28, Def. 4.3].

Definition 7.3 (Conditionally Stable Problem) The problem (7.7) is said to be conditionally stable with respect to a semi-norm $|\cdot|$ on V if

1. For all $\mathfrak{f}$ in the range of $\mathcal{K}$ the solution u of (7.7) is unique.
2. There is a non-decreasing function $C_E : [0, \infty) \to [0, \infty)$ and a modulus of continuity $\Phi : [0, \infty) \to [0, \infty)$ such that for all $\mathfrak{f}$ in the range of $\mathcal{K}$,

$$|u| \le C_E(\|u\|_V)\Phi(\|\mathfrak{f}\|_{\mathcal{X}}).$$

Here Φ being a modulus of continuity means that it is continuous and satisfies $\Phi(0) = 0$.

We restrict our attention to conditionally stable problems where $\mathcal{K}$ and $\mathcal{X}$ consist of two components

$$\mathcal{K} = (\mathcal{L}, \mathcal{R}), \quad \mathcal{X} = W' \times \mathcal{M}.$$

Here, for the Sobolev spaces V and W, W' is the dual of W and $\mathcal{L}$ is a differential operator, mapping V to W' when interpreted in weak form. For the part related to data we let $\mathcal{R} : V \to \mathcal{M}$ denote a restriction operator, possibly composed with a differential operator. To summarize, we will consider problems of the form

$$\mathcal{L}u = \tilde{f}, \quad \mathcal{R}u = \tilde{q} \tag{7.8}$$

where it is assumed that $(\tilde{f}, \tilde{q})$ is in a neighbourhood of the range of $\mathcal{K}$. We will prove estimates that depend on the distance

$$\|\delta f\|_{W'} + \|\delta q\|_{\mathcal{M}}, \quad \delta f = \tilde{f} - f,\ \delta q = \tilde{q} - q,$$

where (f, q) is in the range of $\mathcal{K}$. Observe that this means that we do not assume that the problem (7.8) admits a unique solution, we only assume that it can be solved for some point in a neighbourhood of the data $(\tilde{f}, \tilde{q})$. This allows for perturbed data to be used.

We will now proceed to give examples of problems that are conditionally stable in the above sense.

Example 7.1 (The Elliptic Cauchy Problem and Its Ill-Posedness) Let $\mathcal{L} = -\Delta + \sigma$ where $\sigma \in \mathbb{R}$ and assume that the boundary of Ω consists of two parts Γ and Γ'. Consider the problem of finding $u \in H^1(\Omega)$ such that

$$\mathcal{L}u = f \text{ in } \Omega \tag{7.9}$$

$$u = g \text{ on } \Gamma \tag{7.10}$$

$$\nabla u \cdot n = \psi \text{ on } \Gamma. \tag{7.11}$$

For simplicity, we consider below only the case $g = 0$, and refer to [14] for the case with non-vanishing g. Then

$$\mathcal{R}u = \nabla u \cdot n|_\Gamma, \quad \mathcal{M} = H^{-1/2}(\Gamma). \tag{7.12}$$

Following a classical counter-example by Hadamard, let us exemplify the failure of continuous dependence for this problem. Let $\Omega := \{(x, y) \in \mathbb{R}^2 : x > 0\}$ and $\Gamma = \{(x, y) \in \mathbb{R}^2 : x = 0\}$, $\sigma = 0$, $f = 0$, $g = 0$ and

$$\psi(y) = \frac{1}{n} \sin(ny).$$

It is easy to verify that the solution in that case is

$$u(x, y) = \frac{1}{2n^2} \sin(ny)(e^{nx} - e^{-nx}).$$

Clearly as n becomes large $\|\psi\|_{L^\infty(\Gamma)}$ goes to zero, but $u(x, y)$ blows up for any $x > 0$ and any y outside a countable set, showing the failure of continuous dependence.

Example 7.2 (The Elliptic Data Assimilation Problem and Its Uniqueness) Let $\mathcal{L} = -\Delta$ and assume that measurements u_M of u are available in some open subset of Ω, $\omega \subset \Omega$, then we can formulate the data assimilation problem as

$$\mathcal{L}u = f \text{ in } \Omega \tag{7.13}$$

$$u = u_M \text{ in } \omega. \tag{7.14}$$

Here we choose

$$\mathcal{R}u = u|_\omega, \quad \mathcal{M} = L^2(\omega). \tag{7.15}$$

This problem is often called also a unique continuation problem.

Assume that u_M, f are such that there exists a solution $u \in H^1(\Omega)$ to (7.13)–(7.14). Then this solution is unique which can be proven by using elementary properties of harmonic functions. Indeed, assume that there exists two solutions and let v be their difference. Then

$$\mathcal{L}v = 0 \text{ in } \Omega \tag{7.16}$$

$$v = 0 \text{ in } \omega. \tag{7.17}$$

This means that ν is a harmonic function in Ω and hence real analytic. But ν vanishes in the non-empty open set ω, and hence by analytic continuation, $\nu \equiv 0$ in Ω.

Remark 7.1 For the problem (7.13)–(7.14) to have a solution, it is of course necessary that the compatibility condition $\mathcal{L}u_M|_\omega = f|_\omega$ is satisfied. Using this one may show that, for sufficiently smooth f, (7.13)–(7.14) is equivalent to the Cauchy problem

$$\mathcal{L}u = f \text{ in } \Omega \setminus \omega \tag{7.18}$$

$$u = u_M \text{ on } \partial\omega \tag{7.19}$$

$$\nabla u \cdot n = \nabla u_M \cdot n \text{ on } \partial\omega. \tag{7.20}$$

The conditional stability for the problems in Examples 7.1 and 7.2 is classical, and we discuss it further in Sect. 7.3.2 below. Let us now turn to weak formulation of these problems on which the associated finite element methods will be based.

7.3.1 *Weak Formulations of the Model Problems*

Let us first consider the Cauchy problem in Example 7.1 and introduce the spaces

$$V^\Gamma := \{v \in H^1(\Omega) : v|_\Gamma = 0\} \quad \text{and} \quad W^\Gamma := \{v \in H^1(\Omega) : v|_{\Gamma'} = 0\} (= V^{\Gamma'}).$$

Now observe that the solution of (7.9)–(7.11), with $g = 0$, can be sought in V^Γ. Multiply (7.9) by $v \in W^\Gamma$ and integrate by parts to obtain

$$(\mathcal{L}u, v) = (\nabla u, \nabla v) + (\sigma u, v) - \int_\Gamma \underbrace{\nabla u \cdot n}_{=-\psi} v \, ds - \int_{\Gamma'} \nabla u \cdot n \underbrace{v}_{=0} \, ds$$

By defining

$$a(u, v) := (\nabla u, \nabla v) + (\sigma u, v)$$

we arrive at the weak formulation: find $u \in V^\Gamma$ such that

$$a(u, v) = (f, v) + (\psi, v)_\Gamma, \quad \forall v \in W^\Gamma. \tag{7.21}$$

This weak formulation looks deceptively like the weak formulation for the Poisson problem, but observe that the choice $v = u$ is not allowed since $u \notin W^\Gamma$.

Let us now turn to the data assimilation problem in Example 7.2. Recall from Sect. 7.2 that $V^0 = H_0^1(\Omega)$, and observe that we may multiply (7.13) with $v \in V^0$

to obtain

$$(\mathcal{L}u, v) = (\nabla u, \nabla v) - \int_{\partial\Omega} \nabla u \cdot n \underbrace{v}_{=0} \, ds.$$

This time we define

$$a(u, v) := (\nabla u, \nabla v)$$

and obtain the weak formulation: find $u \in H^1(\Omega)$ such that $u|_\omega = u_M$ and

$$a(u, v) = (f, v) \quad \forall v \in V^0. \tag{7.22}$$

Once again it is not allowed to take $v = u$ due to the different choices of spaces.

7.3.2 Conditional Stability

To unify the treatment of the two examples, we will write V for the primal space and W for the test space. That is, $V = V^\Gamma$ and $W = W^\Gamma$ in the case of Example 7.1, and $V = H^1(\Omega)$ and $W = V^0$ in the case of Example 7.2. Observe that $W' = H^{-1}(\Omega)$ in the case of Example 7.2.

We refer to the review paper [1] for thorough discussion of conditional stability estimates for the two example problems. In particular, the following conditional stability estimate can be deduced from the paper.

Theorem 7.1 *Let $u \in V$ be such that, with $l \in W'$,*

$$a(u, v) = l(v).$$

Let $\mathcal{R} : V \to \mathcal{M}$ be defined by (7.12) *for the Cauchy problem in Example 7.1, and by* (7.15) *for the data assimilation problem in Example 7.2. Write $u_M = \mathcal{R}u$ in both the cases. Then for every open simply connected $\omega' \subset \Omega$ such that dist$(\partial\omega', \partial\Omega) > 0$ there holds*

$$\|u\|_{\omega'} \le C_E(\|u\|)\Phi(|u_M|_{\mathcal{M}} + \|l\|_{W'}),$$

where $C_E(R) = CR^{1-\tau}$ and $\Phi(\eta + \varepsilon) = (\eta + \varepsilon)^\tau$. Here $C > 0$ and $\tau \in (0, 1)$ are constants that depend on ω'.

For a proof of this result with full detail on involved constants see [1, Theorem 1.7] for the Cauchy problem and [1, Theorem 4.4] for the data assimilation case. Let us remark that we state the conditions on ω' in slightly simplified form, for more precise conditions on ω' see [1]. Note that here $\|\cdot\|_{\omega'}$ is viewed as a semi-norm on V.

Remark 7.2 A similar result for global stability of u on the form

$$\|u\|_\Omega \le C_E(\|u\|_V)\Phi(|u_M|_{\mathcal{M}} + \|l\|_{W'}),$$

with $\Phi(\eta+\varepsilon) = |\log(\eta+\varepsilon)|^{-\tau}$, $\tau \in (0,1)$, is also derived in [1] and may be used to derive global error estimates using the techniques below.

Remark 7.3 Conditional stability has been used before to tune the regularisation parameters for Tikhonov regularisation methods see for instance [20]. What is new in the approach that we advocate is that it does not depend on the form of the modulus of continuity Φ, but still allows us to obtain the best possible accuracy with respect to the approximation error and the actual form of Φ.

7.4 Finite Element Approximation of Ill-Posed Problems

The aim of the present section is present a finite element method that draws on our experience of stabilised FEM for convection-diffusion equations. The ideas that are presented below are mainly taken from [13, 19].

We wish to attempt to discretise a conditionally stable ill-posed problem of the form: find $u \in V$ such that

$$a(u, v) = l(w), \quad \forall w \in W \tag{7.23}$$

$$|u - u_M|_{\mathcal{M}} = 0. \tag{7.24}$$

Let us consider, for the moment, the case of Cauchy problem and suppose that l is such that there exists a solution $u \in V$ to (7.23).

Recall the notation defined in Sect. 7.2, and define the finite element spaces

$$V_h^\Gamma := \mathbb{V}_h \cap V^\Gamma \quad \text{and} \quad W_h^\Gamma := \mathbb{V}_h \cap W^\Gamma.$$

We are assuming here that the mesh is fitted to the subsets of the boundary Γ and Γ'. We then have the discrete formulation of the Cauchy problem in Example 7.1: find $u_h \in V_h^\Gamma$ such that

$$a(u_h, w_h) = (f, w_h) + (\psi, w_h)_\Gamma, \quad \forall w_h \in W_h^\Gamma. \tag{7.25}$$

Observe that the corresponding linear system can not be invertible in general, because there is no reason that the system matrix is square. Indeed this only holds in the special case when the number of vertices in Γ is the same as the number of vertices in Γ'. Similarly the matrix corresponding to a naive finite element discretisation of the data assimilation problem in Example 7.2 is not square and in general the system is singular even if we impose $u_h|_\omega = 0$.

The idea is then to reformulate (7.23)–(7.24), on the discrete level, as the problem to *minimise* (7.24) under the *constraint* (7.23). This will allows us also to treat the case of perturbed data that is outside the range of the map $\mathcal{K} = (\mathcal{L}, \mathcal{R})$. In some cases $|\cdot|_{\mathcal{M}}$ may not be the most efficient choice for minimisation purposes and may be replaced by another norm $|\cdot|_{\mathcal{M}_h}$ that is equivalent on the discrete spaces. Then an additional step is required to show that the minimisation with respect to $|\cdot|_{\mathcal{M}_h}$ indeed leads to a bound in $|\cdot|_{\mathcal{M}}$.

Below we will mainly focus on the data assimilation problem in Example 7.2 and use

$$|u_h - \tilde{u}_M|^2_{\mathcal{M}_h} := \int_\omega h^\alpha (u_h - \tilde{u}_M)^2 \, \mathrm{d}x, \tag{7.26}$$

where α is a constant in the interval $[-2, 0]$. Here it is assumed that the mesh is fitted to the domain ω, which can always be easily achieved by replacing ω with a slightly smaller polygonal domain. For the Cauchy problem in Example 7.1, we can take

$$|u_h - \tilde{u}_M|^2_{\mathcal{M}_h} := \int_\Gamma h(\nabla u_h \cdot n - \tilde{\psi})^2 \, \mathrm{d}s. \tag{7.27}$$

In what follows it is important that, in both the cases and for all $\alpha \in [-2, 0]$, there holds for $u \in H^2(\Omega)$ that

$$|u - i_h u|_{\mathcal{M}_h} \leq Ch|u|_{H^2(\Omega)}.$$

We form the tentative Lagrangian

$$Ł(u_h, z_h) := \frac{1}{2}\gamma_M |u_h - \tilde{u}_M|^2_{\mathcal{M}_h} + a(u_h, z_h) - \tilde{l}(z_h),$$

where $\tilde{u}_M = u_M + \delta u$ is the perturbed data available and $\tilde{l}(z_h) = l(z_h) + \delta l(z_h)$ is a perturbed right hand side. Observe that if u is a solution to (7.23) and (7.24) then it will minimise the Lagrangian (if $\delta u = \delta l = 0$) with the associated multiplier $z = 0$. Unfortunately the associated minimisation problem may not be well-posed on the discrete level due to the ill-posedness of $a(\cdot, \cdot)$, even if the data of the continuous problem is in the range of $\mathcal{K}$. It follows that we need some regularisation.

7.4.1 *Regularisation by Stabilisation*

The classical way of obtaining a well-posed optimisation problem is through Tikhonov regularisation. In this case the natural choice would be to add regularising terms in the H^1-semi-norm for both the primal and the dual variable to obtain

$$Ł(u_h, z_h) := \frac{1}{2}\gamma_M |u_h - \tilde{u}_M|^2_{\mathcal{M}_h} + \gamma_1 \|\nabla u_h\|^2 - \gamma_2 \|\nabla z_h\|^2 + a(u_h, z_h) - \tilde{l}(z_h).$$

Computing the Euler-Lagrange equations for this Lagrangian we obtain the system: find $(u_h, z_h) \in V_h \times W_h$ such that

$$a(u_h, w_h) - \gamma_2(\nabla z_h, \nabla w_h) = \tilde{l}(w_h) \quad \forall w_h \in W_h \tag{7.28}$$

$$a(v_h, z_h) + \gamma_1(\nabla u_h, \nabla v_h) + \gamma_M(u_h, v_h)_{\mathcal{M}_h} = \gamma_M(\tilde{u}_M, v_h)_{\mathcal{M}_h} \quad \forall v_h \in V_h \tag{7.29}$$

Here it is assumed that the norm $|\cdot|_{\mathcal{M}_h}$ is associated to an inner product $(\cdot,\cdot)_{\mathcal{M}_h}$. This is of course the case for both (7.26) and (7.27).

Remark 7.4 This system bears a strong resemblance to the quasi-reversibility method for the Cauchy problem in the mixed form as proposed on the continuous level in [7]. Therein it was proven that if the exact solution exists, and the data are unperturbed, then letting $\gamma_1 \to 0$ for bounded γ_2 (that may tend to zero, but at a lower rate than γ_1) the regularised solution converges to the exact solution.

Drawing on our experience from stabilised finite element methods we would like to modify the regularisation terms, so that they vanish at an optimal rate in the limit $u_h \to u \in H^2(\Omega)$, $z_h \to 0$, while keeping the regularisation parameters γ_1 and γ_2 fixed. We therefore introduce the abstract regularisation operators $s : V_h \times V_h \mapsto \mathbb{R}$ and $s^* : W_h \times W_h \mapsto \mathbb{R}$ in the Lagrangian

$$Ł(u_h, z_h) := \frac{1}{2}\gamma_M |u_h - \tilde{u}_M|^2_{\mathcal{M}_h} + \frac{1}{2}s(u_h, u_h) - \frac{1}{2}s^*(z_h, z_h) + a(u_h, z_h) - \tilde{l}(z_h). \tag{7.30}$$

The corresponding Euler-Lagrange equations then reads

$$a(u_h, w_h) - s^*(z_h, w_h) = \tilde{l}(w_h) \tag{7.31}$$

$$a(v_h, z_h) + s(u_h, v_h) + \gamma_M(u_h, v_h)_{\mathcal{M}_h} = \gamma_M(\tilde{u}_M, v_h)_{\mathcal{M}_h}. \tag{7.32}$$

The primal stabilisation operator should be weakly consistent, in the sense that,

$$s(i_h u, i_h u)^{\frac{1}{2}} \le Ch|u|_{H^2(\Omega)}. \tag{7.33}$$

We also require s to be bounded, $s(v_h, v_h) \le C\|v_h\|_V^2$. The dual stabilisation on the other hand must be equivalent with the W norm

$$c_1(h)\|w_h\|_W^2 \le s^*(w_h, w_h) \le C\|w_h\|_W^2,$$

where the lower bound is not required to be uniform in h. No condition analogous to (7.33) is required from s^*, the reason being that $z = 0$ is the solution to the unperturbed problem where data are such that a unique solution $u \in V$ exists. Thus any bilinear form s^* is weakly consistent in the sense that it vanishes in (7.31) when (u_h, z_h) is replaced by the solution to the unperturbed problem.

Anticipating the results in the next section we give the following examples of stabilisation operators,

$$s(v_h, v_h) := \gamma_1 \|h\sigma u_h\|^2 + \gamma_1 \sum_{F \in \mathcal{F}_i} (h_F [\![\nabla v_h]\!], [\![\nabla v_h]\!])_F =: \gamma_1 |v_h|_V^2 \tag{7.34}$$

$$s^*(v_h, v_h) := \gamma_2 (\nabla v_h, \nabla v_h)_\Omega =: \gamma_2 \|v_h\|_W^2. \tag{7.35}$$

We emphasize that, contrary to typical Tikhonov regularisation, the stabilisation parameters $\gamma_1, \gamma_2 > 0$ will not change during computation.

Observe that for $u \in H^2(\Omega)$ there holds $s(u, v_h) = \gamma_1 (h^2\sigma^2 u, v_h)_\Omega$ for all $v_h \in V_h$, since the jump term vanishes when applied to sufficiently smooth functions. The remaining L^2-term, is weakly consistent to the right order for piecewise affine elements. For higher order polynomial approximation of order k, the primal stabilisation operator in the Lagrangian (7.30) must be replaced by a strongly consistent residual based stabilisation of the form

$$s(v_h, v_h) := \|h^k \nabla v_h\|_\Omega^2 + \gamma_1 \|h(f + \Delta v_h - \sigma v_h)\|_h^2 + \gamma_1 \sum_{F \in \mathcal{F}_i} (h_F [\![\nabla v_h]\!], [\![\nabla v_h]\!])_F, \tag{7.36}$$

for details see the discussion in [13]. The weak consistency takes a different form in this case, since the presence of the source term f leads to a contribution on the form $\sum_{K \in \mathcal{T}_h} (f, h^2(-\Delta v_h + \sigma v_h))_K$ in the right hand side of (7.32). Observe also that s defines a semi-norm on $V_h + H^2(\Omega)$ but that s^* defines a norm on W.

Let us now introduce the mesh dependent norm

$$|||(u_h, z_h)|||^2 := \gamma_M |u_h|_{\mathcal{M}_h}^2 + \gamma_1 |u_h|_V^2 + \gamma_2 \|z_h\|_W^2 + \min(\gamma_1, \gamma_M) h^2 \|u_h\|_{H^1(\Omega)}^2. \tag{7.37}$$

As the parameters $\gamma_M, \gamma_1, \gamma_2$ are fixed we could omit including them in the above norm, however, we will keep track of the dependence of the constants in Proposition 7.2 below on these parameters, and for this reason it is convenient to include the parameters in the above norm.

Observe that using (7.4) and (7.5) it is straightforward to prove the interpolation inequality

$$|||(u - i_h u, 0)||| \leq Ch |u|_{h^2(\Omega)}. \tag{7.38}$$

To include the last term in the definition (7.37) we can apply a discrete Poincaré inequality.

Lemma 7.1 (Discrete Poincaré Inequality) *There exists $c_p > 0$ such that for all $v_h \in V_h$ there holds*

$$c_P h \|u_h\|_{H^1(\Omega)} \leq |u_h|_{\mathcal{M}_h} + |u_h|_V.$$

In the case of the Cauchy problem where $|\cdot|_{\mathcal{M}_h}$ is defined by (7.27) and $u_h|_\Gamma = 0$ this is a consequence of the Poincaré inequalities of [9] and for the data assimilation case where $|\cdot|_{\mathcal{M}_h}$ is defined by (7.26) the result was proved in [19].

The system (7.31)–(7.32) can be cast on the compact form, find $(u_h, z_h) \in V_h \times W_h$ such that

$$A_h[(u_h, z_h), (v_h, w_h)] = \tilde{l}(w_h) + \gamma_M(\tilde{u}_M, v_h)_{\mathcal{M}_h}, \quad \forall (v_h, w_h) \in V_h \times W_h, \tag{7.39}$$

where

$$A_h[(u_h, z_h), (v_h, w_h)] := a(u_h, w_h) - s^*(z_h, w_h) + a(v_h, z_h) + s(u_h, v_h) + \gamma_M(u_h, v_h)_{\mathcal{M}_h}.$$

Proposition 7.1 *The system* (7.39) *admits a unique unique solution* $(u_h, z_h) \in V_h \times W_h$.

Proof By construction, for all (v_h, w_h)

$$\gamma_M |v_h|^2_{\mathcal{M}_h} + \gamma_1 |v_h|^2_V + \gamma_2 \|w_h\|^2_W = A_h[(v_h, w_h), (v_h, -w_h)]$$

and therefore by Lemma 7.1 there exists $C > 0$ such that

$$|||(v_h, w_h)|||^2 \leq C\, A_h[(v_h, w_h), (v_h, -w_h)]. \tag{7.40}$$

The linear system (7.39) is square, and by the above positivity there are no zero eigenvalues. We conclude that the system is invertible.

Comparing with the exact problem (7.23)–(7.24) and assuming that $u \in H^2(\Omega)$, we see that the formulation (7.39) satisfies the following consistency relation

$$A_h[(u_h - u, z_h), (v_h, w_h)] = \delta l(w_h) + \gamma_M(\delta u, v_h)_{\mathcal{M}_h}, \quad \forall (v_h, w_h) \in V_h \times W_h. \tag{7.41}$$

7.4.2 Error Analysis Using Conditional Stability

First we will introduce some continuity properties of the bilinear form using the stabilisations. Assume that $u \in H^2(\Omega)$, then there holds

$$a(u - i_h u, v_h) \leq Ch|u|_{H^2(\Omega)} \|v_h\|_W \tag{7.42}$$

and for all $u_h \in V_h$ and all $w \in W$, $i_h w \in W_h$

$$a(u_h, w - i_h w) \leq (Ch\|u\|_{H^2(\Omega)} + |||(u - u_h, 0)|||)\|w\|_W, \tag{7.43}$$

where and the constants are allowed to depend on the parameters γ_1, γ_2 and γ_M.

For the data assimilation problem Eq. (7.42) follows by an application of the Cauchy-Schwarz inequality and (7.4), and (7.43) follows by the integration by parts followed by (7.4) and (7.5) leading to

$$\begin{aligned} a(u_h, w - i_h w) &\leq |(\sigma u_h, w - i_h w)| + \sum_{F \in \mathcal{F}_i} \int_F |h^{\frac{1}{2}} [\![\nabla u_h]\!]| h^{-\frac{1}{2}} |w - i_h w| \, ds \\ &\leq C\gamma_1^{-\frac{1}{2}} (|u - u_h|_V + \|\sigma h u\|)\|w\|_W. \end{aligned}$$

The results for the Cauchy problem are obtained in a similar fashion and we refer to [14] for the details.

We are now ready to prove a first error estimate that holds independently of the stability properties of the continuous model.

Proposition 7.2 *If* (u_h, z_h) *is the solution of* (7.39) *and* $u \in H^2(\Omega)$ *is the solution of* (7.23)–(7.24) *then there holds*

$$|||(u - u_h, z_h)||| \leq C_\gamma h |u|_{H^2(\Omega)} + \delta_\gamma \tag{7.44}$$

where $\delta_\gamma := \gamma_2^{-1/2}\|\delta l\|_{W'} + \gamma_M^{1/2}|\delta u|_{\mathcal{M}_h}$ *and* $C_\gamma := C(1 + \gamma_1^{\frac{1}{2}} + \gamma_2^{-\frac{1}{2}})$.

Proof To prove (7.44) we observe that by (7.38) and the triangle inequality it is enough to consider the discrete error $\xi_h = i_h u - u_h$. By (7.40) we have

$$|||(\xi_h, z_h)|||^2 \leq C\, A_h[(\xi_h, z_h), (\xi_h, -z_h)].$$

Using the Galerkin orthogonality (7.41) we may write

$$A_h[(\xi_h, z_h), (\xi_h, -z_h)] = A_h[(i_h u - u, 0), (\xi_h, -z_h)] - \delta l(z_h) + \gamma_M(\delta u, \xi_h)_{\mathcal{M}_h}.$$

By the continuity (7.42) there holds

$$\begin{aligned} A_h[(i_h u - u, 0), (\xi_h, -z_h)] &= a(u - i_h u, z_h) + s(i_h u - u, \xi_h) + \gamma_M (i_h u - u, \xi_h)_{\mathcal{M}_h} \\ &\leq Ch\gamma_2^{-\frac{1}{2}}|u|_{H^2(\Omega)}\gamma_2^{\frac{1}{2}}\|z_h\|_W + \underbrace{\gamma_1^{\frac{1}{2}}|i_h u - u|_V}_{\leq Ch\gamma_1^{\frac{1}{2}}|u|_{H^2(\Omega)}} \gamma_1^{\frac{1}{2}}|\xi_h|_V + \gamma_M|i_h u - u|_{\mathcal{M}_h}|\xi_h|_{\mathcal{M}_h}. \end{aligned}$$

Bounding also the perturbation terms

$$\delta l(w_h) \leq \gamma_2^{-\frac{1}{2}} \|\delta l\|_{W'} \gamma_2^{\frac{1}{2}} \|z_h\|_W$$

and

$$(\delta u, \xi_h)_{\mathcal{M}_h} \leq |\delta u|_{\mathcal{M}_h} |\xi_h|_{\mathcal{M}_h}$$

we arrive at

$$A_h[(\xi_h, z_h), (-\xi_h, z_h)] \leq C_\gamma h |u|_{H^2(\Omega)} |||(\xi_h, z_h)||| + \delta_\gamma |||(\xi_h, z_h)|||.$$

We conclude by dividing by $|||(\xi_h, z_h)|||$.

This proof is insufficient to show error estimates. However for unperturbed data and $u \in H^2(\Omega)$, it may be used to show that $u_h \to u$ as $h \to 0$, by a compactness argument.

Remark 7.5 Note that δ_γ may depend on h via the quantity $|\delta u|_{\mathcal{M}_h}$. This is the case, for instance, when $|\cdot|_{\mathcal{M}_h}$ is chosen as in (7.26) with $\alpha \neq 0$, and then error in data is amplified for small h.

To prove error estimates we must rely on the conditional stability estimates in Theorem 7.1. The idea behind the argument is to consider the error $e = u - u_h$ and observe that this error satisfies

$$a(e, w) = l(w) - a(u_h, w) =: r(w), \quad \forall w \in W. \tag{7.45}$$

We will then use Proposition 7.2 to get bounds for $\|r\|_{W'}$, $|e|_{\mathcal{M}_h}$ and $\|e\|$, so that the conditional stability can be applied to e.

In the data assimilation case we have $|e|_{\mathcal{M}} = \|e\|_\omega = h^{-\alpha/2}|e|_{\mathcal{M}_h} \leq |e|_{\mathcal{M}_h}$ so this quantity is immediately bounded by (7.44). For the Cauchy problem the continuous and discrete data matching terms are not the same, but one can prove that a suitable bound can be obtained for a perturbed error $\tilde{e}$ by adding a small perturbation to u_h in the interface zone such that

$$|\tilde{e}|_{\mathcal{M}} \leq |||e, 0|||. \tag{7.46}$$

The error analysis then uses the arguments below together with a perturbation argument for $\tilde{e}$, for details see [14]. We will not consider that case here, instead focussing on the data assimilation case.

Theorem 7.2 *Let u be the exact solution to* (7.23)–(7.24), *with $l(w) := (f, w)$, $f \in L^2(\Omega)$, and $|\cdot|_{\mathcal{M}} = \|\cdot\|_\omega$. Let u_h be the solution of* (7.31)–(7.32) *with the stabilisation operators* (7.34)–(7.35)*. Then, for all $\omega' \subset \Omega$ satisfying the* *assumptions in Theorem 7.1 there holds*

$$\|u - u_h\|_{\omega'} \leq Ch^\tau(\|u\|_{H^2(\Omega)} + h^{-1}\delta_\gamma).$$

where the constant depends on the geometry and the constants γ_1, γ_2 *and* γ_M.

Proof As discussed above, the estimate is shown by applying Theorem (7.1) to the problem satisfied by the error. We know that e is a solution to (7.23) with $l(w) = r(w)$ as per Eq. (7.45). By Proposition 7.2 the following bounds hold

$$|e|_{\mathcal{M}_h} = \|e\|_\omega \leq C_\gamma h|u|_{H^2(\Omega)} + \delta_\gamma \tag{7.47}$$

and

$$\|e\|_V \leq C_\gamma |u|_{H^2(\Omega)} + h^{-1}\delta_\gamma. \tag{7.48}$$

Now observe that using Eq. (7.31) we have

$$r(w) = r(w - i_h w) + r(i_h w) = l(w - i_h w) - a(u_h, w - i_h w) - s^*(z_h, i_h w) - \delta l(i_h w).$$

We estimate the terms on the right hand side, assuming that $\|w\|_W = 1$,

$$l(w - i_h w) = (f, w - i_h w) \leq \|f\|\|w - i_h w\| \leq Ch\|f\|,$$

and using the inequality (7.43)

$$a(u_h, w - i_h w) \leq Ch\|u\|_{H^2(\Omega)} + |||(u - u_h, 0)|||.$$

Then applying Proposition 7.2 we obtain the bound

$$a(u_h, w - i_h w) \leq \gamma_1^{-\frac{1}{2}}(C_\gamma h\|u\|_{H^2(\Omega)} + \delta_\gamma).$$

The two remaining terms are handled using the Cauchy-Schwarz inequality in the first case and the duality pairing $H^{-1} \times H^1$ in the second, followed by the stability of the interpolant i_h in the W-norm,

$$s(z_h, i_h w) \leq \gamma_2 \|z_h\|_W \|w\|_W \leq \gamma_2^{\frac{1}{2}}(C_\gamma h|u|_{H^2(\Omega)} + \delta_\gamma)$$

$$\delta l(i_h w) \leq C\|\delta l\|_{W'}$$

Collecting the terms above we have for all $w \in W$ with $\|w\|_W = 1$,

$$r(w) \leq Ch\|f\| + (\gamma_1^{-\frac{1}{2}} + \gamma_2^{\frac{1}{2}})(C_\gamma h\|u\|_{H^2(\Omega)} + \delta_\gamma) + C\|\delta l\|_{W'}. \tag{7.49}$$

But then

$$\|r\|_{W'} = \sup_{w \in W : \|w\|_W = 1} r(w)$$

satisfies the same bound. Note also that $\|f\| \le C\|u\|_{H^2(\Omega)}$. We conclude that e satisfies the assumptions of Theorem 7.1 by with

$$R = \|e\|_V \le C_\gamma |u|_{H^2(\Omega)} + h^{-1}\delta_\gamma, \quad \eta = |e|_{\mathcal{M}_h} \le Ch|u|_{H^2(\Omega)} + \delta_\gamma,$$

$$\varepsilon = \|r\|_{W'} \le C(h\|u\|_{H^2(\Omega)} + \delta_\gamma)$$

c.f. (7.47)–(7.49). In the last step we dropped the dependence on the constants γ_1, γ_2 and γ_M, but it can be traced in the proof.

Remark 7.6 We detailed Theorem 7.2 only in the case of the data assimilation problem, but the same arguments also leads to an analysis for the Cauchy problem, under the assumption (7.46).

Remark 7.7 One may prove Theorem 7.2 for the data assimilation problem if s^* is defined by (7.34). In this case an additional factor h^{-1} multiplies the term measuring perturbations in data.

7.4.3 A Numerical Example

We consider the problem in Example 7.1 on the unit square Ω. The exact solution is $u = 30.0 * x * (1 - x) * y * (1 - y)$, with $f = \mathcal{L}u$, and the data domain ω is defined by

$$\omega := \{(x, y) \in \Omega : |x - 0.5| < 0.25; |y - 0.5| < 0.25\}.$$

We use the formulation (7.31)–(7.32) with $s(\cdot, \cdot)$ given by (7.34) for piecewise affine approximation and (7.36) for piecewise quadratic approximation. The adjoint stabiliser $s^*(\cdot, \cdot)$ was defined by (7.35), and the norm $|\cdot|_{\mathcal{M}_h}$ by (7.26) with $\alpha = 0$ or -2. (Observe that if $\alpha = 0$ then γ_M must have the unit of the square of an inverse length for the equations to be dimensionally correct.)

We chose $\gamma_2 = \gamma_M = 1$ and $\gamma_1 = 10^{-3}$ for all computations. The latter value is similar to that used for computations in the well-posed case. We meshed the domain using structured meshes that were made to fit the boundary of ω. We performed computations on a sequence of meshes with `nele`= 40, 80, 160, 320, elements on each side of the square, using piecewise affine and piecewise quadratic elements. In Fig. 7.1, left graphic, we show a computational mesh and on the right graphic we illustrate the domains ω (the inner square) and ω' (the middle square). In Fig. 7.2, left plot, we show the contourlines of an approximate solution and in the right plot

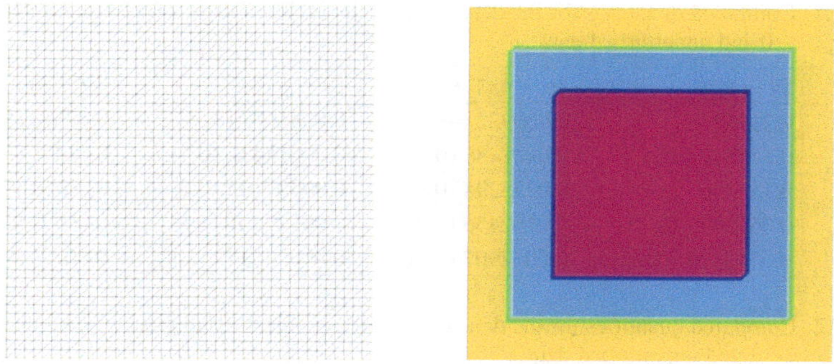

Fig. 7.1 Left: computational mesh, nele=40. Right: the different subdomains ω and ω'

Fig. 7.2 Left: contour lines of approximate solution, nele=40. Right: contour lines of the computational error

the contour lines of the computational error. Observe that the error has a form that is similar to Hadamard's counter-example discussed in Example 7.1, but growing exponentially in the radial direction and oscillating in the direction tangential to the boundary of ω.

In the tables below we report the error in the normalised global L^2-error, the normalised local error for the subset

$$\omega' := \{(x, y) \in \mathbb{R}^2 : |x - 0.5| < 0.375; |y - 0.5| < 0.375\},$$

the data assimilation term, $|u - u_h|_\omega$, and the size of the weakly consistent regularisation

$$|(u - u_h, z)|_s := \sqrt{s(u - u_h, u - u_h) + s^*(z_h, z_h)}. \tag{7.50}$$

The experimental convergence rates are given in parenthesis, where appropriate. We report the results for unperturbed data and $\alpha = 0$ in Tables 7.1 and 7.5 and for $\alpha = -2$ in Tables 7.2 and 7.6. In all cases we observe the expected $O(h^k)$ convergence

Table 7.1 Computed quantities for the data assimilation problem using piecewise affine approximation, $\alpha = 0$ and unperturbed data

nele	$\|u - u_h\|$	$\|u - u_h\|_{\omega'}$	$\|u - u_h\|_{\omega}$	$\|(u - u_h, z)\|_s$
40	0.211594 (–)	0.050922 (–)	0.00816074 (–)	0.0289235 (–)
80	0.175512 (0.3)	0.0407488 (0.3)	0.00618422 (0.4)	0.0147585 (1.0)
160	0.113346 (0.6)	0.0235298 (0.8)	0.00337103 (0.9)	0.00791309 (0.9)
320	0.0672893 (0.75)	0.0102456 (1.2)	0.00119201 (1.5)	0.0042852 (0.9)
640	0.0510429 (0.4)	0.00529074 (1.0)	0.000342379 (1.8)	0.00221974 (0.9)

Table 7.2 Computed quantities for the data assimilation problem using piecewise affine approximation, $\alpha = -2$ and unperturbed data

nele	$\|u - u_h\|$	$\|u - u_h\|_{\omega'}$	$\|u - u_h\|_{\omega}$	$\|(u - u_h, z)\|_s$
40	0.0476335 (–)	0.00481282 (–)	0.000333429 (–)	0.0352793 (–)
80	0.0403148 (0.2)	0.00312934 (0.6)	8.0272e–05 (2.0)	0.0179655 (1.0)
160	0.0304957 (0.4)	0.00188862 (0.7)	1.998e–05 (2.0)	0.00911884 (1.0)
320	0.0227619 (0.4)	0.0009549 (1.0)	4.71016e–06 (2.1)	0.00464924 (1.0)
640	0.0200062 (0.2)	0.000642748 (0.6)	1.15698e–06 (2.0)	0.00234456 (1.0)

Table 7.3 Computed quantities for the data assimilation problem using piecewise affine approximation, $\alpha = 0$ and 2.5% perturbation in data

nele	$\|u - u_h\|$	$\|u - u_h\|_{\omega'}$	$\|u - u_h\|_{\omega}$	$\|(u - u_h, z)\|_s$
40	0.206909	0.0490942	0.0148287	0.0289287 (-)
80	0.176546	0.0409112	0.013946	0.0146984 (1.0)
160	0.119693	0.0267951	0.0131763	0.0077906 (0.9)
320	0.0793605	0.0180773	0.0125264	0.00416117 (0.9)
640	0.0640708	0.0158747	0.0124993	0.00214582 (1.0)

of the stabilising terms (7.50), with $k = 1$ for piecewise affine approximation and $k = 2$ in the quadratic case. We also observe that consistently with theory we have $\|u - u_h\|_\omega = O(h^{k-\alpha/2})$. The convergence of the data term is more even for $\alpha = -2$. For the global and local L^2-norms we see that the error is a factor $5 - 10$ larger when $\alpha = 0$ compared with the case where $\alpha = -2$. Most likely this is due to the fact that the missing length-scale present for $\alpha = 0$ is not well represented when $\gamma_M = 1.0$. Indeed the weak penalty does not impose the data sufficiently well compared to the other terms, when $\alpha = -2$ on the other hand the data penalty term is so strong that the data error is very small already on coarse meshes leading to improved local and global errors. We observe convergence compatible with Hölder stability for all quantities, indicating that possibly we are not yet in the asymptotic regime on these scales. Only on the finest meshes in Table 7.6 we see clearly the decreasing orders characteristic for logarithmic convergence in the global error.

We then make the same sequence of computations but adding a perturbation of 2.5% to the data in ω in the piecewise affine case and 1% in the quadratic case. The results are reported for affine approximation in Tables 7.3 ($\alpha = 0$) and 7.4 ($\alpha = -2$). We observe that although the data assimilation term stagnates, the local

Table 7.4 Computed quantities for the data assimilation problem using piecewise affine approximation, $\alpha = -2$ and 2.5% perturbation in data

nele	$\|u - u_h\|$	$\|u - u_h\|_{\omega'}$	$\|u - u_h\|_{\omega}$	$\|(u - u_h, z)\|_s$
40	0.0520752	0.0145883	0.0124714	0.03529
80	0.0507222	0.014398	0.0125092	0.0186372
160	0.0502568	0.0143645	0.0127194	0.0142032
320	0.0537505	0.0143083	0.0125169	0.0224315
640	0.0427351	0.0138826	0.0125888	0.0434341

Table 7.5 Computed quantities for the data assimilation problem using piecewise quadratic approximation, $\alpha = 0$ and unperturbed data

nele	$\|u - u_h\|$	$\|u - u_h\|_{\omega'}$	$\|u - u_h\|_{\omega}$	$\|(u - u_h, z)\|_s$
20	0.0113854 (–)	0.0020353 (–)	0.000272026 (–)	0.00263335 (–)
40	0.00701791 (0.7)	0.000668735 (1.6)	4.36798e−05 (2.6)	0.00067804 (2.0)
80	0.00630128 (0.16)	0.000458704 (0.54)	1.0293e−05 (2.1)	0.000171095 (2.0)
160	0.00457823 (0.5)	0.000278068 (0.72)	5.50828e−06 (1.0)	4.33632e−05 (2.0)
320	0.00275223 (0.7)	9.14176e−05 (1.6)	7.11806e−07 (2.8)	1.10465e−05 (2.0)

Table 7.6 Computed quantities for the data assimilation problem using piecewise quadratic approximation, $\alpha = -2$ and unperturbed data

nele	$\|u - u_h\|$	$\|u - u_h\|_{\omega'}$	$\|u - u_h\|_{\omega}$	$\|(u - u_h, z)\|_s$
20	0.00594613 (–)	0.000454428 (–)	1.92029e−05 (–)	0.00269387 (–)
40	0.00364274 (0.7)	0.000194766 (1.2)	3.21386e−06 (-2.6)	0.00069238 (–)
80	0.0023773 (0.6)	6.52831e−05 (1.6)	2.95005e−07 (3.4)	0.000176426 (2.0)
160	0.00159176 (0.6)	2.93421e−05 (1.2)	3.91486e−08 (2.9)	4.45628e−05 (2.0)
320	0.00118008 (0.4)	1.27615e−05 (1.2)	4.3179e−09 (3.2)	1.12277e−05 (2.0)

Table 7.7 Computed quantities for the data assimilation problem using piecewise quadratic approximation, $\alpha = 0$ and 1% perturbation in data

nele	$\|u - u_h\|$	$\|u - u_h\|_{\omega'}$	$\|u - u_h\|_{\omega}$	$\|(u - u_h, z)\|_s$
20	0.0146381	0.00619699	0.00510402	0.00260206
40	0.0137215	0.00593519	0.00492976	0.00066236 (2.0)
80	0.0135235	0.00594218	0.00498009	0.000167333 (2.0)
160	0.0110434	0.00593666	0.00497521	4.82896e−05 (1.8)
320	0.00982659	0.0058722	0.00497389	1.23888e−05 (2.0)

and global errors decrease under refinement for $\alpha = 0$. In this case the stabilisation norm also converges to optimal order in spite of the perturbation. When $\alpha = -2$ only the error in the stabilisation semi-norm show any decrease under refinement. On the finest scale we see that both the global error and the error in the stabilisation semi-norm has started to grow. For piecewise affine approximation it appears that the choice $\alpha = -2$ is superior both for perturbed and unperturbed data (at least for the choice $\gamma_M = 1$) (Tables 7.5 and 7.6).

For quadratic approximation the results are reported in Tables 7.7 ($\alpha = 0$) and 7.8 ($\alpha = -2$). Here the effect of the perturbation is present already on the coarsest mesh

Table 7.8 Computed quantities for the data assimilation problem using piecewise quadratic approximation, $\alpha = -2$ and 1% perturbation in data

nele	$\|u - u_h\|$	$\|u - u_h\|_{\omega'}$	$\|u - u_h\|_{\omega}$	$|(u - u_h, z)|_s$
20	0.0177247	0.00638777	0.00513258	0.00275637
40	0.026475	0.00628408	0.00495361	0.00164336
80	0.0503314	0.00644259	0.00500485	0.002676516
160	0.159728	0.0079909	0.0050097	0.00510579
320	0.335852	0.00962178	0.0050035	0.0101055

and the amplification of the error clearly much stronger for $\alpha = -2$. Indeed whereas for $\alpha = 0$ all error quantities still decrease under mesh refinement, the errors for $\alpha = -2$ all stagnate or increase. For the stabilisation norm we clearly see that the error doubles under mesh refinement on finer meshes, which is consistent with theory. In this case it appears that for resolutions where the mesh-size is of similar order as the perturbation it is advantageous to take $\alpha = 0$, also in accordance with theory.

7.5 Time Dependent Problems: Data Assimilation

In this section we consider the extension of the methods in the previous section to the time dependent case, where several interesting new features appear. In particular we can consider a problem which has Lipschitz stability and prove that our method can exploit this in the form of error estimates that are optimal compared to approximation. We consider a data assimilation problem for the heat equation

$$\partial_t u - \Delta u = f, \quad \text{in } (0, T) \times \Omega, \tag{7.51}$$

with homogeneous Dirichlet conditions. Here $T > 0$ and $\Omega \subset \mathbb{R}^n$ is an open convex polyhedral domain. Let $\omega \subset \Omega$ be open and let $0 < T_1 < T$. The data assimilation problems is of the following form: determine the restriction $u|_{(T_1,T)\times\Omega}$ of a solution to the heat equation (7.51) given f and the restriction $u|_{(0,T)\times\omega}$. In this case we have the following stability estimate due to Imanuvilov [23], see also [17, 32, 35] for variations of the estimate.

Theorem 7.3 *Let $\omega \subset \Omega$ be open and non-empty, and let $0 < T_1 < T$. Then there is $C > 0$ such that for all u in the space*

$$H^1(0, T; H^{-1}(\Omega)) \cap L^2(0, T; H_0^1(\Omega)), \tag{7.52}$$

it holds that

$$\|u\|_{C(T_1,T;L^2(\Omega))} + \|u\|_{L^2(T_1,T;H^1(\Omega))} + \|u\|_{H^1(T_1,T;H^{-1}(\Omega))}$$
$$\leq C(\|u\|_{L^2((0,T)\times\omega)} + \|Lu\|_{(0,-1)}),$$

where $L = \partial_t - \Delta$ *and* $\|\cdot\|_{(0,-1)} = \|\cdot\|_{L^2(0,T;H^{-1}(\Omega))}$.

In what follows, we use the shorthand notations

$$H^{(k,m)} = H^k(0,T;H^m(\Omega)), \quad H_0^{(k,m)} = H^{(k,m)} \cap L^2(0,T;H_0^1(\Omega)),$$
$$\|u\|_{(k,m)} = \|u\|_{H^k(0,T;H^m(\Omega))}, \quad \|u\| = \|u\|_{(0,0)},$$

and denote by $\|u\|_\omega$ the norm of $L^2((0,T)\times\omega)$. Moreover, we use the following notation for the data of the problem

$$q = u|_{(0,T)\times\omega}, \quad f = Lu, \tag{7.53}$$

and write

$$a(u,z) = (\nabla u, \nabla z), \quad G_f(u,z) = (\partial_t u, z) + a(u,z) - \langle f, z\rangle, \quad G = G_0,$$

where $(\cdot,\cdot)$ is the inner product of $L^2((0,T)\times\Omega)$ and $\langle\cdot,\cdot\rangle$ is the dual pairing between $L^2(0,T;H^{-1}(\Omega))$ and $L^2(0,T;H_0^1(\Omega))$. Note that for $u \in H^1((0,T)\times\Omega)$, the equations

$$G_f(u,z) = 0, \quad z \in L^2(0,T;H_0^1(\Omega)), \tag{7.54}$$

give the weak formulation of $\partial_t u - \Delta u = f$.

7.5.1 Optimisation Based Finite Element Space Discretisation

We consider only the problem semi-discretised in space, and show that the time continuous dynamical system is well-posed for every fixed h. This section summarizes part of the analysis from [17], where also a problem with weaker stability, similar to that of the data assimilation problem in the previous section was considered. The analysis carries over to the fully discrete case, but the stabilisation operators are not the same. In particular in the fully discrete case, the adjoint stabilisation can be omitted (see reference [18] for details).

Since the problem is time dependent we introduce the spaces $\mathcal{V}_h$ and $\mathcal{W}_h$,

$$\mathcal{V}_h = H^1(0,T;V_h^0), \quad \mathcal{W}_h = L^2(0,T;V_h^0).$$

Observe that contrary to the developments in the previous section both spaces are equipped with Dirichlet conditions in space. The difference between the two spaces here is the regularity in time. Following the development in the previous sections our approach to solve the data assimilation problem is based on minimizing the Lagrangian functional

$$Ł_{q,f}(u,z) = \frac{1}{2}\|u-q\|_\omega^2 + \frac{1}{2}s(u,u) - \frac{1}{2}s^*(z,z), + G_f(u,z), \tag{7.55}$$

where the data q and f are fixed. Here $\|\cdot\|_\omega$ is the norm of $L^2((0,T)\times\omega)$, and s and s^* are the primal and dual stabilizers, respectively. Note that minimizing $Ł_{q,f}$ can be seen as fitting $u|_{(0,T)\times\omega}$ to the data q under the constraint (7.54), z can be interpreted as a Lagrange multiplier, and $s/2$ and $s^*/2$ as regularizing penalty terms.

Let $q \in L^2((0,T)\times\omega)$ and $f \in H^{(0,-1)}$. The Lagrangian $Ł_{q,f}$, defined by (7.55), satisfies

$$\begin{aligned} D_u Ł_{q,f} v &= (u-q,v)_\omega + s(u,v) + G(v,z), \\ D_z Ł_{q,f} w &= -s^*(z,w) + G(u,w) - \langle f, w\rangle, \end{aligned}$$

and therefore the critical points $(u,z) \in \mathcal{V}_h \times \mathcal{W}_h$ of $Ł_{q,f}$ satisfy

$$A[(u,z),(v,w)] = (q,v)_\omega + \langle f,w\rangle, \quad (v,w) \in \mathcal{V}_h \times \mathcal{W}_h, \tag{7.56}$$

where A is the symmetric bilinear form

$$A[(u,z),(v,w)] = (u,v)_\omega + s(u,v) + G(v,z) - s^*(z,w) + G(u,w). \tag{7.57}$$

Note that

$$A[(u,z),(u,-z)] = s(u,u) + \|u\|_\omega^2 + s^*(z,z),$$

Herein we consider only semi-discretisations, that is, we minimize $Ł_{q,f}$ on a scale of spaces that are discrete in the spatial variable but not in the time variable. As before the spatial mesh size $h > 0$ will be the only parameter controlling the convergence of the approximation, and we use piecewise affine finite elements. For simplicity we have set all the auxiliary regularisation parameters $\gamma_1, \gamma_2, \gamma_M$ to one, and we consider only the case of unperturbed data.

7.5.2 A Framework for Stabilisation

Before proceeding to the analysis of the data assimilation problem, we introduce an abstract stabilisation framework.

Let s and s^* be bilinear forms on the spaces $\mathcal{V}_h$ and $\mathcal{W}_h$, respectively. Let $|\cdot|_{\mathcal{V}}$ be a semi-norm on $\mathcal{V}_h$ and let $\|\cdot\|_{\mathcal{W}}$ be a norm on $\mathcal{W}_h$. We relax (7.34) and (7.35) by requiring only that s and s^* are continuous with respect to $|\cdot|_{\mathcal{V}}$ and $\|\cdot\|_{\mathcal{W}}$, that is,

$$s(u,u) \leq C|u|_{\mathcal{V}}^2, \quad s^*(z,z) \leq C\|z\|_{\mathcal{W}}^2, \quad u \in \mathcal{V}_h,\ z \in \mathcal{W}_h,\ h > 0. \tag{7.58}$$

Let $\|\cdot\|_*$ be the norm of a continuously embedded subspace H^* of the energy space (7.52). The space H^* encodes the a priori smoothness. We assume that the stabilizations and norms introduced are such that the following continuities hold

$$G(u, z - \pi_h z) \leq C|u|_{\mathcal{V}}\|z\|_{(0,1)}, \qquad u \in \mathcal{V}_h,\ z \in H_0^{(0,1)}, \tag{7.59}$$

$$G(u - \pi_h u, z) \leq Ch\|z\|_{\mathcal{W}}\|u\|_*, \qquad u \in H^*,\ z \in \mathcal{W}_h, \tag{7.60}$$

where π_h is an interpolator satisfying

$$\pi_h : H_0^1(\Omega) \to V_h^0, \quad h > 0. \tag{7.61}$$

$$\|\pi_h u\|_{H^1(\Omega)} \leq C\|u\|_{H^1(\Omega)}, \qquad u \in H^1(\Omega),\ h > 0, \tag{7.62}$$

$$\|u - \pi_h u\|_{H^m(\Omega)} \leq Ch^{k-m}\|u\|_{H^k(\Omega)}, \qquad u \in H^k(\Omega),\ h > 0, \tag{7.63}$$

where $k = 1, 2$ and $m = 0, k-1$. We assume that the following upper bound holds

$$|\pi_h u|_{\mathcal{V}} \leq Ch\|u\|_*, \qquad u \in H^*, \tag{7.64}$$

and require that analogously to the stationary case

$$\|\pi_h z\|_{\mathcal{W}} \leq C\|z\|_{(0,1)}, \qquad z \in H_0^{(0,1)}. \tag{7.65}$$

We assume that

$$|||(u,z)||| = |u|_{\mathcal{V}} + \|u\|_\omega + \|z\|_{\mathcal{W}},$$

is a norm on $\mathcal{V}_h \times \mathcal{W}_h$. Finally, in the abstract setting, we assume that the s and s^* are sufficiently strong so that the following weak coercivity holds

$$|||(u,z)||| \leq C \sup_{(v,w)\in\mathcal{V}_h\times\mathcal{W}_h} \frac{A[(u,z),(v,w)]}{|||(v,w)|||}, \quad (u,z) \in \mathcal{V}_h \times \mathcal{W}_h \tag{7.66}$$

and for all $(v,w) \in \mathcal{V}_h \times \mathcal{W}_h$,

$$\sup_{\substack{(x,y)\in\mathcal{V}_h\times\mathcal{W}_h \\ x,y\neq 0}} |A[(x,y),(v,w)]| > 0. \tag{7.67}$$

The Babuska-Lax-Milgram theorem implies that Eq. (7.56) has a unique solution in $\mathcal{V}_h \times \mathcal{W}_h$. As we shall see below, these design criteria are sufficient to derive optimal error estimates in the transient case, provided the problem has a conditional stability property.

7.5.3 The Data Assimilation Problem

We will now proceed to a specific case. We choose the stabilizers and semi-norms as follows,

$$s(u,u) = \|h\nabla u(0,\cdot)\|^2_{L^2(\Omega)}, \quad s^* = a, \tag{7.68}$$

$$|u|_{\mathcal{V}} = s(u,u)^{1/2} + \|h\partial_t u\|, \quad \|z\|_{\mathcal{W}} = s^*(z,z)^{1/2}, \tag{7.69}$$

and we define $H^* = H_0^{(1,1)}$. To counter the lack of primal stabilisation on most of the cylinder $(0,T)\times\Omega$, we choose π_h to be the orthogonal projection $\pi_h : H_0^1(\Omega) \to W_h$ as defined in Sect. 7.2.2. As Ω is a convex polyhedron, it is well known that this choice satisfies (7.61)–(7.63), see e.g. [25, Th. 3.12–18].

Lemma 7.2 *The choices (7.68)–(7.69) satisfy the properties (7.58)–(7.64), (7.65) and (7.66). Moreover,* $|||\cdot|||$ *is a norm on* $\mathcal{V}_h \times \mathcal{W}_h$.

Proof It is clear that the continuities (7.58) hold. We begin with the lower bound (7.59). By the orthogonality of π_h,

$$G(u, z-\pi_h z) = (\partial_t u, z-\pi_h z) \le \|h\partial_t u\| h^{-1}\|z-\pi_h z\| \le C\|h\partial_t u\|\,\|z\|_{(0,1)}.$$

Towards the upper bound (7.60), we use the orthogonality as above,

$$G(u-\pi_h u, z) = (\partial_t u - \pi_h \partial_t u, z) \le Ch\|u\|_{(1,1)}\|z\|.$$

The bound (7.60) then follows from an application of the Poincaré inequality on $\|z\|$.

The bound (7.64) follows from the continuity of the trace

$$\|\nabla u(0,\cdot)\|_{L^2(\Omega)} \le \|u\|_{(1,1)}, \tag{7.70}$$

and the continuity of the projection π_h. The bound (7.65) follows immediately from the continuity of π_h.

We turn to the weak coercivity (7.66). The essential difference between the time dependent case and the stationary case is that in the latter case, the choice $w = u$

is prohibited. In this case it is allowed, but due to the time-derivative and the lack of initial condition it does not lead to stability. Instead we observe that $\partial_t u \in \mathcal{W}_h$ when $u \in \mathcal{V}_h$ so that this can be used as a test function $w = \partial_t u$ to obtain

$$A[(u,z),(0,\partial_t u)] = -s^*(z,\partial_t u) + G(u,\partial_t u) = \|\partial_t u\|^2 + a(u,\partial_t u) - a(z,\partial_t u),$$

and thus using bilinearity of A,

$$A[(u,z),(u,\alpha h^2\partial_t u - z)] = s(u,u) + \alpha\|h\partial_t u\|^2 + \|u\|_\omega^2 + s^*(z,z) \qquad (7.71)$$
$$+\alpha h^2 a(u,\partial_t u) - \alpha h^2 a(z,\partial_t u),$$

where $\alpha > 0$. We will establish (7.66) by showing that there is $\alpha \in (0,1)$ such that

$$|||(u, w - z)||| \le C|||(u,z)|||, \qquad (7.72)$$
$$|||(u,z)|||^2 \le CA[(u,z),(u,w-z)], \qquad (7.73)$$

where $w = \alpha h^2 \partial_t u$.

Towards (7.72) we observe that

$$|||(u,w-z)|||^2 = |||(u,z)|||^2 - 2s^*(z,w) + s^*(w,w) \le 2|||(u,z)|||^2 + 2s^*(w,w).$$

We use the discrete inverse inequality (7.1) to bound the second term

$$s^*(w,w) = \alpha^2 h^4 \|\partial_t \nabla u\|^2 \le C\alpha^2 h^2 \|\partial_t u\|^2 \le C\alpha^2 |||(u,z)|||^2, \quad \alpha > 0.$$

It remains to show (7.73). Towards bounding the first cross term in (7.71) we observe that

$$2a(u,\partial_t u) = \int_0^T \partial_t \|\nabla u(t,\cdot)\|^2_{L^2(\Omega)} dt = \|\nabla u(T,\cdot)\|^2_{L^2(\Omega)} - \|\nabla u(0,\cdot)\|^2_{L^2(\Omega)}.$$

Hence $\alpha h^2 a(u,\partial_t u) \ge -\alpha s(u,u)/2$. We use the arithmetic-geometric inequality,

$$ab \le (4\epsilon)^{-1}a^2 + \epsilon b^2, \quad a,b \in \mathbb{R},\ \epsilon > 0,$$

and the discrete inverse inequality (7.1) to bound the second cross term in (7.71),

$$\alpha h^2 a(z,\partial_t u) \le \alpha(4\epsilon)^{-1} a(z,z) + \alpha\epsilon h^4 \|\partial_t \nabla u\|^2 \le \alpha(4\epsilon)^{-1}a(z,z) + C\alpha\epsilon\|h\partial_t u\|^2.$$

Choosing $\epsilon = 1/(2C)$ we obtain

$$A[(u,z),(u,w-z)] \geq (1-\alpha/2)s(u,u)+\alpha\|h\partial_t u\|^2/2+\|u\|_\omega^2+(1-C\alpha/2)s^*(z,z),$$

and therefore (7.73) holds with small enough $\alpha > 0$.

The second condition (7.67) follows using the symmetry of A. Indeed, if $(v,w) \neq 0$, then $A[(x,y),(v,w)] = A[(v,w),(x,y)] > 0$ for some (x,y) by (7.66). Finally, using the Poincaré inequality, we see that $|||(u,z)||| = 0$ implies $z = 0$ and $u(0,\cdot) = 0$. As also $\partial_t u = 0$, we have $u = 0$, and therefore $|||\cdot|||$ is a norm.

7.5.4 Error Estimates

We are now in a situation to prove an error estimate using the abstract theory.

Theorem 7.4 *Let $\omega \subset \Omega$ be open and non-empty and let $0 < T_1 < T$. Suppose that (A2) holds. Let $u \in H^*$ and define $f = \partial_t u - \Delta u$ and $q = u|_\omega$. Suppose that the primal and dual stabilizers satisfy (7.58)–(7.64), (7.65) and (7.66). Then (7.56) has a unique solution $(u_h, z_h) \in \mathcal{V}_h \times \mathcal{W}_h$, and there exists $C > 0$ such that for all $h \in (0,1)$*

$$\begin{aligned}&\|u_h - u\|_{C(T_1,T;L^2(\Omega))} + \|u_h - u\|_{L^2(T_1,T;H^1(\Omega))} + \|u_h - u\|_{H^1(T_1,T;H^{-1}(\Omega))}\\ &\quad \leq Ch(\|u\|_* + \|f\|).\end{aligned}$$

Proof We begin again by showing the estimate

$$|||(u_h - \pi_h u, z_h)||| \leq Ch\|u\|_*. \tag{7.74}$$

The equations $\partial_t u - \Delta u = f$ and $u|_\omega = q$ are equivalent with

$$\begin{aligned} G(u,w) &= \langle f, w\rangle, && w \in L^2(0,T;H_0^1(\Omega)), \\ (q-u,v)_\omega &= 0, && v \in L^2((0,T)\times\omega), \end{aligned} \tag{7.75}$$

and Eqs. (7.56) and (7.75) imply for all $v \in \mathcal{V}_h$ and $w \in \mathcal{W}_h$ that

$$A[(u_h - \pi_h u, z_h),(v,w)] = (u - \pi_h u, v)_\omega + G(u - \pi_h u, w) - s(\pi_h u, v). \tag{7.76}$$

The weak coercivity (7.66) implies that in order to show (7.74) it is enough bound the three terms on the right hand side of (7.76). For the first of them, that is, $(u - \pi_h u, v)_\omega$, we use (7.63). The upper bound (7.60) applies to the second term

$G(u - \pi_h u, w)$, and for the third one we use the continuity (7.58) and the upper bound (7.64),

$$s(\pi_h u, v) \le C|\pi_h u|_{\mathcal{V}}|v|_{\mathcal{V}} \le Ch\|u\|_*|v|_{\mathcal{V}}.$$

We define the residual r as follows. By taking $v = 0$ in (7.56) we get $G(u_h, w) = \langle f, w\rangle + s^*(z_h, w)$, $w \in \mathcal{W}_h$, and therefore

$$\begin{aligned}\langle r, w\rangle &= G(u_h, w) - \langle f, w\rangle - G(u_h, \pi_h w) + G(u_h, \pi_h w) \\ &= G(u_h, w - \pi_h w) - \langle f, w - \pi_h w\rangle + s^*(z_h, \pi_h w), \quad w \in H_0^{(0,1)}.\end{aligned} \tag{7.77}$$

We now wish to arrive to the estimate

$$\|r\|_{(0,-1)} \le C(|u_h|_{\mathcal{V}} + \|z_h\|_{\mathcal{W}} + h\|f\|). \tag{7.78}$$

To show that (7.78) holds, it is enough to bound the three terms on the right hand side of (7.77). The upper bound (7.59) applies to the first term $G(u_h, w - \pi_h w)$, for the second term $\langle f, w - \pi_h w\rangle$ we use (7.63), for the third term we use the continuity (7.58) and the upper bound (7.65)

$$s^*(z_h, \pi_h w) \le C\|z_h\|_{\mathcal{W}}\|\pi_h w\|_{\mathcal{W}} \le C\|z_h\|_{\mathcal{W}}\|w\|_{(0,1)}.$$

The inequalities (7.78), (7.74) and (7.64) imply

$$\|r\|_{(0,-1)} \le C(|u_h - \pi_h u|_{\mathcal{V}} + |\pi_h u|_{\mathcal{V}} + \|z_h\|_{\mathcal{W}} + h\|f\|) \le Ch(\|u\|_* + \|f\|).$$

Finally using the above bound on r, Theorem 7.3 implies that

$$\begin{aligned}&\|u_h - u\|_{C(T_1,T;L^2(\Omega))} + \|u_h - u\|_{L^2(T_1,T;H^1(\Omega))} + \|u_h - u\|_{H^1(T_1,T;H^{-1}(\Omega))} \\ &\quad \le C\|u_h - u\|_\omega + Ch(\|u\|_* + \|f\|).\end{aligned}$$

The claim follows by using (7.74) and (7.63),

$$\|u_h - u\|_\omega \le \|u_h - \pi_h u\|_\omega + \|\pi_h u - u\|_\omega \le Ch\|u\|_*.$$

Here we used also the assumption that H^* is a continuously embedded subspace of the energy space (7.52), namely, this implies that the embedding $H^* \subset H^{(0,1)}$ is continuous.

Remark 7.8 If the data q, f is perturbed in this time-dependent case, the data assimilation problem behaves like a typical well posed problem, that is, the term

$$\|\delta q\|_{L^2(0,T;L^2(\omega))} + \|\delta f\|_{(0,-1)}$$

needs to be added on the right-hand side of the estimate in Theorem 7.4, but this time without any negative power of h. The proof is similar as in the stationary case and we omit it.

7.6 Conclusion

We have shown on some model problems how weakly consistent regularisation may be applied in the context of finite element approximation of ill-posed problems as a means to obtain approximations to the exact solution that are optimal with respect the approximation order of the finite element space and the (conditional) stability of the physical problem. We have only considered piecewise affine approximation here but the extension to high order polynomial approximation (and with associated enhanced accuracy for smooth solutions) is possible using the ideas from [13]. Ongoing work focuses on problems where the stability depends on the parameters of the physical problem in a more intricate way such as for the convection-diffusion equation or the Helmholtz equation. Further work will also address the extension to systems such as the linearised Navier-Stokes' equations.

References

1. Alessandrini, G., Rondi, L., Rosset, E., Vessella, S.: The stability for the Cauchy problem for elliptic equations. Inverse Prob. **25**(12), 123004, 47 (2009). http://dx.doi.org/10.1088/0266-5611/25/12/123004
2. Andrieux, S., Baranger, T.N., Ben Abda, A.: Solving Cauchy problems by minimizing an energy-like functional. Inverse Prob. **22**(1), 115–133 (2006). http://dx.doi.org/10.1088/0266-5611/22/1/007
3. Azaï ez, M., Ben Belgacem, F., El Fekih, H.: On Cauchy's problem. II. Completion, regularization and approximation. Inverse Prob. **22**(4), 1307–1336 (2006). http://dx.doi.org/10.1088/0266-5611/22/4/012
4. Badra, M., Caubet, F., Dardé, J.: Stability estimates for Navier-Stokes equations and application to inverse problems. Discrete Contin. Dyn. Syst. Ser. B **21**(8), 2379–2407 (2016). http://dx.doi.org/10.3934/dcdsb.2016052
5. Baumeister, J.: Stable Solution of Inverse Problems. Advanced Lectures in Mathematics. Friedr. Vieweg & Sohn, Braunschweig (1987). http://dx.doi.org/10.1007/978-3-322-83967-1
6. Ben Belgacem, F., Du, D.T., Jelassi, F.: Extended-domain-Lavrentiev's regularization for the Cauchy problem. Inverse Prob. **27**(4), 045005 (2011). http://dx.doi.org/10.1088/0266-5611/27/4/045005
7. Bourgeois, L.: A mixed formulation of quasi-reversibility to solve the Cauchy problem for Laplace's equation. Inverse Prob. **21**(3), 1087–1104 (2005). http://dx.doi.org/10.1088/0266-5611/21/3/018
8. Bourgeois, L., Dardé, J.: The "exterior approach" to solve the inverse obstacle problem for the Stokes system. Inverse Probl. Imaging **8**(1), 23–51 (2014). http://dx.doi.org/10.3934/ipi.2014.8.23
9. Brenner, S.C.: Poincaré-Friedrichs inequalities for piecewise H^1 functions. SIAM J. Numer. Anal. **41**(1), 306–324 (2003). http://dx.doi.org/10.1137/S0036142902401311

10. Brooks, A.N., Hughes, T.J.R.: Streamline upwind/Petrov-Galerkin formulations for convection dominated flows with particular emphasis on the incompressible Navier-Stokes equations. Comput. Methods Appl. Mech. Eng. **32**(1–3), 199–259 (1982). http://dx.doi.org/10.1016/0045-7825(82)90071-8. FENOMECH '81, Part I (Stuttgart, 1981)
11. Burman, E.: Stabilized finite element methods for nonsymmetric, noncoercive, and ill-posed problems. Part I: Elliptic equations. SIAM J. Sci. Comput. **35**(6), A2752–A2780 (2013). http://dx.doi.org/10.1137/130916862
12. Burman, E.: Error estimates for stabilized finite element methods applied to ill-posed problems. C. R. Math. Acad. Sci. Paris **352**(7–8), 655–659 (2014). http://dx.doi.org/10.1016/j.crma.2014.06.008
13. Burman, E.: Stabilised finite element methods for ill-posed problems with conditional stability. In: Building Bridges: Connections and Challenges in Modern Approaches to Numerical Partial Differential Equations. Lecture Notes in Computational Science and Engineering, vol. 114, pp. 93–127. Springer, Cham (2016)
14. Burman, E.: The elliptic Cauchy problem revisited: control of boundary data in natural norms. C. R. Math. **355**, 479–484 (2017). http://dx.doi.org/10.1016/j.crma.2017.02.014
15. Burman, E., Hansbo, P.: Edge stabilization for Galerkin approximations of convection-diffusion-reaction problems. Comput. Methods Appl. Mech. Eng. **193**(15–16), 1437–1453 (2004). http://dx.doi.org/10.1016/j.cma.2003.12.032
16. Burman, E., Hansbo, P.: Stabilized nonconforming finite element methods for data assimilation in incompressible flows. Math. Comp. **87**(311), 1029–1050 (2018).
17. Burman, E., Oksanen, L.: Data assimilation for the heat equation using stabilized finite element methods. Numer. Math. **139**(3), 505–528 (2018).
18. Burman, E., Oksanen, L., Ish-Horowicz, J.: Fully discrete finite element data assimilation method for the heat equation. ESAIM: Math. Model. Numer. Anal. (2018, in press). https://doi.org/10.1051/m2an/2018030
19. Burman, E., Hansbo, P., Larson, M.: Solving ill-posed control problems by stabilized finite element methods: an alternative to Tikhonov regularization. Inverse Problems **34**(3), (2018).
20. Cheng, J., Yamamoto, M.: One new strategy for a priori choice of regularizing parameters in Tikhonov's regularization. Inverse Prob. **16**(4), L31 (2000). http://stacks.iop.org/0266-5611/16/i=4/a=101
21. Dardé, J., Hannukainen, A., Hyvönen, N.: An H_{div}-based mixed quasi-reversibility method for solving elliptic Cauchy problems. SIAM J. Numer. Anal. **51**(4), 2123–2148 (2013). http://dx.doi.org/10.1137/120895123
22. Di Pietro, D.A., Ern, A.: Mathematical aspects of discontinuous Galerkin methods. In: Mathématiques & Applications (Berlin) [Mathematics & Applications], vol. 69. Springer, Heidelberg (2012). http://dx.doi.org/10.1007/978-3-642-22980-0
23. Èmanuilov, O.Y.: Controllability of parabolic equations. Math. Sb. **186**(6), 109–132 (1995). http://dx.doi.org/10.1070/SM1995v186n06ABEH000047
24. Engl, H.W., Hanke, M., Neubauer, A.: Regularization of inverse problems. In: Mathematics and Its Applications, vol. 375. Kluwer Academic Publishers Group, Dordrecht (1996). http://dx.doi.org/10.1007/978-94-009-1740-8
25. Ern, A., Guermond, J.L.: Theory and practice of finite elements. In: Applied Mathematical Sciences, vol. 159. Springer, New York (2004). http://dx.doi.org/10.1007/978-1-4757-4355-5
26. Guermond, J.L.: Stabilization of Galerkin approximations of transport equations by subgrid modeling. M2AN Math. Model. Numer. Anal. **33**(6), 1293–1316 (1999). http://dx.doi.org/10.1051/m2an:1999145
27. Johnson, C., Nävert, U., Pitkäranta, J.: Finite element methods for linear hyperbolic problems. Comput. Methods Appl. Mech. Eng. **45**(1–3), 285–312 (1984). http://dx.doi.org/10.1016/0045-7825(84)90158-0
28. Kabanikhin, S.I.: Definitions and examples of inverse and ill-posed problems. J. Inverse Ill-Posed Probl. **16**(4), 317–357 (2008). http://dx.doi.org/10.1515/JIIP.2008.019
29. Klibanov, M.V.: Carleman estimates for the regularization of ill-posed Cauchy problems. Appl. Numer. Math. **94**, 46–74 (2015). http://dx.doi.org/10.1016/j.apnum.2015.02.003

30. Kohn, R.V., Vogelius, M.: Relaxation of a variational method for impedance computed tomography. Commun. Pure Appl. Math. **40**(6), 745–777 (1987). http://dx.doi.org/10.1002/cpa.3160400605
31. Lattès, R., Lions, J.L.: Méthode de quasi-réversibilité et applications. Travaux et Recherches Mathématiques, No. 15. Dunod, Paris (1967)
32. Puel, J.P.: A nonstandard approach to a data assimilation problem and Tychonov regularization revisited. SIAM J. Control Optim. **48**(2), 1089–1111 (2009). http://dx.doi.org/10.1137/060670961
33. Scott, L.R., Zhang, S.: Finite element interpolation of nonsmooth functions satisfying boundary conditions. Math. Comput. **54**(190), 483–493 (1990). http://dx.doi.org/10.2307/2008497
34. Tikhonov, A.N., Arsenin, V.Y.: Solutions of Ill-Posed Problems. V. H. Winston & Sons, Washington; Wiley, New York (1977)
35. Yamamoto, M.: Carleman estimates for parabolic equations and applications. Inverse Prob. **25**(12), 123013 (2009). http://dx.doi.org/10.1088/0266-5611/25/12/123013

Chapter 8
Reduced Basis Approximation and A Posteriori Error Estimation: Applications to Elasticity Problems in Several Parametric Settings

Dinh Bao Phuong Huynh, Federico Pichi, and Gianluigi Rozza

Abstract In this work we consider (hierarchical, Lagrange) reduced basis approximation and a posteriori error estimation for elasticity problems in affinely parametrized geometries. The essential ingredients of the methodology are: a Galerkin projection onto a low-dimensional space associated with a smooth "parametric manifold"—dimension reduction; an efficient and effective greedy sampling methods for identification of optimal and numerically stable approximations—rapid convergence; an a posteriori error estimation procedures—rigorous and sharp bounds for the functional outputs related with the underlying solution or related quantities of interest, like stress intensity factor; and Offline-Online computational decomposition strategies—minimum *marginal cost* for high performance in the real-time and many-query (e.g., design and optimization) contexts. We present several illustrative results for linear elasticity problem in parametrized geometries representing 2D Cartesian or 3D axisymmetric configurations like an arc-cantilever beam, a center crack problem, a composite unit cell or a woven composite beam, a multi-material plate, and a closed vessel. We consider different parametrization for the systems: either physical quantities—to model the materials and loads—and geometrical parameters—to model different geometrical configurations—with isotropic and orthotropic materials working in plane stress and plane strain approximation. We would like to underline the versatility of the methodology in very different problems. As last example we provide a nonlinear setting with increased complexity.

D. B. P. Huynh
Akselos SA and Department of Mechanical Engineering, Massachusetts Institute of Technology, Cambridge, MA, USA

F. Pichi · G. Rozza (✉)
mathLab, Mathematics Area, SISSA, International School for Advanced Studies, Trieste, Italy
e-mail: fpichi@sissa.it; grozza@sissa.it

D. A. Di Pietro et al. (eds.), *Numerical Methods for PDEs*,
SEMA SIMAI Springer Series 15, https://doi.org/10.1007/978-3-319-94676-4_8

8.1 Introduction

In several fields, from continuum mechanics to fluid dynamics, we need to solve numerically very complex problems that arise from physics laws. Usually we model these phenomena through partial differential equations (PDEs) and we are interested in finding the field solution and also some other quantities that increase our knowledge on the system we are describing. Almost always we are not able to obtain an analytical solution, so we rely on some discretization techniques, like Finite Element (FE) or Finite Volume (FV), that furnish an approximation of the solution. We refer to this methods as the "truth" ones, because they require very high computational costs, especially in parametrized context. In fact if the problem depends on some physical or geometrical parameter, the *full-order* or *high-fidelity* model has to be solved many times and this might be quite demanding. Examples of typical applications of relevance are optimization, control, design, bifurcation detection and real time query. For this class of problems, we aim to replace the high-fidelity problem by one of much lower numerical complexity, through the *model order reduction* approach [12]. We focus on Reduced Basis (RB) method [3, 4, 17, 34, 35], which provides both *fast* and *reliable* evaluation of an input (parameter)-output relationship. The main features of this methodology are (1) those related to the classic Galerkin projection on which RB method is built upon (2) an a posteriori error estimation which provides sharp and rigorous bounds and (3) offline/online computational strategy which allows rapid computation. The goal of this chapter is to present a very efficient a posteriori error estimation for linear elasticity parametrized problem. We show many different configurations and settings, by applying RB method to approximate problems using plane stress and plane strain formulation and to deal both with isotropic and orthotropic materials. We underline that the setting for very different problems is the same and unique.

This work is organized as follows. In Sect. 8.2, we first present a "unified" linear elasticity formulation; we then briefly introduce the geometric mapping strategy based on domain decomposition; we end the Section with the affine decomposition forms and the definition of the "truth" approximation, which we shall build our RB approximation upon. In Sect. 8.3, we present the RB methodology and the offline-online computational strategy for the RB "compliant" output. In Sect. 8.4, we define our a posteriori error estimators for our RB approach, and provide the computation procedures for the two ingredients of our error estimators, which are the dual norm of the residual and the coercivity lower bound. In Sect. 8.5, we briefly discuss the extension of our RB methodology to the "non-compliant" output. In Sect. 8.6, we show several numerical results to illustrate the capability of this method, with a final subsection devoted to provide an introduction to more complex nonlinear problems. Finally, in Sect. 8.7, we draw discussions and news on future works.

8.2 Preliminaries

In this Section we shall first present a "unified" formulation for all the linear elasticity cases—for isotropic and orthotropic materials, 2D Cartesian and 3D axisymmetric configurations—we consider in this study. We then introduce a domain decomposition and geometric mapping strategy to recast the formulation in the "affine forms", which is a crucial requirement for our RB approximation. Finally, we define the "truth" finite element approximation, upon which we shall build the RB approximation, introduced in the next Section.

8.2.1 Formulation on the "Original" Domain

8.2.1.1 Isotropic/Orthotropic Materials

We first briefly describe our problem formulation based on the original settings (denoted by a superscript ^{o}). We consider a solid body in two dimensions $\Omega^{\mathrm{o}}(\boldsymbol{\mu}) \in \mathbb{R}^2$ with boundary Γ^{o}, where $\boldsymbol{\mu} \in \mathcal{D} \subset \mathbb{R}^P$ is the input parameter and $\mathcal{D}$ is the parameter domain [38, 39]. For the sake of simplicity, in this section, we assume implicitly that any "original" quantities (stress, strain, domains, boundaries, etc.) with superscript ^{o} will depend on the input parameter $\boldsymbol{\mu}$, e.g. $\Omega^{\mathrm{o}} \equiv \Omega^{\mathrm{o}}(\boldsymbol{\mu})$.

We first make the following assumptions: (1) the solid is free of body forces, (2) there are negligible thermal strains; note that the extension to include either or both body forces/thermal strains is straightforward. Let us denote u^{o} as the displacement field, and the spatial coordinate $\mathbf{x}^{\mathrm{o}} = (x_1^{\mathrm{o}}, x_2^{\mathrm{o}})$, the linear elasticity equilibrium reads

$$\frac{\partial \sigma_{ij}^{\mathrm{o}}}{\partial x_j^{\mathrm{o}}} = 0, \quad \text{in } \Omega^{\mathrm{o}} \tag{8.1}$$

where σ^{o} denotes the stresses, which are related to the strains ε^{o} by

$$\sigma_{ij}^{\mathrm{o}} = C_{ijkl}\varepsilon_{kl}^{\mathrm{o}}, \quad 1 \le i, j, k, l \le 2$$

where

$$\varepsilon_{kl}^{\mathrm{o}} = \frac{1}{2}\left(\frac{\partial u_k^{\mathrm{o}}}{\partial x_l^{\mathrm{o}}} + \frac{\partial u_l^{\mathrm{o}}}{\partial x_k^{\mathrm{o}}}\right),$$

$u^o = (u_1^o, u_2^o)$ is the displacement and C_{ijkl} is the elastic tensor, which can be expressed in a matrix form as

$$[\mathbf{C}] = \begin{bmatrix} C_{1111} & C_{1112} & C_{1121} & C_{1122} \\ C_{1211} & C_{1212} & C_{1221} & C_{1222} \\ C_{2111} & C_{2112} & C_{2121} & C_{2122} \\ C_{2211} & C_{2212} & C_{2221} & C_{2222} \end{bmatrix} = [\mathbf{B}]^T[\mathbf{E}][\mathbf{B}],$$

where

$$[\mathbf{B}] = \begin{bmatrix} 1 & 0 & 0 & 0 \\ 0 & 0 & 0 & 1 \\ 0 & 1 & 1 & 0 \end{bmatrix} \quad [\mathbf{E}] = \begin{bmatrix} c_{11} & c_{12} & 0 \\ c_{21} & c_{21} & 0 \\ 0 & 0 & c_{33} \end{bmatrix}.$$

The matrix $[\mathbf{E}]$ varies for different material types and is given in the Appendix.

We next consider Dirichlet boundary conditions for both components of u^o:

$$u_i^o = 0 \quad \text{on} \quad \Gamma_{D,i}^o,$$

and Neumann boundary conditions:

$$\sigma_{ij}^o e_{n,j}^o = \begin{cases} f_n^o e_{n,i}^o & \text{on} \quad \Gamma_N^o \\ 0 & \text{on} \ \Gamma^o \backslash \Gamma_N^o \end{cases}$$

where f_n^o is the specified stress on boundary edge Γ_N^o respectively; and $\mathbf{e}_n^o = [e_{n,1}^o, e_{n,2}^o]$ is the unit normal on Γ_N^o. Zero value of f_n^o indicate free stress (homogeneous Neumann conditions) on a specific boundary. Here we only consider homogeneous Dirichlet boundary conditions, but extensions to non-homogeneous Dirichlet boundary conditions and/or nonzero traction Neumann boundary conditions are simple and straightforward.

We then introduce the functional space

$$X^o = \{v = (v_1, v_2) \in (H^1(\Omega^o))^2 \mid v_i = 0 \text{ on } \Gamma_{D,i}^o, i = 1, 2\},$$

here $H^1(\Omega^o) = \{v \in L^2(\Omega^o) \mid \nabla v \in (L^2(\Omega^o))^2\}$ and $L^2(\Omega^o)$ is the space of square-integrable functions over Ω^o. By multiplying (8.1) by a test function $v \in X^o$ and integrating by part over Ω^o we obtain the weak form

$$\int_{\Omega^o} \frac{\partial v_i}{\partial x_j^o} C_{ijkl} \frac{\partial u_k^o}{\partial x_l^o} d\Omega^o = \int_{\Gamma_N^o} f_n^o e_{n,j}^o v_j d\Gamma^o. \tag{8.2}$$

Finally, we define our output of interest, which usually is a measurement (of our displacement field or even equivalent derived solutions such as stresses, strains) over a boundary segment Γ_L^o or a part of the domain Ω_L^o. Here we just consider a simple case,

$$s^o(\boldsymbol{\mu}) = \int_{\Gamma_L^o} f_{\ell,i}^o u_i^o d\Gamma^o, \tag{8.3}$$

i.e. the measure of the displacement on either or both x_1^o and x_2^o direction along Γ_L^o with multipliers $f_{\ell,i}^o$; more general forms for the output of interest can be extended straightforward. Note that our output of interest is a linear function of the displacement; extension to quadratic function outputs can be found in [20].

We can then now recover our abstract statement: Given a $\boldsymbol{\mu} \in \mathcal{D}$, we evaluate

$$s^o(\boldsymbol{\mu}) = \ell^o(u^o; \boldsymbol{\mu}),$$

where $u^o \in X^o$ satisfies

$$a^o(u^o, v; \boldsymbol{\mu}) = f^o(v; \boldsymbol{\mu}), \quad \forall v \in X^o.$$

Here $a^o(w, v; \boldsymbol{\mu}) : X^o \times X^o \to \mathbb{R}, \forall w, v \in X^o$ is the symmetric and positive bilinear form associated to the left hand side term of (8.2); $f^o(v; \boldsymbol{\mu}) : X^o \to \mathbb{R}$ and $\ell^o(v; \boldsymbol{\mu}) : X^o \to \mathbb{R}, \forall v \in X^o$ are the linear forms associated to the right hand side terms of (8.2) and (8.3), respectively. It shall be proven convenience to recast $a^o(\cdot, \cdot; \boldsymbol{\mu})$, $f^o(\cdot; \boldsymbol{\mu})$ and $\ell^o(\cdot; \boldsymbol{\mu})$ in the following forms

$$a^o(w, v; \boldsymbol{\mu}) = \int_{\Omega^o} \left[\frac{\partial w_1}{\partial x_1^o} \; \frac{\partial w_1}{\partial x_2^o} \; \frac{\partial w_2}{\partial x_1^o} \; \frac{\partial w_2}{\partial x_2^o} \; w_1 \right] [\mathbf{S}^a] \begin{bmatrix} \frac{\partial v_1}{\partial x_1^o} \\ \frac{\partial v_1}{\partial x_2^o} \\ \frac{\partial v_2}{\partial x_1^o} \\ \frac{\partial v_2}{\partial x_2^o} \\ v_1 \end{bmatrix} d\Omega^o, \; \forall w, v \in X^o, \tag{8.4}$$

$$f^o(v; \boldsymbol{\mu}) = \int_{\Gamma_N^o} [\mathbf{S}^f] \begin{bmatrix} v_1 \\ v_2 \end{bmatrix} d\Gamma^o, \quad \forall v \in X^o, \tag{8.5}$$

$$\ell^o(v; \boldsymbol{\mu}) = \int_{\Gamma_L^o} [\mathbf{S}^\ell] \begin{bmatrix} v_1 \\ v_2 \end{bmatrix} d\Gamma^o, \quad \forall v \in X^o, \tag{8.6}$$

where $[\mathbf{S}^a] \in \mathbb{R}^{5\times 5}$; $[\mathbf{S}^f] \in \mathbb{R}^2$ and $[\mathbf{S}^\ell] \in \mathbb{R}^2$ are defined as

$$[\mathbf{S}^a] = \begin{bmatrix} [\mathbf{C}] & [\mathbf{0}]^{4\times 1} \\ [\mathbf{0}]^{1\times 4} & 0 \end{bmatrix}, \quad [\mathbf{S}^f] = \begin{bmatrix} f_n^{\text{o}} e_{n,1}^{\text{o}} & f_n^{\text{o}} e_{n,2}^{\text{o}} \end{bmatrix}, \quad [\mathbf{S}^\ell] = \begin{bmatrix} f_{\ell,1}^{\text{o}} & f_{\ell,2}^{\text{o}} \end{bmatrix}.$$

8.2.1.2 Axisymmetric

Now we shall present the problem formulation for the axisymmetric case. In a cylindrical coordinate system (r, z, θ),[1] the elasticity equilibrium reads

$$\frac{\partial \sigma_{rr}^{\text{o}}}{\partial r} + \frac{\partial \sigma_{zr}^{\text{o}}}{\partial z} + \frac{\sigma_{rr}^{\text{o}} - \sigma_{\theta\theta}^{\text{o}}}{r} = 0, \quad \text{in} \quad \Omega^{\text{o}}$$

$$\frac{\partial \sigma_{rz}^{\text{o}}}{\partial r} + \frac{\partial \sigma_{zz}^{\text{o}}}{\partial z} + \frac{\sigma_{rz}^{\text{o}}}{r} = 0, \quad \text{in} \quad \Omega^{\text{o}}$$

where $\sigma_{rr}^{\text{o}}, \sigma_{zz}^{\text{o}}, \sigma_{rz}^{\text{o}}, \sigma_{\theta\theta}^{\text{o}}$ are the stress components given by

$$\begin{bmatrix} \sigma_{rr}^{\text{o}} \\ \sigma_{zz}^{\text{o}} \\ \sigma_{\theta\theta}^{\text{o}} \\ \sigma_{rz}^{\text{o}} \end{bmatrix} = \underbrace{\frac{E}{(1+\nu)(1-2\nu)} \begin{bmatrix} (1-\nu) & \nu & \nu & 0 \\ \nu & (1-\nu) & \nu & 0 \\ \nu & \nu & (1-\nu) & 0 \\ 0 & 0 & 0 & \dfrac{1-2\nu}{2} \end{bmatrix}}_{[\mathbf{E}]} \begin{bmatrix} \varepsilon_{rr}^{\text{o}} \\ \varepsilon_{zz}^{\text{o}} \\ \varepsilon_{\theta\theta}^{\text{o}} \\ \varepsilon_{rz}^{\text{o}} \end{bmatrix},$$

where E and ν are the axial Young's modulus and Poisson ratio, respectively. We only consider isotropic material, however, extension to general to anisotropic material is possible; as well as axisymmetric plane stress and plane strain [44]. The strain $\varepsilon_{rr}^{\text{o}}, \varepsilon_{zz}^{\text{o}}, \varepsilon_{rz}^{\text{o}}, \varepsilon_{\theta\theta}^{\text{o}}$ are given by

$$\begin{bmatrix} \varepsilon_{rr}^{\text{o}} \\ \varepsilon_{zz}^{\text{o}} \\ \varepsilon_{\theta\theta}^{\text{o}} \\ \varepsilon_{rz}^{\text{o}} \end{bmatrix} = \begin{bmatrix} \dfrac{\partial u_r^{\text{o}}}{\partial r} \\ \dfrac{\partial u_z^{\text{o}}}{\partial z} \\ \dfrac{u_r^{\text{o}}}{r} \\ \dfrac{\partial u_r^{\text{o}}}{\partial z} + \dfrac{\partial u_z^{\text{o}}}{\partial r} \end{bmatrix}, \tag{8.7}$$

where $u_r^{\text{o}}, u_z^{\text{o}}$ are the radial displacement and axial displacement, respectively.

[1] For the sake of simple illustration, we omit the "original" superscript ^{o} on (r, z, θ).

Assuming that the axial axis is $x_2^{\rm o}$, let $[u_1^{\rm o}, u_2^{\rm o}] \equiv [\frac{u_r^{\rm o}}{r}, u_z^{\rm o}]$ and denoting $[x_1^{\rm o}, x_2^{\rm o}, x_3^{\rm o}] \equiv [r, z, \theta]$, we can then express (8.7) as

$$\begin{bmatrix} \varepsilon_{11}^{\rm o} \\ \varepsilon_{22}^{\rm o} \\ \varepsilon_{33}^{\rm o} \\ \varepsilon_{12}^{\rm o} \end{bmatrix} = [\hat{\mathbf{E}}] \underbrace{\begin{bmatrix} x_1^{\rm o} & 0 & 0 & 0 & 1 \\ 0 & 0 & 0 & 1 & 0 \\ 0 & 0 & 0 & 0 & 1 \\ 0 & x_1^{\rm o} & 1 & 0 & 0 \end{bmatrix}}_{[\mathbf{B}_a]} \begin{bmatrix} \frac{\partial u_1^{\rm o}}{\partial x_1^{\rm o}} \\ \frac{\partial u_1^{\rm o}}{\partial x_2^{\rm o}} \\ \frac{\partial u_2^{\rm o}}{\partial x_1^{\rm o}} \\ \frac{\partial u_2^{\rm o}}{\partial x_2^{\rm o}} \\ u_1^{\rm o} \end{bmatrix}.$$

As in the previous case, we consider the usual homogeneous Dirichlet boundary conditions on $\Gamma_{D,i}^{\rm o}$ and Neumann boundary conditions on $\Gamma^{\rm o}$. Then if we consider the output of interest $s^{\rm o}(\boldsymbol{\mu})$ defined upon $\Gamma_L^{\rm o}$, we arrive at the same abstract statement where

$$[\mathbf{S}^a] = x_1^{\rm o}[\mathbf{B}_a]^T[\mathbf{E}][\mathbf{B}_a], \ [\mathbf{S}^f] = \left[(x_1^{\rm o})^2 f_n^{\rm o} e_{n,1}^{\rm o} \ x_1^{\rm o} f_n^{\rm o} e_{n,2}^{\rm o} \right],$$

$$[\mathbf{S}^\ell] = \left[x_1^{\rm o} f_n^{\rm o} e_{n,1}^{\rm o} \ f_n^{\rm o} e_{n,2}^{\rm o} \right].$$

Note that the $x_1^{\rm o}$ multipliers appear in $[\mathbf{S}^f]$ during the weak form derivation, while in $[\mathbf{S}^\ell]$, in order to retrieve the measurement for the axial displacement $u_r^{\rm o}$ rather than $u_1^{\rm o}$ due to the change of variables. Also, the 2π multipliers in both $a^{\rm o}(\cdot, \cdot; \boldsymbol{\mu})$ and $f^{\rm o}(\cdot; \boldsymbol{\mu})$ are disappeared in the weak form during the derivation, and can be included in $\ell^{\rm o}(\cdot; \boldsymbol{\mu})$, i.e. incorporated to $[\mathbf{S}^\ell]$ if measurement is required to be done in truth (rather than in the axisymmetric) domain.

8.2.2 Formulation on Reference Domain

The RB requires that the computational domain must be parameter-independent; however, our "original" domain $\Omega^{\rm o}(\boldsymbol{\mu})$ is obviously parameter-dependent. Hence, to transform $\Omega^{\rm o}(\boldsymbol{\mu})$ into the computational domain, or "reference" (parameter-independent) domain Ω, we must perform geometric transformations in order to express the bilinear and linear forms in our abstract statement in appropriate "affine forms". This "affine forms" formulation allows us to model all possible configurations, corresponding to every $\boldsymbol{\mu} \in \mathcal{D}$, based on a single reference-domain [34, 36].

8.2.2.1 Geometry Mappings

We first assume that, for all $\boldsymbol{\mu} \in \mathcal{D}$, $\Omega^{\mathrm{o}}(\boldsymbol{\mu})$ is expressed as

$$\Omega^{\mathrm{o}}(\boldsymbol{\mu}) = \bigcup_{s=1}^{L_{\mathrm{reg}}} \Omega_s^{\mathrm{o}}(\boldsymbol{\mu}),$$

where the $\Omega_s^{\mathrm{o}}(\boldsymbol{\mu})$, $s = 1, \ldots, L_{\mathrm{reg}}$ are mutually non-overlapping subdomains. In two dimensions, $\Omega_s^{\mathrm{o}}(\boldsymbol{\mu})$, $s = 1, \ldots, L_{\mathrm{reg}}$ is a set of triangles (or in the general case, a set of "curvy triangles"[2] [22].) such that all important domains/edges (those defining different material regions, boundaries, pressures/tractions loaded boundary segments, or boundaries which the output of interests are calculated upon) are included in the set. In practice, such a set is generated by a constrained Delaunay triangulation.

We next assume that there exists a reference domain $\Omega (\equiv \Omega^{\mathrm{o}}(\boldsymbol{\mu}_{\mathrm{ref}}) = \bigcup_{s=1}^{L_{\mathrm{reg}}} \Omega_s$ where, for any $\mathbf{x}^{\mathrm{o}} \in \Omega_s$, $s = 1, \ldots, L_{\mathrm{reg}}$, its image $\mathbf{x}^{\mathrm{o}} \in \Omega_s^{\mathrm{o}}(\boldsymbol{\mu})$ is given by

$$\mathbf{x}^{\mathrm{o}}(\boldsymbol{\mu}) = \mathcal{T}_s^{\mathrm{aff}}(\boldsymbol{\mu}; \mathbf{x}) = [\mathbf{R}_s^{\mathrm{aff}}(\boldsymbol{\mu})]\mathbf{x} + [\mathbf{G}_s^{\mathrm{aff}}(\boldsymbol{\mu})], \tag{8.8}$$

where $[\mathbf{R}_s^{\mathrm{aff}}(\boldsymbol{\mu})] \in \mathbb{R}^{2\times 2}$ and $[\mathbf{G}_s^{\mathrm{aff}}(\boldsymbol{\mu})] \in \mathbb{R}^2$. It thus follows from our definitions that $\mathcal{T}_s(\boldsymbol{\mu}; \mathbf{x}) : \Omega_s \to \Omega_s^{\mathrm{o}}$, $1 \le s \le L_{\mathrm{reg}}$ is an (invertible) affine mapping from Ω_s to $\Omega_s^{\mathrm{o}}(\boldsymbol{\mu})$, hence the Jacobian $|\det([\mathbf{R}_s^{\mathrm{aff}}(\boldsymbol{\mu})])|$ is strictly positive, and that the derivative transformation matrix, $[\mathbf{D}_s^{\mathrm{aff}}(\boldsymbol{\mu})] = [\mathbf{R}_s^{\mathrm{aff}}(\boldsymbol{\mu})]^{-1}$ is well defined. We thus can write

$$\frac{\partial}{\partial x_i^{\mathrm{o}}} = \frac{\partial x_j}{\partial x_i^{\mathrm{o}}} \frac{\partial}{\partial x_j} = D_{s,ij}^{\mathrm{aff}}(\boldsymbol{\mu}) \frac{\partial}{\partial x_j}, \quad 1 \le i, j \le 2. \tag{8.9}$$

As in two dimensions, an affine transformation maps a triangle to a triangle, we can readily calculate $[\mathbf{R}_s^{\mathrm{aff}}(\boldsymbol{\mu})]$ and $[\mathbf{G}_s^{\mathrm{aff}}(\boldsymbol{\mu})]$ for each subdomains s by simply solving a systems of six equations forming from (8.8) by matching parametrized coordinates to reference coordinates for the three triangle vertices.

We further require a mapping continuity condition: for all $\boldsymbol{\mu} \in \mathcal{D}$,

$$\mathcal{T}_s(\boldsymbol{\mu}; \mathbf{x}) = \mathcal{T}_{s'}(\boldsymbol{\mu}; \mathbf{x}), \quad \forall \mathbf{x} \in \Omega_s \cap \Omega_{s'}, \quad 1 \le s, s' \le L_{\mathrm{reg}}.$$

This condition is automatically held if there is no curved edge in the set of $\Omega_s^{\mathrm{o}}(\boldsymbol{\mu})$. If a domain contains one or more "important" curved edge, special "curvy triangles" must be generated appropriately to honour the continuity condition. We refer the readers to [36] for the full discussion and detail algorithm for such cases.

---

[2] In fact, a "curvy triangle" [36] is served as the building block. For its implementation see [22].

The global transformation is for $\mathbf{x} \in \Omega$, the image $\mathbf{x}^{o} \in \Omega^{o}(\boldsymbol{\mu})$ is given by

$$\mathbf{x}^{o}(\boldsymbol{\mu}) = \mathcal{T}(\boldsymbol{\mu}; \mathbf{x}).$$

It thus follows that $\mathcal{T}(\boldsymbol{\mu}; \mathbf{x}) : \Omega \to \Omega^{o}(\boldsymbol{\mu})$ is a piecewise-affine geometric mapping.

8.2.2.2 Affine Forms

We now define our functional space X as

$$X = \{v = (v_1, v_2) \in (H^1(\Omega))^2 | v_i = 0 \text{ on } \Gamma_{D,i}, i = 1, 2\},$$

and recast our bilinear form $a^{o}(w, v; \boldsymbol{\mu})$, by invoking (8.4), (8.8) and (8.9) to obtain $\forall w, v \in X(\Omega)$

$$a(w, v; \boldsymbol{\mu}) = \int_{\bigcup_{s=1}^{L_{\text{reg}}} \Omega_s} \begin{bmatrix} \frac{\partial w_1}{\partial x_1} & \frac{\partial w_1}{\partial x_2} & \frac{\partial w_2}{\partial x_1} & \frac{\partial w_2}{\partial x_2} & w_1 \end{bmatrix} [\mathbf{S}_s^{a,\text{aff}}(\boldsymbol{\mu})] \begin{bmatrix} \frac{\partial v_1}{\partial x_1} \\ \frac{\partial v_1}{\partial x_2} \\ \frac{\partial v_2}{\partial x_1} \\ \frac{\partial v_2}{\partial x_2} \\ v_1 \end{bmatrix} d\Omega.$$

where $[\mathbf{S}_s^{a,\text{aff}}(\boldsymbol{\mu})] = [\mathbf{H}_s(\boldsymbol{\mu})][\mathbf{S}_s^a][\mathbf{H}_s(\boldsymbol{\mu})]^T |\det([\mathbf{R}_s^{\text{aff}}(\boldsymbol{\mu})])|$ is the effective elastic tensor matrix, in which

$$[\mathbf{H}_s(\boldsymbol{\mu})] = \begin{pmatrix} [\mathbf{D}_s(\boldsymbol{\mu})] & [\mathbf{0}]^{2\times 2} & 0 \\ [\mathbf{0}]^{2\times 2} & [\mathbf{D}_s(\boldsymbol{\mu})] & 0 \\ 0 & 0 & 1 \end{pmatrix}.$$

Similarly, the linear form $f^{o}(v; \boldsymbol{\mu}), \forall v \in X$ can be transformed as

$$f(v; \boldsymbol{\mu}) = \int_{\bigcup_{s=1}^{L_{\text{reg}}} \Gamma_{N_s}} [\mathbf{S}_s^{f,\text{aff}}] \begin{bmatrix} v_1 \\ v_2 \end{bmatrix} d\Gamma,$$

where Γ_{N_s} denotes the partial boundary segment of Γ_N of the subdomain Ω_s and $[\mathbf{S}_s^{f,\text{aff}}] = \|([\mathbf{R}_s(\boldsymbol{\mu})]\mathbf{e}_n)\|_2[\mathbf{S}^f]$ is the effective load vector, where $\mathbf{e}_n$ is the normal vector to Γ_{N_s} and $\|\cdot\|_2$ denotes the usual Euclidean norm. The linear form $\ell(v; \boldsymbol{\mu})$ is also transformed in the same manner.

We then replace all "original" x_1^o and x_2^o in the effective elastic tensor matrix $[\mathbf{S}_s^{a,\text{aff}}(\boldsymbol{\mu})]$, effective load/output vectors $[\mathbf{S}_s^{f,\text{aff}}(\boldsymbol{\mu})]$ and $[\mathbf{S}_s^{\ell,\text{aff}}(\boldsymbol{\mu})]$ by (8.8) to obtain a $\mathbf{x}^o$-free effective elastic tensor matrix and effective load/output vectors, respectively,[3] in certain conditions) can be a polynomial function of the spatial coordinates $\mathbf{x}^o$ as well, and we still be able to obtain our affine forms (8.12).

We next expand the bilinear form $a(w, v; \boldsymbol{\mu})$ by treating each entry of the effective elastic tensor matrix for each subdomain separately, namely

$$a(w, v; \boldsymbol{\mu}) = S_{1,11}^{a,\text{aff}}(\boldsymbol{\mu}) \int_{\Omega_1} \frac{\partial w_1}{\partial x_1}\frac{\partial v_1}{\partial x_1} + S_{1,12}^{a,\text{aff}}(\boldsymbol{\mu}) \int_{\Omega_1} \frac{\partial w_1}{\partial x_1}\frac{\partial v_1}{\partial x_2} + \dots \tag{8.10}$$

$$+ S_{L_{\text{reg}},55}^{a,\text{aff}}(\boldsymbol{\mu}) \int_{\Omega_{L_{\text{reg}}}} w_1 w_1. \tag{8.11}$$

Note that here for simplicity, we consider the case where there is no spatial coordinates in $[\mathbf{S}_s^{\ell,\text{aff}}(\boldsymbol{\mu})]$. In general (especially for axisymmetric case), some or most of the integrals may take the form of $\int_{\Omega_s} (x_1)^m (x_2)^n \frac{\partial w_i}{\partial x_j}\frac{\partial v_k}{\partial x_l}$, where $m, n \in \mathbb{R}$.

Taking into account the symmetry of the bilinear form and the effective elastic tensor matrix, there will be at most $Q^a = 7L_{\text{reg}}$ terms in the expansion. However, in practice, most of the terms can be collapsed by noticing that not only there will be a lot of zero entries in $[\mathbf{S}_s^{a,\text{aff}}(\boldsymbol{\mu})]$, $s = 1, \dots, L_{\text{reg}}$, but also there will be a lot of duplicated or "linearly dependent" entries, for example, $S_{1,11}^{a,\text{aff}}(\boldsymbol{\mu}) = [\text{Const}]S_{2,11}^{a,\text{aff}}(\boldsymbol{\mu})$. We can then apply a symbolic manipulation technique [36] to identify, eliminate all zero terms in (8.10) and collapse all "linear dependent" terms to end up with a minimal Q^a expansion. The same procedure is also applied for the linear forms $f(\cdot; \boldsymbol{\mu})$ and $\ell(\cdot; \boldsymbol{\mu})$.

Hence the abstract formulation of the linear elasticity problem in the reference domain Ω reads as follow: given $\boldsymbol{\mu} \in \mathcal{D}$, find

$$s(\boldsymbol{\mu}) = \ell(u(\boldsymbol{\mu}); \boldsymbol{\mu}),$$

where $u(\boldsymbol{\mu}) \in X$ satisfies

$$a(u(\boldsymbol{\mu}), v; \boldsymbol{\mu}) = f(v; \boldsymbol{\mu}), \quad \forall v \in X,$$

[3]Here we note that, the Young's modulus E in the isotropic and axisymmetric cases or E_1, E_2 and E_3 in the orthotropic case only.

where all the bilinear and linear forms are in affine forms,

$$a(w, v; \boldsymbol{\mu}) = \sum_{q=1}^{Q^a} \Theta_q^a(\boldsymbol{\mu}) a_q(w, v),$$

$$f(v; \boldsymbol{\mu}) = \sum_{q=1}^{Q^f} \Theta_q^f(\boldsymbol{\mu}) f_q(v),$$

$$\ell(v; \boldsymbol{\mu}) = \sum_{q=1}^{Q^\ell} \Theta_q^\ell(\boldsymbol{\mu}) \ell_q(v), \quad \forall w, v, \in X. \tag{8.12}$$

Here $\Theta_q^a(\boldsymbol{\mu})$, $a_q(w, v)$, $q = 1, \ldots, Q^a$, $f_q(v)$; $\Theta_q^f(\boldsymbol{\mu})$, $f_q(v)$, $q = 1, \ldots, Q^f$, and $\Theta_q^\ell(\boldsymbol{\mu})$, $\ell_q(v)$, $q = 1, \ldots, Q^\ell$ are parameter-dependent coefficient and parameter-independent bilinear and linear forms, respectively.

We close this section by defining several useful terms. We first define our inner product and energy norm as

$$(w, v)_X = a(w, v; \overline{\boldsymbol{\mu}}) \tag{8.13}$$

and $\|w\|_X = (w, w)^{1/2}$, $\forall w, v \in X$, respectively, where $\overline{\boldsymbol{\mu}} \in \mathcal{D}$ is an arbitrary parameter. Certain other inner norms and associated norms are also possible [36]. We then define our coercivity and continuity constants as

$$\alpha(\boldsymbol{\mu}) = \inf_{w \in X} \frac{a(w, v; \boldsymbol{\mu})}{\|w\|_X^2}, \tag{8.14}$$

$$\gamma(\boldsymbol{\mu}) = \sup_{w \in X} \frac{a(w, v; \boldsymbol{\mu})}{\|w\|_X^2}, \tag{8.15}$$

respectively. We assume that $a(\cdot, \cdot; \boldsymbol{\mu})$ is symmetric, $a(w, v; \boldsymbol{\mu}) = a(v, w; \boldsymbol{\mu})$, $\forall w, v \in X$, coercive, $\alpha(\boldsymbol{\mu}) > \alpha_0 > 0$, and continuous, $\gamma(\boldsymbol{\mu}) < \gamma_0 < \infty$; and also our $f(\cdot; \boldsymbol{\mu})$ and $\ell(\cdot; \boldsymbol{\mu})$ are bounded functionals. It follows that problem which is well-defined and has a unique solution. Those conditions are automatically satisfied given the nature of our considered problems [38, 39].

8.2.3 Truth Approximation

From now on, we shall restrict our attention to the "compliance" case ($f(\cdot; \boldsymbol{\mu}) = \ell(\cdot; \boldsymbol{\mu})$). Extension to the non-compliance case will be discuss in the Sect. 8.5.

We now apply the finite element method and we provide a matrix formulation [37]: given $\boldsymbol{\mu} \in \mathcal{D}$, we evaluate

$$s(\boldsymbol{\mu}) = [\mathbf{F}^{\mathcal{N}}(\boldsymbol{\mu})]^T [\mathbf{u}^{\mathcal{N}}(\boldsymbol{\mu})], \tag{8.16}$$

where $[\mathbf{u}^{\mathcal{N}}(\boldsymbol{\mu})]$ represents a finite element solution $u^{\mathcal{N}}(\boldsymbol{\mu}) \in X^{\mathcal{N}} \in X$ of size $\mathcal{N}$ which satisfies

$$[\mathbf{K}^{\mathcal{N}}(\boldsymbol{\mu})][\mathbf{u}^{\mathcal{N}}(\boldsymbol{\mu})] = [\mathbf{F}^{\mathcal{N}}(\boldsymbol{\mu})]; \tag{8.17}$$

here $[\mathbf{K}^{\mathcal{N}}(\boldsymbol{\mu})]$, and $[\mathbf{F}^{\mathcal{N}}(\boldsymbol{\mu})]$ and the (discrete forms) stiffness matrix and load vector of $a(\cdot, \cdot; \boldsymbol{\mu})$, and $f(\cdot; \boldsymbol{\mu})$, respectively. Note that the stiffness matrix $[\mathbf{K}^{\mathcal{N}}(\boldsymbol{\mu})]$ is symmetric positive definite (SPD). By invoking the affine forms (8.12), we can express $[\mathbf{K}^{\mathcal{N}}(\boldsymbol{\mu})]$, and $[\mathbf{F}^{\mathcal{N}}(\boldsymbol{\mu})]$ as

$$[\mathbf{K}^{\mathcal{N}}(\boldsymbol{\mu})] = \sum_{q=1}^{Q^a} \Theta_q^a(\boldsymbol{\mu})[\mathbf{K}_q^{\mathcal{N}}],$$

$$\left[\mathbf{F}^{\mathcal{N}}(\boldsymbol{\mu})\right] = \sum_{q=1}^{Q^f} \Theta_q^f(\boldsymbol{\mu})[\mathbf{F}_q^{\mathcal{N}}], \tag{8.18}$$

where $[\mathbf{K}_q^{\mathcal{N}}]$, $[\mathbf{F}_q^{\mathcal{N}}]$ and are the discrete forms of the parameter-independent bilinear and linear forms $a_q(\cdot, \cdot)$ and $f_q(\cdot)$, respectively. We also denote (the SPD matrix) $[\mathbf{Y}^{\mathcal{N}}]$ as the discrete form of our inner product (8.13). We also assume that the size of our FE approximation, $\mathcal{N}$ is large enough such that our FE solution is an accurate approximation of the exact solution.

8.3 Reduced Basis Method

In this Section we shall restrict our attention by recalling the RB method for the "compliant" output. We shall first define the RB spaces and the Galerkin projection. We then describe an Offline-Online computational strategy, which allows us to obtain $\mathcal{N}$-independent calculation of the RB output approximation [17, 26].

8.3.1 RB Spaces and the Greedy Algorithm

To define the RB approximation we first introduce a (nested) Lagrangian parameter sample for $1 \leq N \leq N_{\max}$,

$$S_N = \{\boldsymbol{\mu}_1, \boldsymbol{\mu}_2, \ldots, \boldsymbol{\mu}_N\},$$

and associated hierarchical reduced basis spaces $(X_N^{\mathcal{N}} =) W_N^{\mathcal{N}}$, $1 \leq N \leq N_{\max}$,

$$W_N^{\mathcal{N}} = \text{span}\{u^{\mathcal{N}}(\boldsymbol{\mu}_n), 1 \leq n \leq N\},$$

where $\boldsymbol{\mu}_n \in \mathcal{D}$ are determined by the means of a Greedy sampling algorithm [35, 36]; this is an iterative procedure where at each step a new basis function is added in order to improve the precision of the basis set.

The key point of this methodology is the availability of an estimate of the error induced by replacing the full space $X^{\mathcal{N}}$ with the reduced order one $W_N^{\mathcal{N}}$ in the variational formulation. More specifically we assume that for all $\boldsymbol{\mu} \in \mathcal{D}$ there exist an estimator $\eta(\boldsymbol{\mu})$ such that

$$||u^{\mathcal{N}}(\boldsymbol{\mu}) - u^{\mathcal{N}}_{\text{RB},N}(\boldsymbol{\mu})|| \leq \eta(\boldsymbol{\mu}),$$

where $u^{\mathcal{N}}(\boldsymbol{\mu}) \in X^{\mathcal{N}} \in X$ represents the finite element solution, $u^{\mathcal{N}}_{\text{RB},N}(\boldsymbol{\mu}) \in X_N^{\mathcal{N}} \subset X^{\mathcal{N}}$ the reduced basis one and we can choose either the induced or the energy norm.

During this iterative basis selection process and if at the j-th step a j-dimensional reduced basis space $W_j^{\mathcal{N}}$ is given, the next basis function is the one that maximizes the estimated model order reduction error given the j-dimensional space $W_j^{\mathcal{N}}$ over $\mathcal{D}$. So at the $n+1$ iteration we select

$$\boldsymbol{\mu}_{n+1} = arg \max_{\boldsymbol{\mu} \in \mathcal{D}} \eta(\boldsymbol{\mu})$$

and compute $u^{\mathcal{N}}(\boldsymbol{\mu}_{n+1})$ to enrich the reduced space. This is repeated until the maximal estimated error is below a required error tolerance. With this choice the greedy algorithm always selects the next parameter sample point as the one for which the model error is the maximum as estimated by $\eta(\boldsymbol{\mu})$ and this yields a basis that aims to be optimal in the maximum norm over $\mathcal{D}$.

Furthermore we can rewrite the reduced space as

$$W_N^{\mathcal{N}} = \text{span}\{\zeta_n^{\mathcal{N}}, 1 \leq n \leq N\},$$

where the basis functions $\left\{\zeta^{\mathcal{N}}\right\}$ are computed from the snapshots $u^{\mathcal{N}}(\boldsymbol{\mu})$ by a Gram-Schmidt orthonormalization process such that $[\boldsymbol{\zeta}_m^{\mathcal{N}}]^T[\mathbf{Y}^{\mathcal{N}}][\boldsymbol{\zeta}_n^{\mathcal{N}}] = \delta_{mn}$, where δ_{mn} is the Kronecker-delta symbol. We then define our orthonormalized-snapshot matrix $[\mathbf{Z}_N] \equiv [\mathbf{Z}_N^{\mathcal{N}}] = [[\boldsymbol{\zeta}_1^{\mathcal{N}}]| \cdots |[\boldsymbol{\zeta}_n^{\mathcal{N}}]]$ of dimension $\mathcal{N} \times N$.

8.3.2 Galerkin Projection

We then apply a Galerkin projection on our "truth" problem [1, 27–29, 36]: given $\boldsymbol{\mu} \in \mathcal{D}$, we could evaluate the RB output as

$$s_N(\boldsymbol{\mu}) = [\mathbf{F}^{\mathcal{N}}(\boldsymbol{\mu})]^T[\mathbf{u}^{\mathcal{N}}_{\text{RB},N}(\boldsymbol{\mu})],$$

where

$$[\mathbf{u}^{\mathcal{N}}_{\text{RB},N}(\boldsymbol{\mu})] = [\mathbf{Z}_N][\mathbf{u}_N(\boldsymbol{\mu})] \tag{8.19}$$

represents the RB solution $\mathbf{u}^{\mathcal{N}}_{\text{RB},N}(\boldsymbol{\mu}) \in X^{\mathcal{N}}_N \subset X^{\mathcal{N}}$ of size $\mathcal{N}$. Here $[\mathbf{u}_N(\boldsymbol{\mu})]$ is the RB coefficient vector of dimension N satisfies the RB "stiffness" equations

$$[\mathbf{K}_N(\boldsymbol{\mu})][\mathbf{u}_N(\boldsymbol{\mu})] = [\mathbf{F}_N(\boldsymbol{\mu})], \tag{8.20}$$

where

$$\begin{aligned} [\mathbf{K}_N(\boldsymbol{\mu})] &= [\mathbf{Z}_N]^T[\mathbf{K}^{\mathcal{N}}(\boldsymbol{\mu})][\mathbf{Z}_N], \\ [\mathbf{F}_N(\boldsymbol{\mu})] &= [\mathbf{Z}_N]^T[\mathbf{F}^{\mathcal{N}}(\boldsymbol{\mu})]. \end{aligned} \tag{8.21}$$

Note that the system (8.20) is of small size: it is just a set of N linear algebraic equations, in this way we can now evaluate our output as

$$s_N(\boldsymbol{\mu}) = [\mathbf{F}_N(\boldsymbol{\mu})]^T[\mathbf{u}_N(\boldsymbol{\mu})]. \tag{8.22}$$

It can be shown [31] that the condition number of the RB "stiffness" matrix $[\mathbf{Z}_N]^T[\mathbf{K}^{\mathcal{N}}(\boldsymbol{\mu})][\mathbf{Z}_N]$ is bounded by $\gamma_0(\boldsymbol{\mu})/\alpha_0(\boldsymbol{\mu})$, and independent of both N and $\mathcal{N}$.

8.3.3 Offline-Online Procedure

Although the system (8.20) is of small size, the computational cost for assembling the RB "stiffness" matrix (and the RB "output" vector $[\mathbf{F}^{\mathcal{N}}(\boldsymbol{\mu})]^T[\mathbf{Z}_N]$) is still involves $\mathcal{N}$ and costly, $O(N\mathcal{N}^2 + N^2\mathcal{N})$ (and $O(N\mathcal{N})$, respectively). However, we can use our affine forms (8.12) to construct very efficient Offline-Online procedures, as we shall discuss below.

We first insert our affine forms (8.18) into the expansion (8.20) and (8.22), by using (8.21) we obtain

$$\sum_{q=1}^{Q^a} \Theta_q^a(\boldsymbol{\mu})[\mathbf{K}_{qN}][\mathbf{u}_N(\boldsymbol{\mu})] = \sum_{q=1}^{Q^f} \Theta_q^f(\boldsymbol{\mu})[\mathbf{F}_{qN}]$$

and

$$s_N(\boldsymbol{\mu}) = \sum_{q=1}^{Q^f} \Theta_q^f(\boldsymbol{\mu})[\mathbf{F}_{qN}][\mathbf{u}_N(\boldsymbol{\mu})],$$

respectively. Here

$$[\mathbf{K}_{qN}] = [\mathbf{Z}_N]^T[\mathbf{K}_q^{\mathcal{N}}][\mathbf{Z}_N], \quad 1 \le q \le Q^a$$
$$[\mathbf{F}_{qN}] = [\mathbf{Z}_N]^T[\mathbf{F}_q^{\mathcal{N}}], \quad 1 \le q \le Q^f,$$

are parameter independent quantities that can be computed just once and than stored for all the subsequent $\boldsymbol{\mu}$-dependent queries. We then observe that all the "expensive" matrices $[\mathbf{K}_{qN}]$, $1 \le q \le Q^a$, $1 \le N \le N_{\max}$ and vectors $[\mathbf{F}_{qN}]$, $1 \le q \le Q^f$, $1 \le N \le N_{\max}$, are now separated and parameter-independent, hence those can be *pre-computed* in an Offline-Online procedure.

In the Offline stage, we first compute the $[\mathbf{u}^{\mathcal{N}}(\boldsymbol{\mu}^n)]$, $1 \le n \le N_{\max}$, form the matrix $[\mathbf{Z}_{N_{\max}}]$ and then form and store $[\mathbf{F}_{N_{\max}}]$ and $[\mathbf{K}_{qN_{\max}}]$. The Offline operation count depends on $N_{\max}$, Q^a and $\mathcal{N}$ but requires only $O(Q^a N_{\max}^2 + Q^f N_{\max} + Q^\ell N_{\max})$ permanent storage.

In the Online stage, for a given $\boldsymbol{\mu}$ and N ($1 \le N \le N_{\max}$), we retrieve the pre-computed $[\mathbf{K}_{qN}]$ and $[\mathbf{F}_N]$ (subarrays of $[\mathbf{K}_{qN_{\max}}]$, $[\mathbf{F}_{N_{\max}}]$), form $[\mathbf{K}_N(\boldsymbol{\mu})]$, solve the resulting $N \times N$ system (8.20) to obtain $\{\mathbf{u}_N(\boldsymbol{\mu})\}$, and finally evaluate the output $s_N(\boldsymbol{\mu})$ from (8.22). The Online operation count is thus $O(N^3)$ and independent of $\mathcal{N}$. The implication of the latter is twofold: first, we will achieve very fast response in the many-query and real-time contexts, as N is typically very small, $N \ll \mathcal{N}$; and second, we can choose $\mathcal{N}$ arbitrary large—to obtain as accurate FE predictions as we wish—without adversely affecting the Online (marginal) cost.

8.4 A Posteriori Error Estimation

In this Section we recall the a posteriori error estimator for our RB approximation. We shall discuss in details the computation procedures for the two ingredients of the error estimator: the dual norm of the residual and the coercivity lower bound. We first present the Offline-Online strategy for the computation of the dual norm of the residual; we then briefly discuss the Successive Constraint Method [21] in order to compute the coercivity lower bound.

8.4.1 Definitions

We first introduce the error $e^{\mathcal{N}}(\boldsymbol{\mu}) \equiv u^{\mathcal{N}}(\boldsymbol{\mu}) - u^{\mathcal{N}}_{\mathrm{RB},N}(\boldsymbol{\mu}) \in X^{\mathcal{N}}$ and the residual $r^{\mathcal{N}}(v;\boldsymbol{\mu}) \in (X^{\mathcal{N}})'$ (the dual space to $X^{\mathcal{N}}$), $\forall v \in X^{\mathcal{N}}$,

$$r^{\mathcal{N}}(v;\boldsymbol{\mu}) = f(v) - a(u^{\mathcal{N}}(\boldsymbol{\mu}), v; \boldsymbol{\mu}), \tag{8.23}$$

which can be given in the discrete form as

$$[\mathbf{r}^{\mathcal{N}}(\boldsymbol{\mu})] = [\mathbf{F}^{\mathcal{N}}(\boldsymbol{\mu})] - [\mathbf{K}^{\mathcal{N}}(\boldsymbol{\mu})][\mathbf{u}^{\mathcal{N}}_{\mathrm{RB},N}(\boldsymbol{\mu})]. \tag{8.24}$$

We then introduce the Riesz representation of $r^{\mathcal{N}}(v;\boldsymbol{\mu})$: $\hat{e}(\boldsymbol{\mu}) \in X^{\mathcal{N}}$ defined by $(\hat{e}(\boldsymbol{\mu}), v)_{X^{\mathcal{N}}} = r^{\mathcal{N}}(v;\boldsymbol{\mu})$, $\forall v \in X^{\mathcal{N}}$. In vector form, $\hat{e}(\boldsymbol{\mu})$ can be expressed as

$$[\mathbf{Y}^{\mathcal{N}}][\hat{\boldsymbol{e}}(\boldsymbol{\mu})] = [\mathbf{r}^{\mathcal{N}}(\boldsymbol{\mu})]. \tag{8.25}$$

We also require a lower bound to the coercivity constant

$$\alpha^{\mathcal{N}}(\boldsymbol{\mu}) = \inf_{w \in X^{\mathcal{N}}} \frac{a(w, w; \boldsymbol{\mu})}{\|w\|^2_{X^{\mathcal{N}}}}, \tag{8.26}$$

such that $0 < \alpha^{\mathcal{N}}_{\mathrm{LB}}(\boldsymbol{\mu}) \le \alpha^{\mathcal{N}}(\boldsymbol{\mu})$, $\forall \boldsymbol{\mu} \in \mathcal{D}$.

We may now define our error estimator for our output as

$$\Delta^s_N(\boldsymbol{\mu}) \equiv \frac{\|\hat{e}(\boldsymbol{\mu})\|^2_{X^{\mathcal{N}}}}{\alpha^{\mathcal{N}}_{\mathrm{LB}}}, \tag{8.27}$$

where $\|\hat{e}(\boldsymbol{\mu})\|_{X^{\mathcal{N}}}$ is the dual norm of the residual. We can also equip the error estimator with an effectivity defined by

$$\eta^s_N(\boldsymbol{\mu}) \equiv \frac{\Delta^s_N(\boldsymbol{\mu})}{|s^{\mathcal{N}}(\boldsymbol{\mu}) - s_N(\boldsymbol{\mu})|}. \tag{8.28}$$

We can readily demonstrate [31, 36] that

$$1 \leq \eta_N^s(\boldsymbol{\mu}) \leq \frac{\gamma_0(\boldsymbol{\mu})}{\alpha_{\text{LB}}^{\mathcal{N}}(\boldsymbol{\mu})}, \quad \forall \boldsymbol{\mu} \in \mathcal{D};$$

so that the error estimator is both *rigorous* and *sharp*. Note that here we can only claim the *sharp* property for this current "compliant" case.

We shall next provide procedures for the computation of the two ingredients of our error estimator: we shall first discuss the Offline-Online strategy to compute the dual norm of the residual $\|\hat{e}(\boldsymbol{\mu})\|_{X^{\mathcal{N}}}$, and then provide the construction for the lower bound of the coercivity constant $\alpha^{\mathcal{N}}(\boldsymbol{\mu})$.

8.4.2 Dual Norm of the Residual

In discrete form, the dual norm of the residual $\varepsilon(\boldsymbol{\mu}) = \|\hat{e}(\boldsymbol{\mu})\|_{X^{\mathcal{N}}}$ is given by

$$\varepsilon^2(\boldsymbol{\mu}) = [\hat{\boldsymbol{e}}(\boldsymbol{\mu})]^T [\mathbf{Y}^{\mathcal{N}}][\hat{\boldsymbol{e}}(\boldsymbol{\mu})]. \tag{8.29}$$

We next invoke (8.24), (8.25) and (8.29) to arrive at

$$\begin{aligned}
\varepsilon^2(\boldsymbol{\mu}) &= \Big([\mathbf{F}^{\mathcal{N}}(\boldsymbol{\mu})] - [\mathbf{K}^{\mathcal{N}}(\boldsymbol{\mu})][\mathbf{u}_{\text{RB},N}^{\mathcal{N}}(\boldsymbol{\mu})]\Big)^T [\mathbf{Y}^{\mathcal{N}}]^{-1} \\
&\quad \Big([\mathbf{F}^{\mathcal{N}}(\boldsymbol{\mu})] - [\mathbf{K}^{\mathcal{N}}(\boldsymbol{\mu})][\mathbf{u}_{\text{RB},N}^{\mathcal{N}}(\boldsymbol{\mu})]\Big) \\
&= [\mathbf{F}^{\mathcal{N}}(\boldsymbol{\mu})]^T [\mathbf{Y}^{\mathcal{N}}]^{-1} [\mathbf{F}^{\mathcal{N}}(\boldsymbol{\mu})] - 2[\mathbf{F}^{\mathcal{N}}(\boldsymbol{\mu})]^T [\mathbf{Y}^{\mathcal{N}}]^{-1} [\mathbf{K}^{\mathcal{N}}(\boldsymbol{\mu})] \\
&\quad + [\mathbf{K}^{\mathcal{N}}(\boldsymbol{\mu})]^T [\mathbf{Y}^{\mathcal{N}}]^{-1} [\mathbf{K}^{\mathcal{N}}(\boldsymbol{\mu})].
\end{aligned} \tag{8.30}$$

We next defines the "pseudo"-solutions $[\mathbf{P}_q^f] = [\mathbf{Y}^{\mathcal{N}}]^{-1}[\mathbf{F}_q^{\mathcal{N}}]$, $1 \leq q \leq Q^f$ and $[\mathbf{P}_{qN}^a] = [\mathbf{Y}^{\mathcal{N}}]^{-1}[\mathbf{K}_q^{\mathcal{N}}][\mathbf{Z}_N]$, $1 \leq q \leq Q^a$, then apply the affine form (8.18) and (8.19) into (8.30) to obtain

$$\begin{aligned}
\varepsilon^2(\boldsymbol{\mu}) &= \sum_{q=1}^{Q^f} \sum_{q'=1}^{Q^f} \Theta_q^f(\boldsymbol{\mu}) \Theta_{q'}^f(\boldsymbol{\mu}) \Big([\mathbf{P}_q^f]^T [\mathbf{Y}^{\mathcal{N}}][\mathbf{P}_{q'}^f]\Big) \\
&\quad -2 \sum_{q=1}^{Q^a} \sum_{q'=1}^{Q^f} \Theta_q^a(\boldsymbol{\mu}) \Theta_{q'}^f(\boldsymbol{\mu}) \Big([\mathbf{P}_q^f]^T [\mathbf{Y}^{\mathcal{N}}][\mathbf{P}_{q'N}^a]\Big) [\mathbf{u}_N^{RB}(\boldsymbol{\mu})] \\
&\quad + \sum_{q=1}^{Q^a} \sum_{q'=1}^{Q^a} \Theta_q^a(\boldsymbol{\mu}) \Theta_{q'}^a(\boldsymbol{\mu}) [\mathbf{u}_N^{RB}(\boldsymbol{\mu})]^T \Big([\mathbf{P}_{qN}^a]^T [\mathbf{Y}^{\mathcal{N}}][\mathbf{P}_{q'N}^a]\Big) [\mathbf{u}_N^{RB}(\boldsymbol{\mu})].
\end{aligned} \tag{8.31}$$

It is observed that all the terms in bracket in (8.31) are all parameter-independent, hence they can be *pre-computed* in the Offline stage. The Offline-Online strategy is now clear.

In the Offline stage we form the parameter-independent quantities. We first compute the "pseudo"-solutions $[\mathbf{P}_q^f] = [\mathbf{Y}^{\mathcal{N}}]^{-1}[\mathbf{F}_q^{\mathcal{N}}]$, $1 \le q \le Q^f$ and $[\mathbf{P}_{qN}^a] = [\mathbf{Y}^{\mathcal{N}}]^{-1}[\mathbf{K}_q^{\mathcal{N}}][\mathbf{Z}_N]$, $1 \le q \le Q^a$, $1 \le N \le N_{\max}$; and form/store $[\mathbf{P}_q^f]^T[\mathbf{Y}^{\mathcal{N}}][\mathbf{P}_{q'}^f]$, $1 \le q, q' \le Q^f$, $[\mathbf{P}_q^f]^T[\mathbf{Y}^{\mathcal{N}}][\mathbf{P}_{q'N}^a]$, $1 \le q \le Q^f$, $1 \le q \le Q^a$, $1 \le N \le N_{\max}$, $[\mathbf{P}_{qN}^a][\mathbf{Y}^{\mathcal{N}}][\mathbf{P}_{q'N}^a]$, $1 \le q, q' \le Q^a$, $1 \le N \le N_{\max}$. The Offline operation count depends on $N_{\max}$, Q^a, Q^f, and $\mathcal{N}$.

In the Online stage, for a given $\boldsymbol{\mu}$ and N $(1 \le N \le N_{\max})$, we retrieve the pre-computed quantities $[\mathbf{P}_q^f]^T[\mathbf{Y}^{\mathcal{N}}][\mathbf{P}_{q'}^f]$, $1 \le q, q' \le Q^f$, $[\mathbf{P}_q^f]^T[\mathbf{Y}^{\mathcal{N}}][\mathbf{P}_{q'N}^a]$, $1 \le q \le Q^f$, $1 \le q \le Q^a$, and $[\mathbf{P}_{qN}^a]^T[\mathbf{Y}^{\mathcal{N}}][\mathbf{P}_{q'N}^a]$, $1 \le q, q' \le Q^a$, and then evaluate the sum (8.31). The Online operation count is dominated by $O(((Q^a)^2+(Q^f)^2)N^2)$ and independent of $\mathcal{N}$.

8.4.3 Lower Bound of the Coercivity Constant

We now briefly address some elements for the computation of the lower bound in the coercive case. In order to derive the discrete form of the coercivity constant (8.26) we introduce the discrete eigenvalue problem: given $\boldsymbol{\mu} \in \mathcal{D}$, find the minimum set $([\boldsymbol{\chi}_{\min}(\boldsymbol{\mu})], \lambda_{\min}(\boldsymbol{\mu}))$ such that

$$\begin{aligned} [\mathbf{K}^{\mathcal{N}}(\boldsymbol{\mu})][\boldsymbol{\chi}(\boldsymbol{\mu})] &= \lambda_{\min}[\mathbf{Y}^{\mathcal{N}}][\boldsymbol{\chi}(\boldsymbol{\mu})], \\ [\boldsymbol{\chi}(\boldsymbol{\mu})]^T\,[\mathbf{Y}^{\mathcal{N}}][\boldsymbol{\chi}(\boldsymbol{\mu})] &= 1. \end{aligned} \tag{8.32}$$

We can then recover

$$\alpha^{\mathcal{N}}(\boldsymbol{\mu}) = \sqrt{\lambda_{\min}(\boldsymbol{\mu})}. \tag{8.33}$$

However, the eigenproblem (8.32) is of size $\mathcal{N}$, so using direct solution as an ingredient for our error estimator is very expensive. Hence, we will construct an inexpensive yet of good quality lower bound $\alpha_{\mathrm{LB}}^{\mathcal{N}}(\boldsymbol{\mu})$ and use this lower bound instead of the truth (direct) expensive coercivity constant $\alpha^{\mathcal{N}}(\boldsymbol{\mu})$ in our error estimator.

For our current target problems, our bilinear form is coercive and symmetric. We shall construct our coercivity lower bound by the Successive Constraint Method (SCM) [21]. It is noted that the SCM method can be readily extended to non-symmetric as well as non-coercive bilinear forms [21, 23, 31, 36].

We first introduce an alternative (albeit not very computation-friendly) discrete form for our coercivity constant as

$$\text{minimum} \sum_{q=1}^{Q^a} \Theta_q^a(\boldsymbol{\mu}) y_q, \tag{8.34}$$

$$\text{subject to } y_q = \frac{[\mathbf{w}_q]^T [\mathbf{K}_q^{\mathcal{N}}][\mathbf{w}_q]}{[\mathbf{w}_q]^T [\mathbf{Y}^{\mathcal{N}}][\mathbf{w}_q]}, \quad 1 \le q \le Q^a,$$

where $[\mathbf{w}_q]$ is the discrete vector of any arbitrary $w_y \in X^{\mathcal{N}}$.

We shall now "relax" the constraint in (8.34) by defining the "continuity constraint box" associated with $y_{q,\min}$ and $y_{q,\max}$, $1 \le q \le Q^a$ obtained from the minimum set $([\mathbf{y}_-(\boldsymbol{\mu})], y_{q,\min})$ and maximum set $([\mathbf{y}_+(\boldsymbol{\mu})], y_{q,\max})$ solutions of the eigenproblems

$$[\mathbf{K}_q^{\mathcal{N}}][\mathbf{y}_-(\boldsymbol{\mu})] = y_{q,\min}[\mathbf{Y}^{\mathcal{N}}][\mathbf{y}_-(\boldsymbol{\mu})],$$
$$[\mathbf{y}_-(\boldsymbol{\mu})]\,[\mathbf{Y}^{\mathcal{N}}][\mathbf{y}_-(\boldsymbol{\mu})] = 1,$$

and

$$[\mathbf{K}_q^{\mathcal{N}}][\mathbf{y}_+(\boldsymbol{\mu})] = y_{q,\max}[\mathbf{Y}^{\mathcal{N}}][\mathbf{y}_+(\boldsymbol{\mu})],$$
$$[\mathbf{y}_+(\boldsymbol{\mu})]\,[\mathbf{Y}^{\mathcal{N}}][\mathbf{y}_+(\boldsymbol{\mu})] = 1,$$

respectively, for $1 \le q \le Q^a$. We next define a "coercivity constraint" sample

$$C_J = \{\boldsymbol{\mu}_1^{\mathrm{SCM}} \in \mathcal{D}, \ldots, \boldsymbol{\mu}_J^{\mathrm{SCM}} \in \mathcal{D}\},$$

and denote $C_J^{M,\boldsymbol{\mu}}$ the set of M $(1 \le M \le J)$ points in C_J closest (in the usual Euclidean norm) to a given $\boldsymbol{\mu} \in \mathcal{D}$. The construction of the set C_J is done by means of a Greedy procedure [21, 31, 36]. The Greedy selection of C_J can be called the "Offline stage", which involves the solutions of J eigenproblems (8.32) to obtain $\alpha^{\mathcal{N}}(\boldsymbol{\mu})$, $\forall \boldsymbol{\mu} \in C_J$.

We may now define our lower bound $\alpha_{\mathrm{LB}}^{\mathcal{N}}(\boldsymbol{\mu})$ as the solution of

$$\text{minimum} \sum_{q=1}^{Q^a} \Theta_q^a(\boldsymbol{\mu}) y_q, \tag{8.35}$$

$$\text{subject to } y_{q,\min} \le y_q \le y_{q,\max}, \quad 1 \le q \le Q^a,$$

$$\sum_{q=1}^{Q^a} \Theta_q^a(\boldsymbol{\mu}') y_q \ge \alpha^{\mathcal{N}}(\boldsymbol{\mu}'), \quad \forall \boldsymbol{\mu}' \in C_J^{M,\boldsymbol{\mu}}.$$

We then "restrict" the constraint in (8.34) and define our upper bound $\alpha_{\mathrm{UB}}^{\mathcal{N}}(\boldsymbol{\mu})$ as the solution of

$$\text{mininum} \sum_{q=1}^{Q^a} \Theta_q^a(\boldsymbol{\mu}) y_{q,*}(\boldsymbol{\mu}'), \tag{8.36}$$

$$\text{subject to } y_{q,*}(\boldsymbol{\mu}') = [\boldsymbol{\chi}(\boldsymbol{\mu}')]^T [\mathbf{K}_q^{\mathcal{N}}][\boldsymbol{\chi}(\boldsymbol{\mu}')], \quad 1 \le q \le Q^a, \quad \forall \boldsymbol{\mu}' \in C_J^{M,\boldsymbol{\mu}},$$

where $[\boldsymbol{\chi}(\boldsymbol{\mu})]$ is defined by (8.32). It can be shown [21, 31, 36] that the feasible region of (8.36) is a subset of that of (8.34), which in turn, is a subset of that of (8.35): hence $\alpha_{\mathrm{LB}}^{\mathcal{N}}(\boldsymbol{\mu}) \le \alpha^{\mathcal{N}}(\boldsymbol{\mu}) \le \alpha_{\mathrm{UB}}^{\mathcal{N}}(\boldsymbol{\mu})$.

We note that the lower bound (8.35) is a linear optimization problem (or Linear Program (LP)) which contains Q^a design variables and $2Q^a + M$ inequality constraints. Given a value of the parameter $\boldsymbol{\mu}$, the Online evaluation $\boldsymbol{\mu} \to \alpha_{\mathrm{LB}}^{\mathcal{N}}(\boldsymbol{\mu})$ is thus as follows: we find the subset $C_J^{M,\boldsymbol{\mu}}$ of C_J for a given M, we then calculate $\alpha_{\mathrm{LB}}^{\mathcal{N}}(\boldsymbol{\mu})$ by solving the LP (8.35). The crucial point here is that the online evaluation $\boldsymbol{\mu} \to \alpha_{\mathrm{LB}}^{\mathcal{N}}(\boldsymbol{\mu})$ is totally independent of $\mathcal{N}$. The upper bound (8.35), however, can be obtained as the solution of just a simple enumeration problem; the online evaluation of $\alpha_{\mathrm{UB}}^{\mathcal{N}}(\boldsymbol{\mu})$ is also independent of $\mathcal{N}$. In general, the upper bound $\alpha_{\mathrm{UB}}^{\mathcal{N}}(\boldsymbol{\mu})$ is not used in the calculation of the error estimator, however, it is used in the Greedy construction of the set C_J [21]. In practice, when the set C_J does not guarantee to produce a positive $\alpha_{\mathrm{LB}}^{\mathcal{N}}(\boldsymbol{\mu})$, the upper bound $\alpha_{\mathrm{UB}}^{\mathcal{N}}(\boldsymbol{\mu})$ can be used as a substitution for $\alpha_{\mathrm{UB}}^{\mathcal{N}}(\boldsymbol{\mu})$ since it approximates the "truth" $\alpha^{\mathcal{N}}(\boldsymbol{\mu})$ in a very way; however we will lose the rigorous property of the error estimators.

8.5 Extension of the RB Method to Non-compliant Output

We shall briefly provide the extension of our RB methodology for the "non-compliant" case in this Section. We first present a suitable primal-dual formulation for the "non-compliant" output; we then briefly provide the extension to the RB methodology, including the RB approximation and its a posteriori error estimation.

8.5.1 Adjoint Problem

We shall briefly discuss the extension of our methodology to the non-compliant problems. We still require that both f and ℓ are bounded functionals, but now $(f(\cdot;\boldsymbol{\mu}) \neq \ell(\cdot;\boldsymbol{\mu}))$. We still use the previous abstract statement in Sect. 8.2. We begin with the definition of the dual problem associated to ℓ: find $\psi(\boldsymbol{\mu}) \in X$ (our "adjoint" or "dual" field) such that

$$a(v, \psi(\boldsymbol{\mu}); \boldsymbol{\mu}) = -\ell(\boldsymbol{\mu}), \quad \forall v \in X.$$

8.5.2 Truth Approximation

We now again apply the finite element method to the dual formulation: given $\boldsymbol{\mu} \in \mathcal{D}$, we evaluate

$$s(\boldsymbol{\mu}) = [\mathbf{L}^{\mathcal{N}}(\boldsymbol{\mu})]^T [\mathbf{u}^{\mathcal{N}}(\boldsymbol{\mu})],$$

where $[\mathbf{u}^{\mathcal{N}}(\boldsymbol{\mu})]$ is the finite element solution of size $\mathcal{N}$ satisfying (8.17). The discrete form of the dual solution $\psi^{\mathcal{N}}(\boldsymbol{\mu}) \in X^{\mathcal{N}}$ is given

$$[\mathbf{K}^{\mathcal{N}}(\boldsymbol{\mu})][\boldsymbol{\psi}^{\mathcal{N}}(\boldsymbol{\mu})] = -[\mathbf{L}^{\mathcal{N}}(\boldsymbol{\mu})];$$

here $[\mathbf{L}^{\mathcal{N}}(\boldsymbol{\mu})]$ is the discrete load vector of $\ell(\cdot; \boldsymbol{\mu})$. We also invoke the affine forms (8.12) to express $[\mathbf{L}^{\mathcal{N}}(\boldsymbol{\mu})]$ as

$$[\mathbf{L}^{\mathcal{N}}(\boldsymbol{\mu})] = \sum_{q=1}^{Q^\ell} \Theta_q^\ell(\boldsymbol{\mu})[\mathbf{L}_q^{\mathcal{N}}], \tag{8.37}$$

where all the $[\mathbf{L}_q^{\mathcal{N}}]$ are the discrete forms of the parameter-independent linear forms $\ell_q(\cdot)$, $1 \le q \le Q^\ell$.

8.5.3 Reduced Basis Approximation

We now define our RB spaces: we shall need to define two Lagrangian parameter samples set, $S_{N^{\mathrm{pr}}} = \{\boldsymbol{\mu}_1, \boldsymbol{\mu}_2, \ldots, \boldsymbol{\mu}_{N^{\mathrm{pr}}}\}$ and $S_{N^{\mathrm{du}}} = \{\boldsymbol{\mu}_1, \boldsymbol{\mu}_2, \ldots, \boldsymbol{\mu}_{N^{\mathrm{du}}}\}$ corresponding to the set of our primal and dual parameter samples set, respectively. We also associate the primal and dual reduced basis spaces $(X_{N^{\mathrm{pr}}}^{\mathcal{N}} =) W_{N^{\mathrm{pr}}}^{\mathcal{N}}$, $1 \le N \le N_{\max}^{\mathrm{pr}}$ and $(X_{N^{\mathrm{du}}}^{\mathcal{N}} =) W_{N^{\mathrm{du}}}^{\mathcal{N}}$, $1 \le N \le N_{\max}^{\mathrm{du}}$ to our $S_{N^{\mathrm{pr}}}$ and $S_{N^{\mathrm{du}}}$ set, respectively, which are constructed from the primal $u^{\mathcal{N}}(\boldsymbol{\mu})$ and dual $\psi^{\mathcal{N}}(\boldsymbol{\mu})$ snapshots by a Gram-Schmidt process as in Sect. 8.3. Finally, we denote our primal and dual orthonormalized-snapshot as $[\mathbf{Z}_{N^{\mathrm{pr}}}^{\mathrm{pr}}]$ and $[\mathbf{Z}_{N^{\mathrm{du}}}^{\mathrm{du}}]$ basis matrices, respectively.

8.5.4 Galerkin Projection

We first denote the RB primal approximation to the primal "truth" approximation $u^{\mathcal{N}}(\boldsymbol{\mu})$ as $u^{\mathcal{N}}_{\mathrm{RB},N}(\boldsymbol{\mu})$ and the RB dual approximation to the primal "truth" dual approximation $\psi^{\mathcal{N}}(\boldsymbol{\mu})$ as $\psi^{\mathcal{N}}_{\mathrm{RB},N}(\boldsymbol{\mu})$: their discrete forms are given by $[\mathbf{u}^{\mathcal{N}}_{RB,N^{\mathrm{pr}}}(\boldsymbol{\mu})]=[\mathbf{Z}^{\mathrm{pr}}_{N^{\mathrm{pr}}}][\mathbf{u}_{N^{\mathrm{pr}}}(\boldsymbol{\mu})]$ and $[\boldsymbol{\psi}^{\mathcal{N}}_{RB,N^{\mathrm{du}}}(\boldsymbol{\mu})]=[\mathbf{Z}^{\mathrm{du}}_{N^{\mathrm{du}}}][\boldsymbol{\psi}_{N^{\mathrm{du}}}(\boldsymbol{\mu})]$, respectively.

We then apply a Galerkin projection (note that in this case, a Galerkin-Petrov projection is also possible [2, 31, 36]). given a $\boldsymbol{\mu} \in \mathcal{D}$, we evaluate the RB output

$$s_{N^{\mathrm{pr}},N^{\mathrm{du}}}(\boldsymbol{\mu}) = [\mathbf{L}^{\mathcal{N}}(\boldsymbol{\mu})]^T[\mathbf{u}^{\mathcal{N}}_{\mathrm{RB},N^{\mathrm{pr}}}(\boldsymbol{\mu})] - [\mathbf{r}^{\mathcal{N}}_{\mathrm{pr}}(\boldsymbol{\mu})]^T[\boldsymbol{\psi}^{\mathcal{N}}_{\mathrm{RB},N^{\mathrm{du}}}(\boldsymbol{\mu})],$$

recall that $[\mathbf{r}^{\mathcal{N}}_{\mathrm{pr}}(\boldsymbol{\mu})]$ is the discrete form of the RB primal residual defined in (8.23). The RB coefficient primal and dual are given by

$$\begin{aligned}\sum_{q=1}^{Q^a} \Theta^a_q(\boldsymbol{\mu})[\mathbf{K}_{qN^{\mathrm{pr}}N^{\mathrm{pr}}}][\mathbf{u}_{N^{\mathrm{pr}}}(\boldsymbol{\mu})] &= \sum_{q=1}^{Q^f} \Theta^f_q(\boldsymbol{\mu})[\mathbf{F}_{qN^{\mathrm{pr}}}],\\ \sum_{q=1}^{Q^a} \Theta^a_q(\boldsymbol{\mu})[\mathbf{K}_{qN^{\mathrm{du}}N^{\mathrm{du}}}[\boldsymbol{\psi}_{N^{\mathrm{du}}}(\boldsymbol{\mu})] &= -\sum_{q=1}^{Q^\ell} \Theta^\ell_q(\boldsymbol{\mu})[\mathbf{L}_{qN^{\mathrm{du}}}].\end{aligned} \tag{8.38}$$

Note that the two systems (8.38) are also of small size: their sizes are of N^{pr} and N^{du}, respectively. We can now evaluate our output as

$$\begin{aligned}s_{N^{\mathrm{pr}},N^{\mathrm{du}}}(\boldsymbol{\mu}) = &\sum_{q=1}^{Q^\ell} \Theta^\ell_q(\boldsymbol{\mu})[\mathbf{L}_{qN^{\mathrm{pr}}}][\mathbf{u}_{N^{\mathrm{pr}}}(\boldsymbol{\mu})] - \sum_{q=1}^{Q^f} \Theta^f_q(\boldsymbol{\mu})[\mathbf{F}_{qN^{\mathrm{du}}}][\boldsymbol{\psi}_{N^{\mathrm{du}}}(\boldsymbol{\mu})]\\ &+\sum_{q=1}^{Q^a} \Theta^a_q(\boldsymbol{\mu})[\boldsymbol{\psi}_{N^{\mathrm{du}}}(\boldsymbol{\mu})]^T[\mathbf{K}_{qN^{\mathrm{du}}N^{\mathrm{pr}}}][\mathbf{u}_{N^{\mathrm{pr}}}(\boldsymbol{\mu})].\end{aligned} \tag{8.39}$$

All the quantities in (8.38) and (8.39) are given by

$$\begin{aligned}[\mathbf{K}_{qN^{\mathrm{pr}}N^{\mathrm{pr}}}] &= [\mathbf{Z}^{\mathrm{pr}}_{N^{\mathrm{pr}}}]^T[\mathbf{K}_q][\mathbf{Z}^{\mathrm{pr}}_{N^{\mathrm{pr}}}], \quad 1 \le q \le Q^a,\ 1 \le N^{\mathrm{pr}} \le N^{\mathrm{pr}}_{\max},\\ \left[\mathbf{K}_{qN^{\mathrm{du}}N^{\mathrm{du}}}\right] &= [\mathbf{Z}^{\mathrm{du}}_{N^{\mathrm{du}}}]^T[\mathbf{K}_q][\mathbf{Z}^{\mathrm{du}}_{N^{\mathrm{du}}}], \quad 1 \le q \le Q^a,\ 1 \le N^{\mathrm{du}} \le N^{\mathrm{du}}_{\max},\\ \left[\mathbf{K}_{qN^{\mathrm{du}}N^{\mathrm{pr}}}\right] &= [\mathbf{Z}^{\mathrm{du}}_{N^{\mathrm{du}}}]^T[\mathbf{K}_q][\mathbf{Z}^{\mathrm{pr}}_{N^{\mathrm{pr}}}], \quad 1 \le q \le Q^a,\ 1 \le N^{\mathrm{pr}} \le N^{\mathrm{pr}}_{\max},\ 1 \le N^{\mathrm{du}} \le N^{\mathrm{du}}_{\max}\\ \left[\mathbf{F}_{qN^{\mathrm{pr}}}\right] &= [\mathbf{Z}^{\mathrm{pr}}_{N^{\mathrm{pr}}}]^T[\mathbf{F}_q], \quad 1 \le q \le Q^f,\ 1 \le N^{\mathrm{pr}} \le N^{\mathrm{pr}}_{\max},\\ \left[\mathbf{F}_{qN^{\mathrm{du}}}\right] &= [\mathbf{Z}^{\mathrm{du}}_{N^{\mathrm{du}}}]^T[\mathbf{F}_q], \quad 1 \le q \le Q^f,\ 1 \le N^{\mathrm{du}} \le N^{\mathrm{du}}_{\max},\end{aligned}$$

$$[\mathbf{L}_{qN^{\rm pr}}] = [\mathbf{Z}^{\rm pr}_{N^{\rm pr}}]^T[\mathbf{L}_q], \quad 1 \le q \le Q^\ell, 1 \le N^{\rm pr} \le N^{\rm pr}_{\rm max},$$

$$\left[\mathbf{L}_{qN^{\rm du}}\right] = [\mathbf{Z}^{\rm du}_{N^{\rm du}}]^T[\mathbf{L}_q], \quad 1 \le q \le Q^\ell, 1 \le N^{\rm du} \le N^{\rm du}_{\rm max}.$$

The computation of the output $s_{N^{\rm pr},N^{\rm du}}(\boldsymbol{\mu})$ clearly admits an Offline-Online computational strategy similar to the one we discuss previously in Sect. 8.3.

8.5.5 A Posteriori Error Estimation

We now introduce the dual residual $r^{\mathcal{N}}_{\rm du}(v; \boldsymbol{\mu})$,

$$r^{\mathcal{N}}_{\rm du}(v; \boldsymbol{\mu}) = -\ell(v) - a(v, \psi^{\mathcal{N}}_{N^{\rm du}}(\boldsymbol{\mu}); \boldsymbol{\mu}), \quad \forall v \in X^{\mathcal{N}}.$$

and its Riesz representation of $r^{\mathcal{N}}_{\rm du}(v; \boldsymbol{\mu})$: $\hat{e}^{\rm du}(\boldsymbol{\mu}) \in X^{\mathcal{N}}$ defined by $(\hat{e}^{\rm du}(\boldsymbol{\mu}), v)_{X^{\mathcal{N}}} = r^{\mathcal{N}}_{\rm du}(v; \boldsymbol{\mu})$, $\forall v \in X^{\mathcal{N}}$.

We may now define our error estimator for our output as

$$\Delta^s_{N^{\rm pr}N^{\rm du}}(\boldsymbol{\mu}) \equiv \frac{\|\hat{e}^{\rm pr}(\boldsymbol{\mu})\|_{X^{\mathcal{N}}}}{(\alpha^{\mathcal{N}}_{\rm LB})^{1/2}} \frac{\|\hat{e}^{\rm du}(\boldsymbol{\mu})\|_{X^{\mathcal{N}}}}{(\alpha^{\mathcal{N}}_{\rm LB})^{1/2}}, \tag{8.40}$$

where $\hat{e}^{\rm pr}(\boldsymbol{\mu})$ is the Riesz representation of the primal residual. We then define the effectivity associated with our error bound

$$\eta^s_{N^{\rm pr}N^{\rm du}}(\boldsymbol{\mu}) \equiv \frac{\Delta^s_{N^{\rm pr}N^{\rm du}}(\boldsymbol{\mu})}{|s^{\mathcal{N}}(\boldsymbol{\mu}) - s_{N^{\rm pr}N^{\rm du}}(\boldsymbol{\mu})|}. \tag{8.41}$$

We can readily demonstrate [15, 31, 36] that

$$1 \le \eta^s_{N^{\rm pr}N^{\rm du}}(\boldsymbol{\mu}), \quad \forall \boldsymbol{\mu} \in \mathcal{D};$$

note that the error estimator is still *rigorous*, however it is less *sharp* than that in the "compliant" case since in this case we could not provide an upper bound to $\eta^s_{N^{\rm pr}N^{\rm du}}(\boldsymbol{\mu})$.

The computation of the dual norm of the primal/dual residual also follows an Offline-Online computation strategy: the dual norm of the primal residual is in fact, the same as in Sect. 8.4.2; the same procedure can be applied to compute the dual norm of the dual residual.

8.6 Numerical Results

In this sections we shall consider several "model problems" to demonstrate the feasibility of our methodology. We note that in all cases, these model problems are presented in non-dimensional form unless stated otherwise. In all problems below, displacement is, in fact, in non-dimensional form $u = \tilde{u}\tilde{E}/\tilde{\sigma}_0$, where $\tilde{u}$, $\tilde{E}$, $\tilde{\sigma}_0$ are the dimensional displacement, Young's modulus and load strength, respectively, while E and σ_0 are our non-dimensional Young's modulus and load strength and usually are around unity.

We shall not provide any details for $\Theta^a_q(\boldsymbol{\mu})$, $\Theta^f_q(\boldsymbol{\mu})$ and $\Theta^\ell_q(\boldsymbol{\mu})$ and their associated bilinear and linear forms $a_q(\cdot,\cdot)$, $f_q(\cdot)$ and $\ell_q(\cdot)$ for any of the below examples as they are usually quite complex, due to the complicated structure of the effective elastic tensor and our symbolic manipulation technique. We refer the users to [20, 24, 31, 40], in which all the above terms are provided in details for some simple model problems.

In the below, the timing t_{FE} for an evaluation of the FE solution $\boldsymbol{\mu} \rightarrow s^{\mathcal{N}}(\boldsymbol{\mu})$ is the computation time taken by solving (8.17) and evaluating (8.16) by using (8.18) and (8.37), in which all the stiffness matrix components, $[\mathbf{K}_q]$, $1 \leq q \leq Q^a$, load and output vector components, $[\mathbf{F}_q]$, $1 \leq q \leq Q^f$ and $[\mathbf{L}_q]$, $1 \leq q \leq Q^\ell$, respectively, are pre-computed and pre-stored. We do not include the computation time of forming those components (or alternatively, calculate the stiffness matrix, load and output vector directly) in t_{FE}.

Finally, for the sake of simplicity, we shall denote the number of basis N defined as $N = N^{\mathrm{pr}} = N^{\mathrm{du}}$ in all of our model problems in this Section.

8.6.1 The Arc-Cantilever Beam

We consider a thick arc cantilever beam correspond to the domain $\Omega^{\mathrm{o}}(\boldsymbol{\mu})$ representing the shape of a quarter of an annulus as shown in Fig. 8.1. We apply (clamped) homogeneous Dirichlet conditions on Γ^{o}_D and non-homogeneous Neumann boundary conditions corresponding to a unit tension on Γ^{o}_N. The width of the cantilever beam is $2d$, and the material is isotropic with $(E, \nu) = (1, 0.3)$ under plane stress assumption. Our output of interest is the integral of the tangential displacement (u_2) over Γ^{o}_N, which can be interpreted as the average tangential displacement on Γ^{o}_N.[4] Note that our output of interest is "non-compliant".

[4]The average tangential displacement on Γ^{o}_N is not exactly $s(\boldsymbol{\mu})$ but rather $s(\boldsymbol{\mu})/l_{\Gamma^{\mathrm{o}}_N}$, where $l_{\Gamma^{\mathrm{o}}_N}$ is the length of Γ^{o}_N. It is obviously that the two descriptions of the two outputs, "integral of" and "average of", are pretty much equivalent to each other.

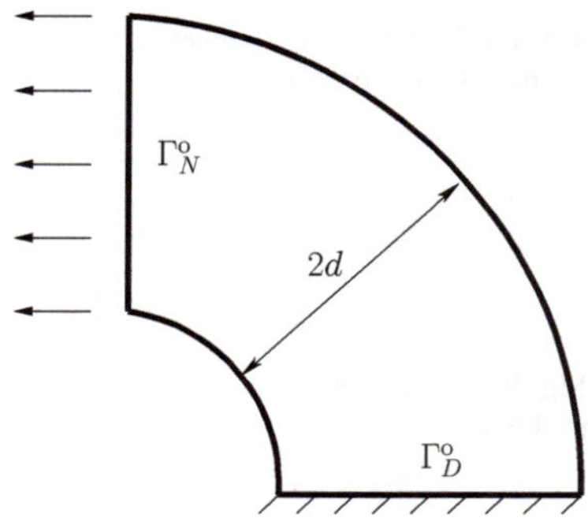

Fig. 8.1 The arc-cantilever beam

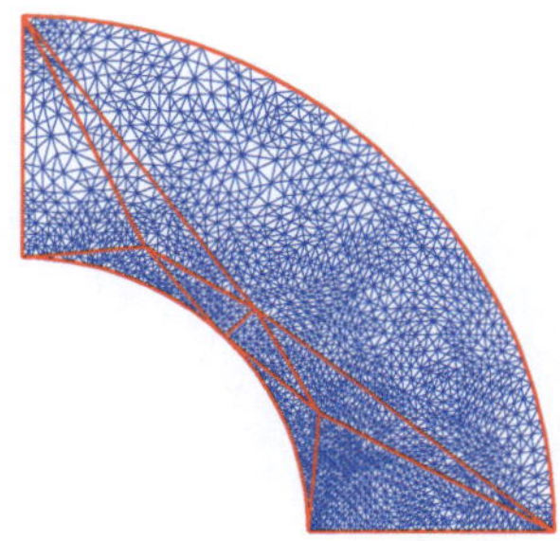

Fig. 8.2 The arc-cantilever beam problem: domain composition and FE mesh

The parameter is the half-width of the cantilever beam $\boldsymbol{\mu} = [\mu_1] \equiv [d]$. The parameter domain is chosen as $\mathcal{D} = [0.3, 0.9]$, which can model a moderately thick beam to a very thick beam. We then choose $\boldsymbol{\mu}_{\text{ref}} = 0.3$ and apply the domain decomposition and obtain $L_{\text{reg}} = 9$ subdomains as shown in Fig. 8.2, in which three subdomains are the general "curvy triangles", generated by our computer automatic procedure [36]. Note that geometric transformations are relatively complicated, due to the appearances of the "curvy triangles" and all subdomains transformations are classified as the "general transformation case" [19, 31]. We then recover our affine forms with $Q^a = 54$, $Q^f = 1$ and $Q^l = 1$.

We next consider a FE approximation where the mesh contains $n_{\text{node}} = 2747$ nodes and $n_{\text{elem}} = 5322$ P_1 elements, which corresponds to $\mathcal{N} = 5426$ degrees of freedoms[5] as shown in Fig. 8.2. To verify our FE approximation, we compare our FE results with the approximated solution for thick arc cantilever beam by Roark [41] for a 100 uniformly distributed test points in $\mathcal{D}$: the maximum difference between our results and Roark's one is just 2.9%.

We then apply our RB approximation. We present in Table 8.1 our convergence results: the RB error bounds and effectivities as a function of $N (= N^{\text{pr}} = N^{\text{du}})$. The error bound reported, $\mathcal{E}_N = \Delta_N^s(\boldsymbol{\mu})/|s_N(\boldsymbol{\mu})|$ is the maximum of the relative error bound over a random test sample Ξ_{test} of size $n_{\text{test}} = 100$. We denote by $\overline{\eta}_N^s$ the average of the effectivity $\eta_N^s(\boldsymbol{\mu})$ over Ξ_{test}. We observe that average effectivity

[5]Note that $\mathcal{N} \neq 2n_{\text{node}}$ since Dirichlet boundary nodes are eliminated from the FE system.

Table 8.1 The arc-cantilever beam: RB convergence

N	ε_N	$\overline{\eta}_N^s$
2	3.57E+00	86.37
4	3.70E−03	18.82
6	4.07E−05	35.72
8	6.55E−07	41.58
10	1.99E−08	40.99

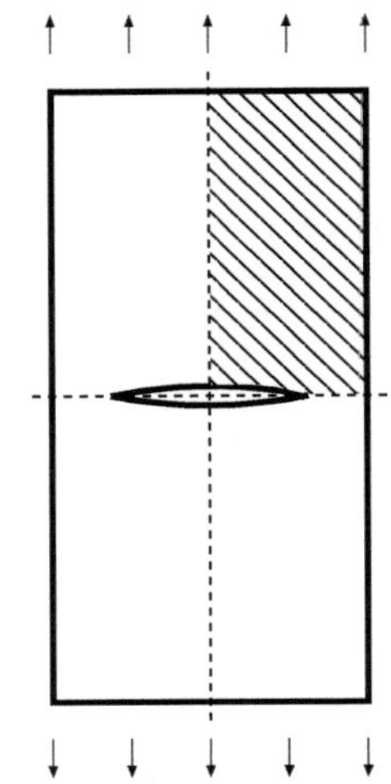

Fig. 8.3 The center crack problem

is of order $O(20-90)$, not very *sharp*, but this is expected due to the fact that the output is "non-compliant".

As regards computational times, a RB online evaluation $\boldsymbol{\mu} \rightarrow (s_N(\boldsymbol{\mu}), \Delta_N^s(\boldsymbol{\mu}))$ requires just $t_{\mathrm{RB}} = 115$(ms) for $N = 10$; while the FE solution $\boldsymbol{\mu} \rightarrow s^{\mathcal{N}}(\boldsymbol{\mu})$ requires $t_{\mathrm{FE}} = 9$(s): thus our RB online evaluation is just 1.28% of the FEM computational cost.

8.6.2 The Center Crack Problem

We next consider a fracture model corresponds to a center crack in a plate under tension at both sides as shown in Fig. 8.3.

Due to the symmetry of the geometry and loading, we only consider one quarter of the physical domain, as shown in Fig. 8.3, note that the crack corresponds to the boundary segment Γ_C^{o}. The crack (in our "quarter" model) is of size d, and the plate is of height h (and of fixed width $w = 1$). We consider plane strain isotropic material with $(E, \nu) = (1, 0.3)$. We consider (symmetric about the x_1^{o} direction and x_2^{o} direction) Dirichlet boundary conditions on the left and bottom boundaries Γ_L^{o} and Γ_B^{o}, respectively; and non-homogeneous Neumann boundary conditions (tension)

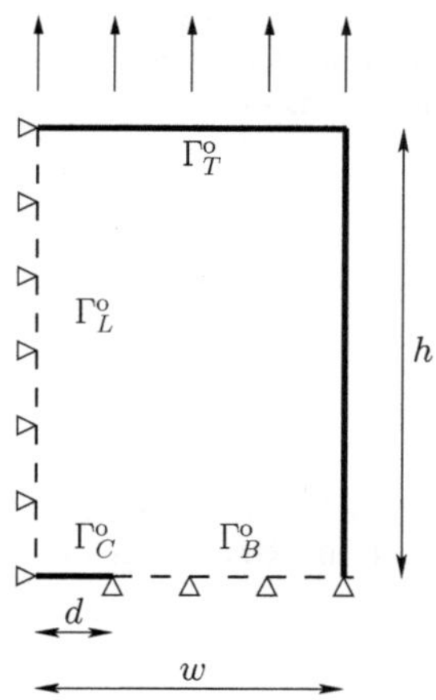

Fig. 8.4 The center crack problem

on the top boundary Γ_T^o. Our ultimate output of interest is the stress intensity factor (SIF) for the crack, which will be derived from an intermediate (compliant) energy output by application of the virtual crack extension approach [30]. The SIF plays an important role in the field of fracture mechanics, for examples, if we have to estimate the propagation path of cracks in structures [18]. We further note that analytical result for SIF of a center-crack in a plate under tension is only available for the infinite plate [25], which can be compared with our solutions for small crack length d and large plate height h values (Fig. 8.4).

Our parameters are the crack length and the plate height $\boldsymbol{\mu} = [\mu_1, \mu_2] \equiv [d, h]$, and the parameter domain is given by $\mathcal{D} = [0.3, 0.7] \times [0.5, 2.0]$. We then choose $\boldsymbol{\mu}_{\mathrm{ref}} = [0.5, 1.0]$ and apply a domain decomposition: the final setting contains $L_{\mathrm{reg}} = 3$ subdomains, which in turn gives us $Q^a = 10$ and $Q^f = 1$. Note that our "compliant" output $s(\boldsymbol{\mu})$ is just an intermediate result for the calculation of the SIF. In particular, the virtual crack extension method (VCE) [30] allows us to extract the "Mode-I" SIF though the energy $s(\boldsymbol{\mu})$ though the Energy Release Rate (ERR), $G(\boldsymbol{\mu})$, defined by

$$G(\boldsymbol{\mu}) = -\left(\frac{\partial s(\boldsymbol{\mu})}{\partial \mu_1}\right).$$

In practice, the ERR is approximated by a finite-difference (FD) approach for a suitable small value $\delta\mu_1$ as

$$\widehat{G}(\boldsymbol{\mu}) = -\left(\frac{s(\boldsymbol{\mu} + \delta\mu_1) - s(\boldsymbol{\mu})}{\delta\mu_1}\right),$$

which then give the SIF approximation $\widehat{\mathrm{SIF}}(\boldsymbol{\mu}) = \sqrt{\widehat{G}(\boldsymbol{\mu})/(1 - \nu^2)}$.

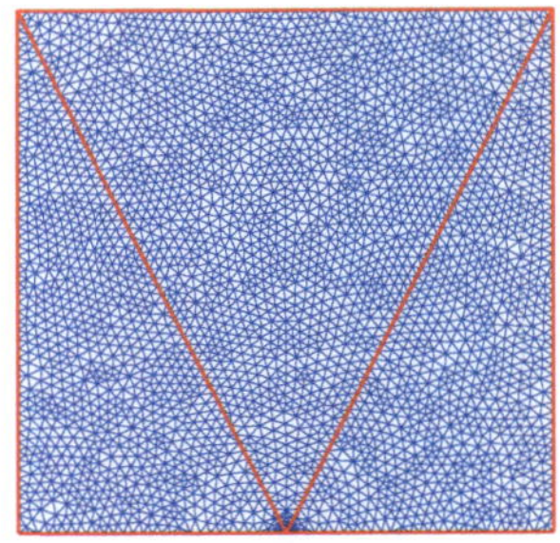

Fig. 8.5 The center crack problem: domain composition and FE mesh

Table 8.2 The center crack problem: RB convergence

N	$\mathcal{E}_N$	$\overline{\eta}^s_N$
5	2.73E−02	6.16
10	9.48E−04	8.47
20	5.71E−06	7.39
30	5.59E−08	7.01
40	8.91E−10	7.54
50	6.26E−11	8.32

We then consider a FE approximation with a mesh contains $n_{\text{node}} = 3257$ nodes and $n_{\text{elem}} = 6276$ P_1 elements, which corresponds to $\mathcal{N} = 6422$ degrees of freedoms; the mesh is refined around the crack tip in order to give a good approximation for the (singular) solution near this region as shown in Fig. 8.5.

We present in Table 8.2 the convergence results for the "compliant" output $s(\boldsymbol{\mu})$: the RB error bounds and effectivities as a function of N. The error bound reported, $\mathcal{E}_N = \Delta^s_N(\boldsymbol{\mu})/|s_N(\boldsymbol{\mu})|$ is the maximum of the relative error bound over a random test sample Ξ_{test} of size $n_{\text{test}} = 200$. We denote by $\overline{\eta}^s_N$ the average of the effectivity $\eta^s_N(\boldsymbol{\mu})$ over Ξ_{test}. We observe that the effectivity average is very sharp, and of order $O(10)$.

We next define the ERR RB approximation $\widehat{G}_N(\boldsymbol{\mu})$ to our "truth" (FE) $\widehat{G}^{\mathcal{N}}_{\text{FE}}(\boldsymbol{\mu})$ and its associated ERR RB error $\Delta^{\widehat{G}}_N(\boldsymbol{\mu})$ by

$$
\begin{aligned}
\widehat{G}_N(\boldsymbol{\mu}) &= \frac{s_N(\boldsymbol{\mu}) - \Delta^s_N(\boldsymbol{\mu} + \delta\mu_1)}{\delta\mu_1}, \\
\Delta^{\widehat{G}}_N(\boldsymbol{\mu}) &= \frac{\Delta^s_N(\boldsymbol{\mu} + \delta\mu_1) + \Delta^s_N(\boldsymbol{\mu})}{\delta\mu_1}.
\end{aligned}
\tag{8.42}
$$

It can be readily proven [36] that our SIF RB error is a rigorous bound for the ERR RB prediction $\widehat{G}_N(\boldsymbol{\mu})$: $|\widehat{G}_N(\boldsymbol{\mu}) - \widehat{G}^{\mathcal{N}}_{\text{FE}}(\boldsymbol{\mu})| \le \Delta^{\widehat{G}}_N(\boldsymbol{\mu})$. It is note that the choice of $\delta\mu_1$ is not arbitrary: $\delta\mu_1$ needed to be small enough to provide a good FD approximation, while still provide a good ERR RB error bound (8.42). Here we choose $\delta\mu_1 = 1\text{E} - 03$.

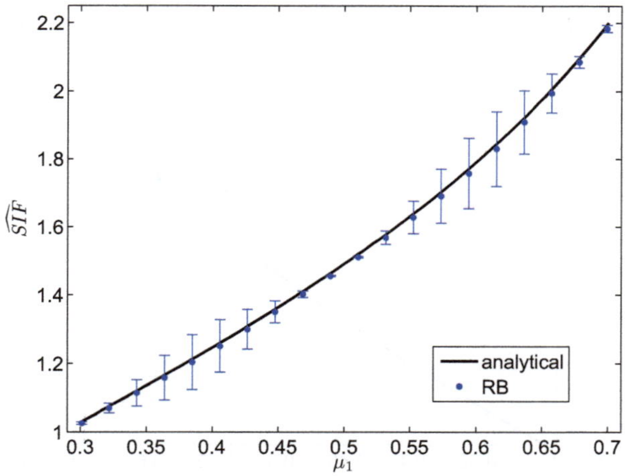

Fig. 8.6 The center crack problem: SIF solution for $N = 15$

We then can define the SIF RB approximation $\widehat{\mathrm{SIF}}_N(\boldsymbol{\mu})$ to our "truth" (FE) $\widehat{\mathrm{SIF}}^{\mathcal{N}}_{\mathrm{FE}}(\boldsymbol{\mu})$ and its associated SIF RB error estimation $\Delta_N^{\widehat{\mathrm{SIF}}}(\boldsymbol{\mu})$ as

$$\widehat{\mathrm{SIF}}_N(\boldsymbol{\mu}) = \frac{1}{2\sqrt{1-\nu^2}}\left\{\sqrt{\widehat{G}_N(\boldsymbol{\mu}) + \Delta_N^{\widehat{G}}(\boldsymbol{\mu})} + \sqrt{\widehat{G}_N(\boldsymbol{\mu}) - \Delta_N^{\widehat{G}}(\boldsymbol{\mu})}\right\},$$

$$\Delta_N^{\widehat{\mathrm{SIF}}}(\boldsymbol{\mu}) = \frac{1}{2\sqrt{1-\nu^2}}\left\{\sqrt{\widehat{G}_N(\boldsymbol{\mu}) + \Delta_N^{\widehat{G}}(\boldsymbol{\mu})} - \sqrt{\widehat{G}_N(\boldsymbol{\mu}) - \Delta_N^{\widehat{G}}(\boldsymbol{\mu})}\right\}.$$

It is readily proven in [20] that $|\widehat{\mathrm{SIF}}_N(\boldsymbol{\mu}) - \widehat{\mathrm{SIF}}^{\mathcal{N}}_{\mathrm{FE}}(\boldsymbol{\mu})| \leq \Delta_N^{\widehat{\mathrm{SIF}}}(\boldsymbol{\mu})$.

We plot the SIF RB results $\widehat{\mathrm{SIF}}(\boldsymbol{\mu})$ with error bars correspond to $\Delta_N^{\widehat{\mathrm{SIF}}}(\boldsymbol{\mu})$, and the analytical results $\widehat{\mathrm{SIF}}(\boldsymbol{\mu})$ [25] in Fig. 8.6 for the case $\mu_1 \in [0.3, 0.7]$, $\mu_2 = 2.0$ for $N = 15$. It is observed that the RB error is large since the small number of basis $N = 15$ does not compromise the small $\delta\mu_1 = 1\mathrm{E}-03$ value. We next plot, in Fig. 8.7, SIF RB results and error for the same $\boldsymbol{\mu}$ range as in Fig. 8.6, but for $N = 30$. It is observed now that the SIF RB error is significantly improved—thanks to the better RB approximation that compensates the small value $\delta\mu_1$. We also want to point out that, in both Figs. 8.6 and 8.7, it is clearly shown that our RB SIF error is not a *rigorous* bound for the *exact* SIF values $\widehat{\mathrm{SIF}}(\boldsymbol{\mu})$ but rather is a *rigorous* bound for the "truth" (FE) approximation $\widehat{\mathrm{SIF}}^{\mathcal{N}}_{\mathrm{FE}}(\boldsymbol{\mu})$. It is shown, however, that FE SIF approximation (which is considered in Fig. 8.7 thanks to the negligible RB error) are of good quality compared with the exact SIF. The VCE in this case works quite well, however it is not suitable for complicate crack settings. In such cases, other SIF calculation methods and appropriate RB approximations might be preferable [19, 20].

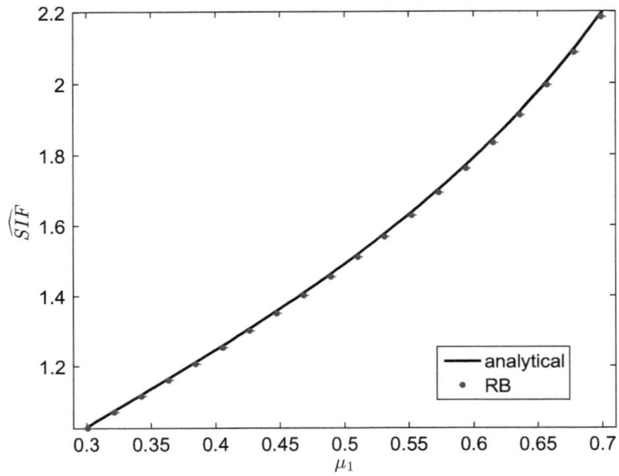

Fig. 8.7 The center crack problem: SIF solution for $N = 30$

As regards computational times, a RB online evaluation $\boldsymbol{\mu} \rightarrow (\widehat{\mathrm{SIF}}_N(\boldsymbol{\mu}), \Delta_N^{\widehat{\mathrm{SIF}}}(\boldsymbol{\mu}))$ requires just $t_{\mathrm{RB}}(= \; 25\times) \; = \; 50$(ms) for $N \; = \; 40$; while the FE solution $\boldsymbol{\mu} \rightarrow \widehat{\mathrm{SIF}}_{\mathrm{FE}}^{\mathcal{N}}(\boldsymbol{\mu})$ requires $t_{\mathrm{FE}}(= 7 \times 2) = 14$(s): thus our RB online evaluation takes only 0.36% of the FEM computational cost.

8.6.3 The Composite Unit Cell Problem

We consider a unit cell contains an ellipse region as shown in Fig. 8.8. We apply (clamped) Dirichlet boundary conditions on the bottom of the cell Γ_B^{o} and (unit tension) non-homogeneous Neumann boundary conditions on Γ_T^{o}. We denote the two semimajor axis and semiminor axis of the ellipse region as d_1 and d_2, respectively. We assume plane stress isotropic materials: the material properties of the matrix (outside of the region) is given by $(E_m, \nu_m) = (1, 0.3)$, and the material properties of the ellipse region is given by $(E_f, \nu_f) = (E_f, 0.3)$. Our output of interest is the integral of normal displacement (u_1) over Γ_T^{o}. We note our output of interest is thus "compliant".

We consider $P = 3$ parameters $\boldsymbol{\mu} = [\mu_1, \mu_2, \mu_3] \equiv [d_1, d_2, E_f]$. The parameter domain is chosen as $\mathcal{D} = [0.8, 1.2] \times [0.8, 1.2] \times [0.2, 5]$. Note that the third parameter (the Young modulus of the ellipse region) can represent the ellipse region from an "inclusion" (with softer Young's modulus $E_f < E_m(= 1)$) to a "fiber" (with stiffer Young's modulus $E_f > E_m(= 1)$).

We then choose $\boldsymbol{\mu}_{\mathrm{ref}} = [1.0, 1.0, 1.0]$ and apply the domain decomposition [36] and obtain $L_{\mathrm{reg}} = 34$ subdomains, in which 16 subdomains are the general "curvy triangles" (8 inward "curvy triangles" and 8 outward curvy "triangles") as shown

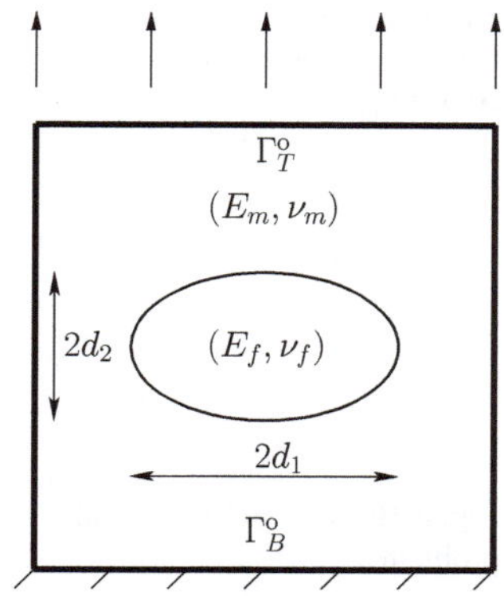

Fig. 8.8 The composite unit cell problem

Fig. 8.9 The composite unite cell problem: domain composition and FE mesh

in Fig. 8.9. However, despite the large number of "curvy triangles" in the domain decomposition, it is observed that almost all transformations are congruent, hence we expected a small number of Q^a than (says), that of the "arc-cantilever beam" example, in which all the subdomains transformations are different. Indeed, we recover our affine forms with $Q^a = 30$ and $Q^f = 1$, note that Q^a is relatively small for such a complex domain decomposition thanks to our efficient symbolic manipulation "collapsing" technique and those congruent "curvy triangles".

We next consider a FE approximation where the mesh contains $n_{\text{node}} = 3906$ nodes and $n_{\text{elem}} = 7650$ P_1 elements, which corresponds to $\mathcal{N} = 7730$ degrees of freedoms. The mesh is refined around the interface of the matrix and the inclusion/fiber.

We then apply the RB approximation. We present in Table 8.3 our convergence results: the RB error bounds and effectivities as a function of N. The error bound reported, $\mathcal{E}_N = \Delta^s_N(\boldsymbol{\mu})/|s_N(\boldsymbol{\mu})|$ is the maximum of the relative error bound over a random test sample Ξ_{test} of size $n_{\text{test}} = 200$. We denote by $\overline{\eta}^s_N$ the average of the effectivity $\eta^s_N(\boldsymbol{\mu})$ over Ξ_{test}. We observe that our effectivity average is of order $O(10)$.

As regards computational times, a RB online evaluation $\boldsymbol{\mu} \rightarrow (s_N(\boldsymbol{\mu}), \Delta^s_N(\boldsymbol{\mu}))$ requires just $t_{\text{RB}} = 66$(ms) for $N = 30$; while the FE solution $\boldsymbol{\mu} \rightarrow s^{\mathcal{N}}(\boldsymbol{\mu})$ requires approximately $t_{\text{FE}} = 8$(s): thus our RB online evaluation is just 0.83% of the FEM computational cost.

Table 8.3 The composite unit cell problem: RB convergence

N	$\mathcal{E}_N$	$\overline{\eta}^s_N$
5	9.38E−03	8.86
10	2.54E−04	7.18
15	1.37E−05	5.11
20	3.91E−06	9.74
25	9.09E−07	6.05
30	2.73E−07	10.64
35	9.00E−08	10.17
40	2.66E−08	10.35

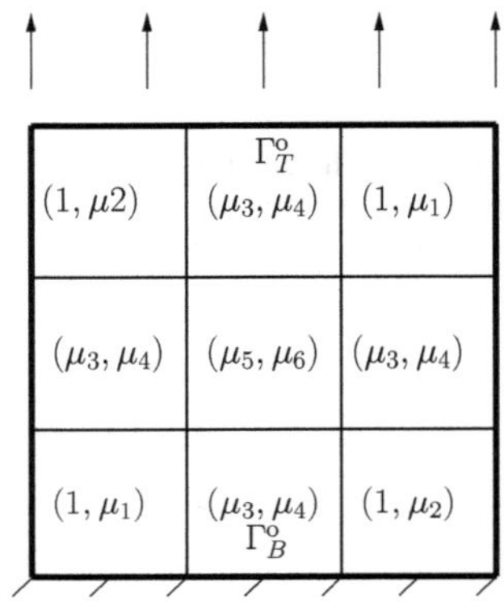

Fig. 8.10 The multi-material problem

8.6.4 The Multi-Material Plate Problem

We consider a unit cell divided into 9 square subdomains of equal size as shown in Fig. 8.10. We apply (clamped) Dirichlet boundary conditions on the bottom of the cell Γ_B^o and (unit tension) non-homogeneous Neumann boundary conditions on Γ_T^o. We consider orthotropic plane stress materials: the Young's modulus properties for all 9 subdomains are given in Figure, the Poisson's ratio is chosen as $\nu_{12,i} = 0.3$, $i = 1, \ldots, 9$ and $\nu_{21,i}$ is determined by (8.44). The shear modulus is chosen as a function of the two Young's moduli as in (8.45) for all 9 subdomains. All material axes are aligned with the coordinate system (and loading). Our output of interest is the integral of normal displacement (u_1) over Γ_T^o, which represents the average normal displacement on Γ_T^o. We note our output of interest is thus "compliant".

We consider $P = 6$ parameters $\boldsymbol{\mu} = [\mu_1, \ldots, \mu_6]$, correspond to the six Young's moduli values as shown in Fig. 8.10 (the two Young's moduli for each subdomain are shown in those brackets). The parameter domain is chosen as $\mathcal{D} = [0.5, 2.0]^6$.

We then apply the domain decomposition [36] and obtain $L_{\text{reg}} = 18$ subdomains. Despite the large L_{reg} number of domains, there is no geometric transformation in this case. We recover our affine forms with $Q^a = 12$, $Q^f = 1$, note that all Q^a are contributed from all the Young's moduli since there is no geometric transformation

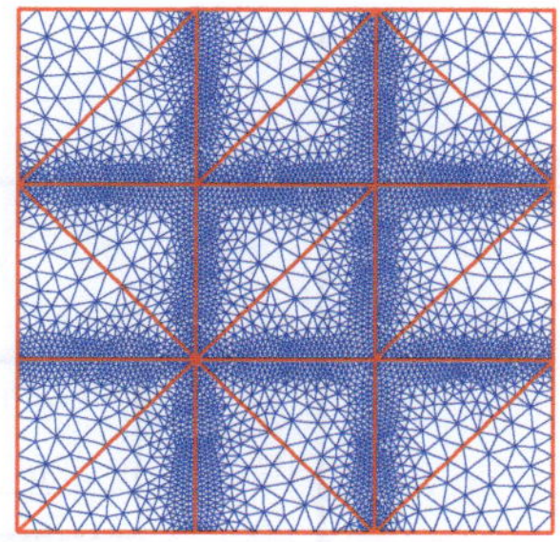

Fig. 8.11 The multi-material problem: domain composition and FE mesh

Table 8.4 The multi-material problem: RB convergence

N	$\mathcal{E}_N$	$\overline{\eta}^s_N$
5	1.01E−02	8.11
10	1.45E−03	11.16
20	3.30E−04	11.47
30	1.12E−04	12.59
40	2.34E−05	11.33
50	9.85E−06	12.90

involved. Moreover, it is observed that the bilinear form can be, in fact, classified as a "parametrically coercive" one [31].

We next consider a FE approximation where the mesh contains $n_{\text{node}} = 4098$ nodes and $n_{\text{elem}} = 8032$ P_1 elements, which corresponds to $\mathcal{N} = 8112$ degrees of freedoms. The mesh is refined around all the interfaces between different subdomains as shown in Fig. 8.11.

We then apply the RB approximation. We present in Table 8.4 our convergence results: the RB error bounds and effectivities as a function of N. The error bound reported, $\mathcal{E}_N = \Delta^s_N(\boldsymbol{\mu})/|s_N(\boldsymbol{\mu})|$ is the maximum of the relative error bound over a random test sample Ξ_{test} of size $n_{\text{test}} = 200$. We denote by $\overline{\eta}^s_N$ the average of the effectivity $\eta^s_N(\boldsymbol{\mu})$ over Ξ_{test}. We observe that our effectivity average is of order $O(10)$.

As regards computational times, a RB online evaluation $\boldsymbol{\mu} \to (s_N(\boldsymbol{\mu}), \Delta^s_N(\boldsymbol{\mu}))$ requires just $t_{\text{RB}} = 33$(ms) for $N = 40$; while the FE solution $\boldsymbol{\mu} \to s^{\mathcal{N}}(\boldsymbol{\mu})$ requires $t_{\text{FE}} = 8.1$(s): thus the RB online evaluation is just 0.41% of the FEM computational cost.

8.6.5 The Woven Composite Beam Problem

We consider a composite cantilever beam as shown in Fig. 8.12. The beam is divided into two regions, each with a square hole in the center of (equal) size $2w$. We apply (clamped) Dirichlet boundary conditions on the left side of the beam Γ^{o}_L, (symmetric about the x^{o}_1 direction) Dirichlet boundary conditions on the right side of

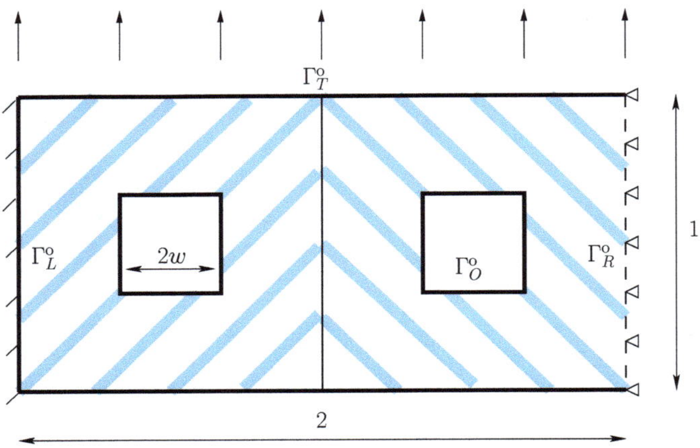

Fig. 8.12 The woven composite beam problem

the beam Γ_R^o, and (unit tension) non-homogeneous Neumann boundary conditions on the top side Γ_T^o. We consider the same orthotropic plane stress materials for both regions: $(E_1, E_2) = (1, E_2)$, $\nu_{12} = 0.3$, ν_{21} is determined by (8.44) and the shear modulus G_{12} is given by (8.45). The material axes of both regions are not aligned with the coordinate system and loading: the angles of the material axes and the coordinate system of the first and second region are θ and $-\theta$, respectively. The setting represents a "woven" composite material across the beam horizontally. Our output of interest is the integral of the normal displacement (u_1) over the boundary Γ_O^o. We note our output of interest is thus "non-compliant".

We consider $P = 3$ parameters $\boldsymbol{\mu} = [\mu_1, \mu_2, \mu_3] \equiv [w, E_2, \theta]$. The parameter domain is chosen as $\mathcal{D} = [1/6, 1/12] \times [1/2, 2] \times [-\pi/4, \pi/4]$.

We then apply the domain decomposition [36] and obtain $L_{\text{reg}} = 32$ subdomains, note that all subdomains transformations are just simply translations due to the "added control points" along the external (and interface) boundaries strategy [36]. We recover the affine forms with $Q^a = 19$, $Q^f = 2$, and $Q^\ell = 1$.

We next consider a FE approximation where the mesh contains $n_{\text{node}} = 3569$ nodes and $n_{\text{elem}} = 6607$ P_1 elements, which corresponds to $\mathcal{N} = 6865$ degrees of freedoms. The mesh is refined around the holes, the interfaces between the two regions, and the clamped boundary as shown in Fig. 8.13.

We then apply the RB approximation. We present in Table 8.5 our convergence results: the RB error bounds and effectivities as a function of N. The error bound reported, $\mathcal{E}_N = \Delta_N^s(\boldsymbol{\mu})/|s_N(\boldsymbol{\mu})|$ is the maximum of the relative error bound over a random test sample Ξ_{test} of size $n_{\text{test}} = 200$. We denote by $\overline{\eta}_N^s$ the average of the effectivity $\eta_N^s(\boldsymbol{\mu})$ over Ξ_{test}. We observe that our effectivity average is of order $O(5-25)$.

As regards computational times, a RB online evaluation $\boldsymbol{\mu} \to (s_N(\boldsymbol{\mu}), \Delta_N^s(\boldsymbol{\mu}))$ requires just $t_{\text{RB}} = 40$(ms) for $N = 20$; while the FE solution $\boldsymbol{\mu} \to s^{\mathcal{N}}(\boldsymbol{\mu})$ requires

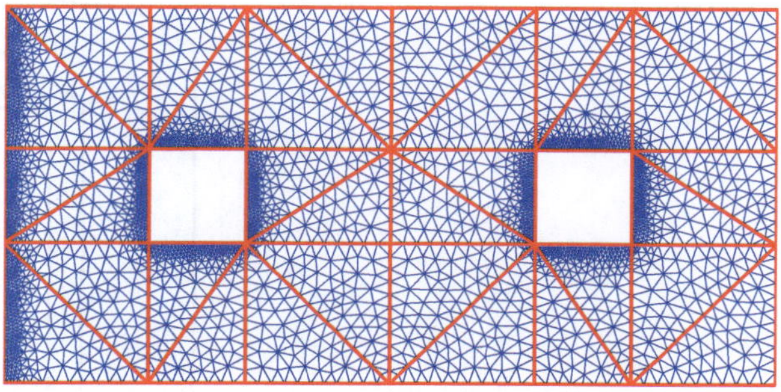

Fig. 8.13 The woven composite beam problem: domain composition and FE mesh

Table 8.5 The woven composite beam problem: RB convergence

N	$\mathcal{E}_N$	$\overline{\eta}^s_N$
4	4.64E−02	22.66
8	1.47E−03	7.39
12	2.35E−04	9.44
16	6.69E−05	14.29
20	1.31E−05	11.41

$t_{\text{FE}} = 7.5$(s): thus our RB online evaluation is just 0.53% of the FEM computational cost.

8.6.6 The Closed Vessel Problem

We consider a closed vessel under tension at both ends as shown in Fig. 8.14. The vessel is axial symmetric about the x_2^{o} axis, and symmetric about the x_1^{o} axis, hence we only consider a representation "slice" by our axisymmetric formulation as shown in Fig. 8.15. The vessel is consists of two layered, the outer layer is of fixed width $w^{\text{out}} = 1$, while the inner layer is of width $w^{\text{in}} = w$. The material properties of the inner layer and outer layer are given by $(E^{\text{in}}, \nu) = (E^{\text{in}}, 0.3)$ and $(E^{\text{out}}, \nu) = (1, 0.3)$, respectively. We apply (symmetric about the x_2^{o} direction) Dirichlet boundary conditions on the bottom boundary of the model Γ_B^{o}, (symmetric about the x_1^{o} direction) Dirichlet boundary conditions on the left boundary of the model Γ_L^{o} and (unit tension) non-homogeneous Neumann boundary conditions on the top boundary Γ_T^{o}. Our output of interest is the integral of the axial displacement (u_r) over the right boundary Γ_R^{o}. We note our output of interest is thus "non-compliant".

We consider $P = 2$ parameters $\boldsymbol{\mu} = [\mu_1, \mu_2] \equiv [w, E^{\text{in}}]$. The parameter domain is chosen as $\mathcal{D} = [0.1, 1.9] \times [0.1, 10]$.

We then apply the domain decomposition [36] and obtain $L_{\text{reg}} = 12$ subdomains as shown in Fig. 8.16. We recover our affine forms with $Q^a = 47$, $Q^f = 1$,

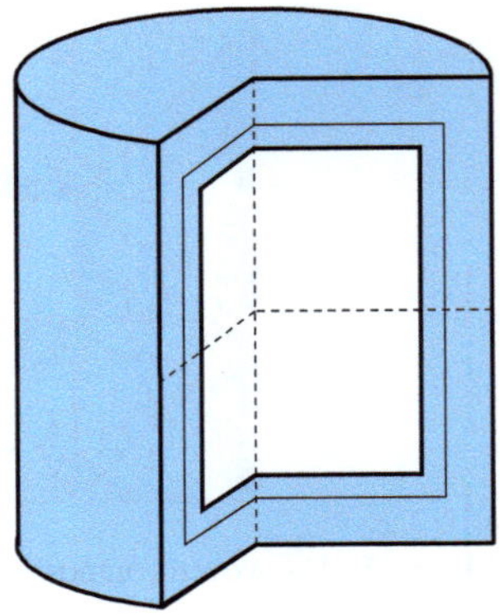

Fig. 8.14 The closed vessel problem

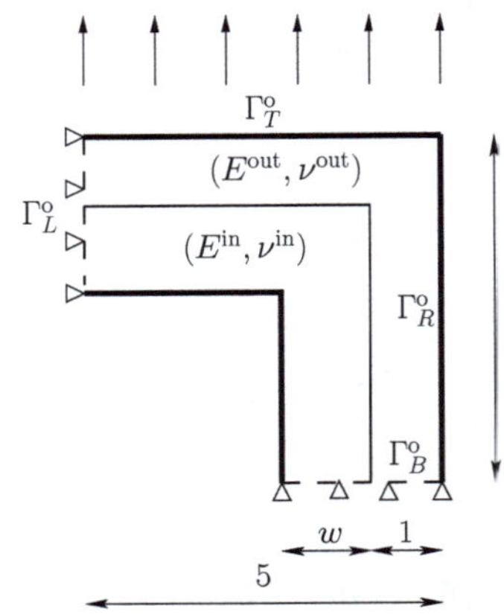

Fig. 8.15 The closed vessel problem

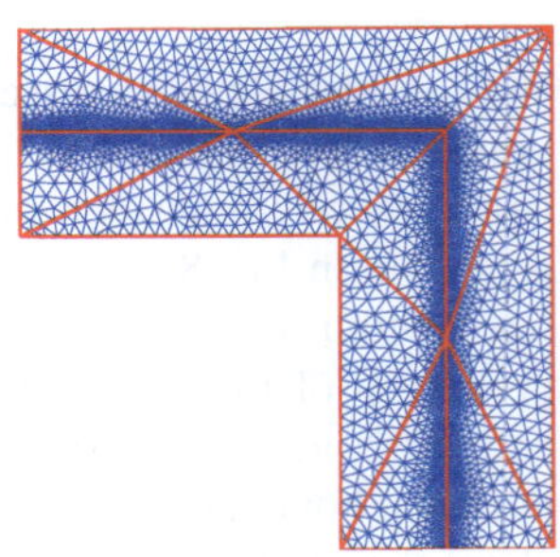

Fig. 8.16 The closed vessel problem: domain composition and FE mesh

and $Q^\ell = 1$. Despite the small number of parameter (and seemingly simple transformations), Q^a is large in this case. A major contribution to Q^a come from the expansion of the terms x_1^{o} in the effective elastic tensor [**S**], which appeared due to the geometric transformation of the inner layer.

We next consider a FE approximation where the mesh contains $n_{\text{node}} = 3737$ nodes and $n_{\text{elem}} = 7285$ P_1 elements, which corresponds to $\mathcal{N} = 7423$ degrees of freedoms. The mesh is refined around the interfaces between the two layers.

Table 8.6 The closed vessel problem: RB convergence

N	$\mathcal{E}_N$	$\overline{\eta}^s_N$
10	7.12E−02	56.01
20	1.20E−03	111.28
30	3.96E−05	49.62
40	2.55E−06	59.96
50	5.70E−07	113.86
60	5.90E−08	111.23
70	6.95E−09	77.12

We then apply the RB approximation. We present in Table 8.6 convergence results: the RB error bounds and effectivities as a function of N. The error bound reported, $\mathcal{E}_N = \Delta^s_N(\boldsymbol{\mu})/|s_N(\boldsymbol{\mu})|$ is the maximum of the relative error bound over a random test sample Ξ_{test} of size $n_{\text{test}} = 200$. We denote by $\overline{\eta}^s_N$ the average of the effectivity $\eta^s_N(\boldsymbol{\mu})$ over Ξ_{test}. We observe that our effectivity average is of order $O(50-120)$, which is quite large, however it is not surprising since our output is "non-compliant".

As regards computational times, a RB online evaluation $\boldsymbol{\mu} \to (s_N(\boldsymbol{\mu}), \Delta^s_N(\boldsymbol{\mu}))$ requires just $t_{\text{RB}} = 167$(ms) for $N = 40$; while the FE solution $\boldsymbol{\mu} \to s^{\mathcal{N}}(\boldsymbol{\mu})$ requires $t_{\text{FE}} = 8.2$(s): thus our RB online evaluation is just 2.04% of the FEM computational cost.

8.6.7 The Von Kármán Plate Problem

We consider now a different problem that can be derived from the classical elasticity equations [13, 14]. It turns out to be nonlinear and brings with it a lot of technical difficulties. Let us consider an elastic, bidimensional and rectangular plate $\Omega = [0, l] \times [0, 1]$ in its undeformed state, subjected to a μ-parametrized external load acting on its edge, then the *Airy stress potential* and the deformation from its flat state, respectively ϕ and u are defined by the Von Kármán equations

$$\begin{cases} \Delta^2 u + \mu u_{xx} = [\phi, u] + f\ , & \text{in } \Omega \\ \Delta^2 \phi = -\,[u, u]\ , & \text{in } \Omega \end{cases} \tag{8.43}$$

where

$$\Delta^2 := \Delta\Delta = \left(\frac{\partial^2}{\partial x^2} + \frac{\partial^2}{\partial y^2}\right)^2 ,$$

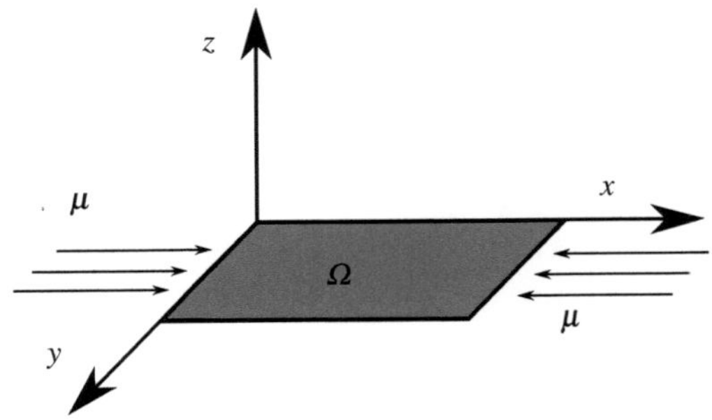

Fig. 8.17 A rectangular bidimensional elastic plate compressed on its edges

is the biharmonic operator and

$$[u,\phi] := \frac{\partial^2 u}{\partial x^2}\frac{\partial^2 \phi}{\partial y^2} - 2\frac{\partial^2 u}{\partial x \partial y}\frac{\partial^2 \phi}{\partial x \partial y} + \frac{\partial^2 u}{\partial y^2}\frac{\partial^2 \phi}{\partial x^2},$$

is the *bracket of Monge-Ampére*. So we have a system of two nonlinear and parametrized equations of the fourth order with μ the parameter that measures the compression along the sides of the plate (Fig. 8.17).

From the mathematical point of view, we will suppose the plate is simply supported, i.e. that holds boundary conditions

$$u = \Delta u = 0, \qquad \phi = \Delta\phi = 0, \qquad \text{on } \partial\Omega.$$

In this model problem we are interested in the study of stability and uniqueness of the solution for a given parameter. In fact due to the nonlinearity of the bracket we obtain the so called *buckling phenomena* [43], that is the main feature studied in bifurcations theory. What we seek is the critical value of μ for which the stable (initial configuration) solution become unstable while there are two new stable and symmetric solutions.

To detect this value we need a very complex algorithm that mixes a *continuation method*, a nonlinear solver and finally a full-order method to find the buckled state. At the end for every $\mu \in \mathcal{D}_{train}$ (a fine discretization of the parameter domain $\mathcal{D}$) we have a loop due to the nonlinearity, for which at each iteration we have to solve the Finite Element method applied to the weak formulation of the problem.

Here we consider $P = 1$ parameter μ and its domain is suitably chosen[6] as $\mathcal{D} = [30, 70]$.

Also in this case we can simply recover the affine forms with $Q^a = 3$. For the rectangular plate test case with $l = 2$ we applied the Finite Element method, with $n_{node} = 441$ nodes and $n_{elem} = 800$ P_2 elements, which corresponds to $\mathcal{N} = 6724$ degrees of freedom. We stress on the fact that the linear system obtained by the

[6]It is possible to show that the bifurcation point is related to the eigenvalue of the linearized model [5], so we are able to set in a proper way the range of the parameter domain.

Galerkin projection has to be solved at each step of the nonlinear solver, here we chose the classic *Newton method* [33].

Moreover, for a given parameter, we have to solve a FE system until Newton method converges just to obtain one of the possible solutions of our model; keeping in mind that we do not know a priori where is the bifurcation point and we have to investigate the whole parameter domain. It is clear that despite the simple geometry and the quite coarse mesh, the reduction strategies are fundamental in this kind of applications.

For example, in order to plot a *bifurcation diagram* like the one in Fig. 8.18, the full order code running on a standard computer takes approximately 1 h.

Once selected a specific parameter, $\lambda = 70$, we can see in Fig. 8.19 the two solutions that belong to the different branches of the plot reported in Fig. 8.18.

We then applied RB approximation and present in Table 8.7 a convergence results: the error between the truth approximation and the reduced one as a function of N. The error reported, $\mathcal{E}_N = \max_{\mu \in \mathcal{D}} ||\mathbf{u}^{\mathcal{N}}(\mu) - \mathbf{u}^{\mathcal{N}}_{\mathrm{RB},N}(\mu)||_X$ is the maximum of the approximation error over a uniformly chosen test sample.

As we can see in Fig. 8.20 e obtain very good results with a low number of snapshots due to the strong properties of the underlying biharmonic operator.

A suitable extension for the a posteriori error estimate of the solution can be obtained by applying Brezzi-Rappaz-Raviart (BRR) theory on the numerical approximation of nonlinear problems [6–9, 16]. However, the adaptation of BRR theory to RB methods in bifurcating problems is not straightforward, and we leave it for further future investigation [32].

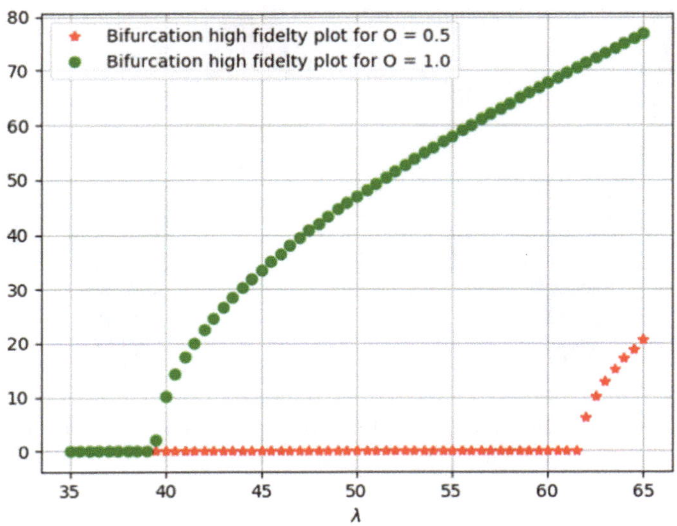

Fig. 8.18 Bifurcation diagram for a square plate and different initial guess for Newton method, on y-axis is represented the infinite norm of the solution

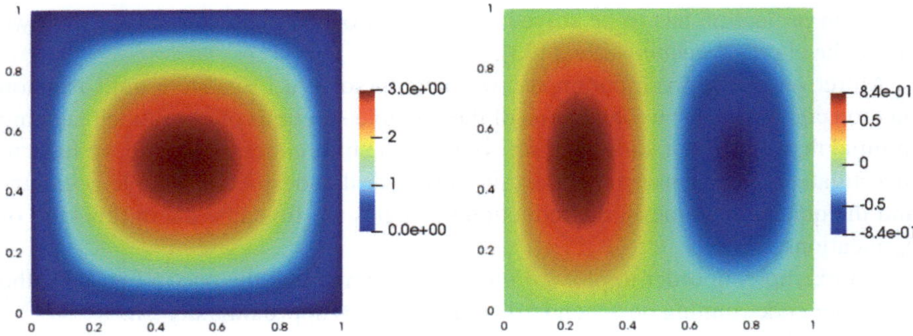

Fig. 8.19 Contour plot of the two solutions belonging to the green and red branches of the bifurcation diagram for $\lambda = 70$, respectively

Table 8.7 The Von Kármán plate problem: RB convergence

N	$\mathcal{E}_N$
1	6.61E+00
2	6.90E−01
3	7.81E−02
4	2.53E−02
5	1.88E−02
6	1.24E−02
7	9.02E−03
8	8.46E−03

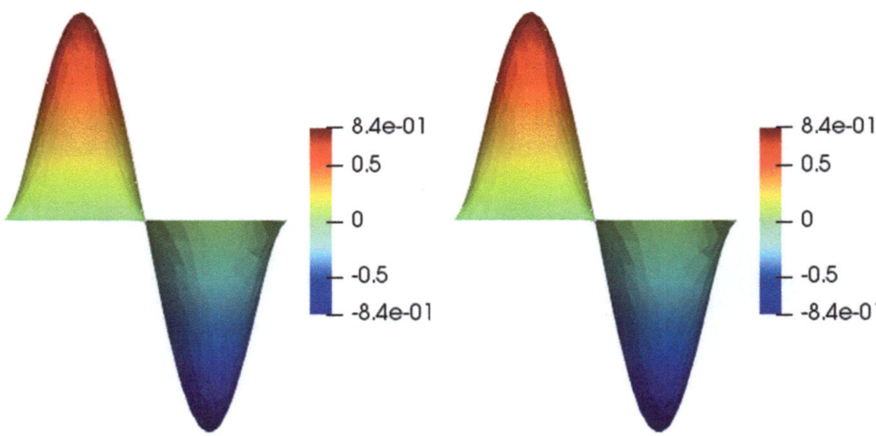

Fig. 8.20 Comparison between the full order solution (left) and reduced order one (right) for $\lambda = 65$

As regards computational times, a RB online evaluation $\boldsymbol{\mu} \to \mathbf{u}^{\mathcal{N}}_{\mathrm{RB},N}(\boldsymbol{\mu})$ requires just $t_{\mathrm{RB}} = 100(\mathrm{ms})$ for $N = 8$; while the FE solution $\boldsymbol{\mu} \to \mathbf{u}^{\mathcal{N}}(\boldsymbol{\mu})$ requires $t_{\mathrm{FE}} = 8.17(\mathrm{s})$: thus our RB online evaluation is just 1.22% of the FEM computational cost.

8.7 Conclusions

We have provided some examples of applications of reduced basis methods in linear elasticity problems depending also on many parameters of different kind (geometrical, physical, engineering) using different linear elasticity approximations, a 2D Cartesian setting or a 3D axisymmetric one, different material models (isotropic and orthotropic), as well as an overview on nonlinear problems. Reduced basis methods have confirmed a very good computational performance with respect to a classical finite element formulation, not very suitable to solve parametrized problems in the real-time and many-query contexts. We have extended and generalized previous work [24] with the possibility to treat with more complex outputs by introducing a dual problem [36]. Another very important aspect addressed in this work is the certification of the errors in the reduced basis approximation by means of a posteriori error estimators, see for example [21]. This work looks also at more complex 3D parametrized applications (not only in the special axisymmetric case) as quite promising problem to be solved with the same certified methodology [11, 42].

Acknowledgements We are sincerely grateful to Prof. A.T. Patera (MIT) and Dr. C.N. Nguyen (MIT) for important suggestions, remarks, insights, and codevelopers of the `rbMIT` and `RBniCS` (http://mathlab.sissa.it/rbnics) software libraries used for the numerical tests. We acknowledge the European Research Council consolidator grant H2020 ERC CoG 2015 AROMA-CFD GA 681447 (PI Prof. G. Rozza).

Appendix

Stress-Strain Matrices

In this section, we denote E_i, $i = 1, 3$ as the Young's moduli, ν_{ij}; $i, j = 1, 2, 3$ as the Poisson ratios; and G_{12} as the shear modulus of the material.

Isotropic Cases

For both of the following cases, $E = E_1 = E_2$, and $\nu = \nu_{12} = \nu_{21}$.

Isotropic plane stress:

$$[\mathbf{E}] = \frac{E}{(1-\nu^2)} \begin{bmatrix} 1 & \nu & 0 \\ \nu & 1 & 0 \\ 0 & 0 & 2(1+\nu) \end{bmatrix}.$$

Isotropic plane strain:

$$[\mathbf{E}] = \frac{E}{(1+\nu)(1-2\nu)} \begin{bmatrix} 1 & \nu & 0 \\ \nu & 1 & 0 \\ 0 & 0 & 2(1+\nu) \end{bmatrix}.$$

Orthotropic Cases

Here we assume that the orthotropic material axes are aligned with the axes used for the analysis of the structure. If the structural axes are not aligned with the orthotropic material axes, orthotropic material rotation must be rotated by with respect to the structural axes. Assuming the angle between the orthogonal material axes and the structural axes is θ, the stress-strain matrix is given by $[\mathbf{E}] = [\boldsymbol{T}(\theta)][\hat{\mathbf{E}}][\boldsymbol{T}(\theta)]^T$, where

$$[\boldsymbol{T}(\theta)] = \begin{bmatrix} \cos^2\theta & \sin^2\theta & -2\sin\theta\cos\theta \\ \sin^2\theta & \cos^2\theta & 2\sin\theta\cos\theta \\ \sin\theta\cos\theta & -sin\theta\cos\theta & \cos^2\theta - \sin^2\theta \end{bmatrix}.$$

Orthotropic plane stress:

$$[\hat{\mathbf{E}}] = \frac{1}{(1-\nu_{12}\nu_{21})} \begin{bmatrix} E_1 & \nu_{12}E_1 & 0 \\ \nu_{21}E_2 & E_2 & 0 \\ 0 & 0 & (1-\nu_{12}\nu_{21})G_{12} \end{bmatrix}.$$

Note here that the condition

$$\nu_{12}E_1 = \nu_{21}E_2 \tag{8.44}$$

must be required in order to yield a symmetric [**E**].

Orthotropic plane strain:

$$[\hat{\mathbf{E}}] = \frac{1}{\Lambda} \begin{bmatrix} (1-\nu_{23}\nu_{32})E_1 & (\nu_{12}+\nu_{13}\nu_{32})E_1 & 0 \\ (\nu_{21}+\nu_{23}\nu_{31})E_2 & (1-\nu_{13}\nu_{31})E_2 & 0 \\ 0 & 0 & \Lambda G_{12} \end{bmatrix}.$$

Here $\Lambda = (1-\nu_{13}\nu_{31})(1-\nu_{23}\nu_{32}) - (\nu_{12}+\nu_{13}\nu_{32})(\nu_{21}+\nu_{23}\nu_{31})$. Furthermore, the following conditions,

$$\nu_{12}E_1 = \nu_{21}E_2, \quad \nu_{13}E_1 = \nu_{31}E_3, \quad \nu_{23}E_2 = \nu_{32}E_3,$$

must be satisfied, which leads to a symmetric [**E**].

An reasonable good approximation for the shear modulus G_{12} in orthotropic case is given by [10] as

$$\frac{1}{G_{12}} \approx \frac{(1+\nu_{21})}{E_1} + \frac{(1+\nu_{12})}{E_2}. \tag{8.45}$$

References

1. Almroth, B.O., Stern, P., Brogan, F.A.: Automatic choice of global shape functions in structural analysis. AIAA J. **16**, 525–528 (1978)
2. Benner, P., Gugercin, S., Willcox, K.: A survey of projection-based model reduction methods for parametric dynamical systems. SIAM Rev. **57**(4), 483–531 (2015)
3. Benner, P., Cohen, A., Ohlberger, M., Willcox, K.: Model Reduction and Approximation: Theory and Algorithms. Computational Science and Engineering Series. Society for Industrial and Applied Mathematics, Philadelphia (2017)
4. Benner, P., Ohlberger, M., Patera, A., Rozza, G., Urban, K. (eds.): Model Reduction of Parametrized Systems. Springer, Berlin (2017)
5. Berger, M.: On Von Kármán equations and the buckling of a thin elastic plate, I the clamped plate. Commun. Pure Appl. Math. **20**, 687–719 (1967)
6. Brezzi, F., Rappaz, J., Raviart, P.A.: Finite dimensional approximation of nonlinear problems. part I: branches of non singular solutions. Numer. Math. **36**(1), 1–25 (1980)
7. Brezzi, F., Rappaz, J., Raviart, P.A.: Finite dimensional approximation of nonlinear problems. part II: limit points. Numer. Math. **37**(1), 1–28 (1981)
8. Brezzi, F., Rappaz, J., Raviart, P.A.: Finite dimensional approximation of nonlinear problems. part III: simple bifurcation points. Numer. Math. **38**(1), 1–30 (1982)
9. Canuto, C., Tonn, T., Urban, K.: A posteriori error analysis of the reduced basis method for nonaffine parametrized nonlinear PDEs. SIAM J. Numer. Anal. **47**(3), 2001–2022 (2009)
10. Carroll, W.F.: A Primer for Finite Elements in Elastic Structures. Wiley, Hoboken (1998)
11. Chinesta, F., Ladevèze, P.: Separated Representations and PGD-Based Model Reduction: Fundamentals and Applications. CISM International Centre for Mechanical Sciences. Springer, Vienna (2014)
12. Chinesta, F., Huerta, A., Rozza, G., Willcox, K.: Model order reduction: a survey. In: Wiley Encyclopedia of Computational Mechanics. Wiley, Hoboken (2016)
13. Ciarlet, P.G.: Mathematical Elasticity, Volume I: Three-Dimensional Elasticity. Elsevier, Amsterdam (1988)
14. Ciarlet, P.G.: Mathematical Elasticity: Volume II: Theory of Plates. Studies in Mathematics and its Applications. Elsevier, Amsterdam (1997)
15. Grepl, M.A., Patera, A.T.: *A posteriori* error bounds for reduced-basis approximations of parametrized parabolic partial differential equations. Math. Model. Numer. Anal. **39**(1), 157–181 (2005)
16. Grepl, M., Maday, Y., Nguyen, N., Patera, A.: Efficient reduced-basis treatment of nonaffine and nonlinear partial differential equations. Math. Model. Numer. Anal. **41**(3), 575–605 (2007)
17. Hesthaven, J., Rozza, G., Stamm, B.: Certified Reduced Basis Methods for Parametrized Partial Differential Equations. Springer Briefs in Mathematics. Springer, Berlin (2015)
18. Hutchingson, J.: Nonlinear Fracture Mechanics. Monograph, Department of Solid Mechanics, Technical University, Lyngby (1979)
19. Huynh, D.B.P.: Reduced-basis approximation and applications in fracture mechanics. Ph.D. thesis, Singapore-MIT Alliance, National University of Singapore (2007)
20. Huynh, D.B.P., Patera, A.T.: Reduced basis approximation and a posteriori error estimation for stress intensity factors. Int. J. Numer. Methods Eng. **72**(10), 1219–1259 (2007)

21. Huynh, D.B.P., Rozza, G., Sen, S., Patera, A.T.: A successive constraint linear optimization method for lower bounds of parametric coercivity and inf-sup stability constants. CR Acad. Sci. Paris Ser. I **345**, 473–478 (2007)
22. Huynh, D.B.P., Nguyen, N.C., Rozza, G., Patera, A.T.: rbMIT software http://augustine.mit.edu/methodology/methodology_rbMIT_System.htm. Copyright MIT, Technology Licensing Office, case 12600, Cambridge, MA (2007–2009)
23. Huynh, D., Knezevic, D.J., Chen, Y., Hesthaven, J.S., Patera, A.T.: A natural-norm Successive Constraint Method for inf-sup lower bounds. Comput. Methods Appl. Mech. Eng. **199**(29–32), 1963–1975 (2010)
24. Milani, R., Quarteroni, A., Rozza, G.: Reduced basis method for linear elasticity problems with many parameters. Comput. Methods Appl. Mech. Eng. **197**, 4812–4829 (2008)
25. Murakami, Y.: Stress Intensity Factors Handbook. Elsevier, Amsterdam (2001)
26. Nguyen, N., Veroy, K., Patera, A.: Certified real-time solution of parametrized partial differential equations. In: Yip, S. (ed.) Handbook of Materials Modeling: Methods, pp. 1529–1564. Springer, Dordrecht (2005)
27. Noor, A.K.: Recent advances in reduction methods for nonlinear problems. Comput. Struct. **13**, 31–44 (1981)
28. Noor, A.K.: On making large nonlinear problems small. Comput. Methods Appl. Mech. Eng. **34**, 955–985 (1982)
29. Noor, A.K., Peters, J.M.: Reduced basis technique for nonlinear analysis of structures. AIAA J. **18**(4), 455–462 (1980)
30. Parks, D.M.: A stiffness derivative finite element technique for determination of crack tip stress intensity factors. Int. J. Fract. **10**(4), 487–502 (1974)
31. Patera, A., Rozza, G.: Reduced basis approximation and A posteriori error estimation for Parametrized Partial Differential Equation. MIT Pappalardo Monographs in Mechanical Engineering (Copyright MIT (2007–2010)). Http://augustine.mit.edu
32. Pichi, F., Rozza, G.: Reduced basis approaches for parametrized bifurcation problems held by non-linear Von Kármán equations (2018). arXiv:1804.02014
33. Quateroni, A., Valli, A.: Numerical Approximation of Partial Differential Equations, 2nd edn. Springer, Berlin (1997)
34. Quarteroni, A., Rozza, G., Manzoni, A.: Certified reduced basis approximation for parametrized partial differential equations and applications. J. Math. Ind. **1**(1), 3 (2011)
35. Quarteroni, A., Manzoni, A., Negri, F.: Reduced Basis Methods for Partial Differential Equations: An Introduction. UNITEXT. Springer, Berlin (2015)
36. Rozza, G., Huynh, D., Patera, A.: Reduced basis approximation and a posteriori error estimation for affinely parametrized elliptic coercive partial differential equations: application to transport and continuum mechanics. Arch. Comput. Meth. Eng. **15**, 229–275 (2008)
37. Rozza, G., Nguyen, N.C., Patera, A., Deparis, S.: Reduced basis methods and a posteriori error estimators for heat transfer problems. In: Proceedings of the ASME HT 2009 Summer Conference, Heat Transfer Conference. ASME, New York (2009). Paper No. HT2009-88211, pp. 753–762
38. Sneddon, I., Dautray, R., Benilan, P., Lions, J., Cessenat, M., Gervat, A., Kavenoky, A., Lanchon, H.: Mathematical Analysis and Numerical Methods for Science and Technology. Volume 1: Physical Origins and Classical Methods. Mathematical Analysis and Numerical Methods for Science and Technology. Springer, Berlin (1999)
39. Sneddon, I., Dautray, R., Benilan, P., Lions, J., Cessenat, M., Gervat, A., Kavenoky, A., Lanchon, H.: Mathematical Analysis and Numerical Methods for Science and Technology. Volume 2: Functional and Variational Methods. Mathematical Analysis and Numerical Methods for Science and Technology. Springer, Berlin (2000)
40. Veroy, K.: Reduced-basis methods applied to problems in elasticity: analysis and applications. Ph.D. thesis, Massachusetts Institute of Technology (2003)
41. Young, W., Budynas, R.: Roark's Formulas for Stress and Strain, 7th edn. McGraw, New York (2001)

42. Zanon, L.: Model order reduction for nonlinear elasticity: applications of the reduced basis method to geometrical nonlinearity and finite deformation. Ph.D. thesis, RWTH Aachen University (2017)
43. Zanon, L., Veroy-Grepl, K.: The reduced basis method for an elastic buckling problem. Proc. Appl. Math. Mech. **13**(1), 439–440 (2013)
44. Zienkiewicz, O.C., Taylor, R.L., Zhu, J.Z.: The Finite Element Method: Its Basis and Fundamentals, 6th edn. Butterworth-Heinemann, Oxford (2005)

Chapter 9
Adaptive Tree Approximation with Finite Element Functions: A First Look

Andreas Veeser

Abstract We provide an introduction to adaptive tree approximation with finite element functions over meshes that are generated by bisection. This approximation technique can be seen as a benchmark for adaptive finite element methods, but may be also used therein for the approximation of data and coarsening. Correspondingly, we focus on approximation problems related to adaptive finite element methods, the design and performance of algorithms, and the resulting convergence rates, together with the involved regularity. For simplicity and clarity, these issues are presented and discussed in detail in the univariate case. The additional technicalities and difficulties of the multivariate case are briefly outlined.

9.1 Introduction and Motivation

The approximate numerical solution of partial differential equations (PDEs) is ubiquitous in applications. Of course, the balance of quality and cost of the approximate solutions is of primary interest. Adaptive techniques tailoring the discretization to a given solution often improve this balance, in certain cases even dramatically.

The adaptive solution of PDEs in particular replaces a possibly complicated target function (the PDE solution) by a simple one (the numerical solution). If the target function is known, this corresponds to adaptive approximation. Apart from being of interest by its own, adaptive approximation is of great interest for the adaptive solution of PDEs, because it

- allows to study aspects of adaptivity without the difficulty that the target function is unknown,
- can be viewed as a benchmark for adaptive solution,

A. Veeser (✉)
Dipartimento di Matematica, Università degli Studi di Milano, Milano, Italy
e-mail: andreas.veeser@unimi.it

D. A. Di Pietro et al. (eds.), *Numerical Methods for PDEs*,
SEMA SIMAI Springer Series 15, https://doi.org/10.1007/978-3-319-94676-4_9

- may be used for data approximation and coarsening and, thus, may be a part of adaptive solution,
- should appear as a special case of adaptive solution.

Finite element methods are a well-established and successful technique for the numerical solution of PDEs. Here we shall consider adaptive tree approximation, which is the counterpart of adaptive finite element methods that are based upon mesh refinement with a tree structure. There are several surveys for such adaptive finite element methods, e.g. [9, 20, 21, 25]. However, they do not provide much information about adaptive tree approximation. This contribution intends to fill this gap, in the spirit of the primer [20] and focusing on aspects related to the adaptive solution of PDEs.

9.2 Finite Elements and Bisection

Our interest in adaptive tree approximation is motivated by adaptive finite element methods with bisection. In this section we illustrate this motivation by presenting, in a very simple framework, a finite element method, mesh refinement by bisection, and related approximation problems.

9.2.1 A Finite Element Method

Let us consider the following simple boundary value problem:

$$-u'' = f \text{ in } (0,1), \quad u(0) = 0, \quad u'(1) = 0. \tag{9.1}$$

In order to choose a weak formulation, let φ be a test function of a space V to be determined. Multiply the differential equation in (9.1) by φ, integrate over the domain $(0, 1)$, and integrate by parts the left-hand side. Assuming $V \subset \{v \mid v(0) = 0\}$, we arrive at the problem

$$\text{find } u \in V \text{ such that } \forall \varphi \in V \quad \int_0^1 u'\varphi' = \langle f, \varphi \rangle. \tag{9.2}$$

We let $H^1(0, 1)$ denote the Sobolev space of functions whose derivative are square-integrable and set

$$V = \{v \in H^1(0,1) \mid v(0) = 0\}.$$

Here the point value $v(0)$ is well-defined due to the embedding $H^1(0, 1) \subset C^0[0, 1]$, where $C^0[0, 1]$ stands for the set of all functions that are continuous in $[0, 1]$. It is well known that the left-hand side of the variational equation in (9.2) is a scalar product on V and its induced norm is

$$\|v\|_V := \|v'\|_{(0,1)} \quad \text{where} \quad \|w\|_{(0,1)} := \left(\int_0^1 |w|^2\right)^{\frac{1}{2}}. \tag{9.3}$$

Consequently, the Riesz representation theorem implies that problem (9.2) is well-posed according to Hadamard for any functional $f \in V^*$ in the (topological) dual space of V.

We are interested in the adaptive approximate solution of (9.2) and the adaptive approximation of some function in V, where, in both cases, the approximation quality is measured with (9.3). The approximants are constructed by means of meshes and adaptivity is based upon a particular form of mesh refinement. Before discussing the refinement technique, it is useful to consider the approximants over a fixed mesh. Let

$$0 = x_0 < x_1 < \cdots < x_{n-1} < x_n = 1,$$

which induces the mesh

$$M := \{[x_{i-1}, x_i] \mid i = 1, \ldots, n\}$$

and the finite-dimensional linear spaces

$$\mathbb{P}_1(M) := \{v \in L^\infty(0,1) \mid \forall K \in M \ v_{|K} \in \mathbb{P}_1\},$$
$$V(M) := \{v \in \mathbb{P}_1(M) \mid v \in C^0[0,1], \ v(0) = 0\}.$$

Since $V(M) \subset V$, we can associate with M the Galerkin approximation

$$U_M \in V(M) \text{ such that } \forall \varphi \in V(M) \int_0^1 U_M' \varphi' = \langle f, \varphi \rangle. \tag{9.4}$$

Céa's lemma then shows that the error of the approximate solution U_M coincides with the best error when approximating $u \in V$ with functions from $V(M)$:

$$\|u' - U_M'\|_V = \inf_{v \in V(M)} \|u' - v'\|_V. \tag{9.5}$$

9.2.2 Error Localizations

Adaptive or local mesh refinement aims at tailoring the mesh to a specific target function, thereby changing the local approximation properties of the discrete space. The following result links global and local best errors.

Lemma 9.1 (Best Error Localization) *For any function $u \in V$ and any mesh M of the interval $(0, 1)$, we have*

$$\inf_{v \in V(M)} \|u' - v'\|_{(0,1)} = \left(\sum_{K \in M} \inf_{p \in \mathbb{P}_1} \|u' - p'\|_K^2 \right)^{\frac{1}{2}}.$$

Proof Let us start by deriving a necessary condition on local best approximations. Given any element $K \in M$, suppose that $p_K \in \mathbb{P}_1$ is a best approximation on K, i.e. $\|u' - p_K'\|_K = \inf_{p \in \mathbb{P}_1} \|u' - p'\|_K$. If we square this equality, the objective function is quadratic in p and therefore $p_K \in \mathbb{P}_1$ is a best approximation on K if and only if

$$\forall p \in \mathbb{P}_1 \quad \int_K (u' - p_K')p' = 0. \tag{9.6}$$

Since p' is constant for any $p \in \mathbb{P}_1$, the fundamental theorem of calculus shows that this is in turn equivalent to

$$0 = \big[u(x_i) - p_K(x_i)\big] - \big[u(x_{i-1}) - p_K(x_{i-1})\big]$$

for $K = [x_{i-1}, x_i]$. Consequently, the best approximations on the element K are given by

$$p_K(x_{i-1}) = u(x_{i-1}) + c_K, \quad \text{and} \quad p_K(x_i) = u(x_i) + c_K,$$

where $c_K \in \mathbb{R}$ is a free constant. Remarkably, choosing $c_K = 0$ for all elements $K \in M$ corresponds to the Lagrange interpolant I_M which is characterized by

$$\forall i = 0, \ldots, n \quad I_M u(x_i) = u(x_i)$$

and is in $V(M)$ for $u \in V$. Moreover, summing (9.6) over all mesh elements yields that $I_M u$ is the best approximation in $V(M)$ to $u \in V$. We thus conclude by the identities

$$\begin{aligned} \inf_{v \in V(M)} \|u' - v'\|_{(0,1)} = \|u' - (I_M u)'\|_{(0,1)} &= \left(\sum_{K \in M} \|u' - (I_M u)'\|_K^2 \right)^{\frac{1}{2}} \\ &= \left(\sum_{K \in M} \inf_{p \in \mathbb{P}_1} \|u' - p'\|_K^2 \right)^{\frac{1}{2}}. \end{aligned}$$

□

Since the best approximation in $V(M)$ is unique, the preceding proof and Céa's lemma (9.5) show that the approximate solution U_M coincides with the Lagrange interpolant $I_M u$ of the exact solution. An alternative proof of this fact can be based upon the Green function; see, e.g., Brenner/Scott [8, Section 0.7]. That proof however does not reveal that the local pieces of the Lagrange interpolant are also best approximations.

Let us now see that we have a result of the same flavor without referring to the exact solution u in the localization.

Lemma 9.2 (A Posteriori Error Analysis) *For any functional* $f \in V^*$ *and any mesh* M *of the interval* $(0, 1)$, *the distance between the solutions* u *and* U_M *to* (9.2) *and* (9.4) *satisfies*

$$\|u' - U'_M\|_V = \left(\sum_{K\in M} \|f\|^2_{H^{-1}(K)}\right)^{\frac{1}{2}},$$

where

$$\|f\|_{H^{-1}(K)} := \sup_{\varphi\in H^1_0(K), \|\varphi'\|_K=1} \langle f, \varphi\rangle$$

is a local dual norm.

Proof We deliberately present a proof that does not start with Lemma 9.1 but connects to a posteriori error analysis. Given any test function $\varphi \in V$, the definition of the solution u yields

$$\int_0^1 (u' - U'_M)\varphi' = \langle f, \varphi\rangle - \int_0^1 U'_M\varphi' =: \langle R_M, \varphi\rangle, \tag{9.7}$$

where the residual R_M does not involve the exact solution u. Moreover, the residual R_M simplifies to

$$\begin{aligned}\langle R_M, \varphi\rangle &= \langle f, \varphi\rangle - \int_0^1 U'_M\varphi' = \langle f, \varphi - I_M\varphi\rangle - \int_0^1 U'_M(\varphi - I_M\varphi)' \\ &= \langle f, \varphi - I_M\varphi\rangle - \sum_{i=1}^n \left(\left[U'_M(\varphi - I_M\varphi)\right]_{x_{i-1}}^{x_i} - \int_{x_{i-1}}^{x_i} U''_M(\varphi - I_M\varphi)\right) \\ &= \langle f, \varphi - I_M\varphi\rangle\end{aligned} \tag{9.8}$$

in view of the definition of U_M, elementwise integration by parts, $U_M \in \mathbb{P}_1(M)$, and $\varphi(x_i) - I_M\varphi(x_i) = 0$ for all $i = 0, \ldots, n$.

Fix any mesh element $K \in M$, denote by $\langle\cdot, \cdot\rangle_K$ the duality pairing associated with $H^1_0(K)$ and let $\varphi_K \in H^1_0(K) \subset V$ such that $\|\varphi'_K\|_K = 1$. The identity $I_M\varphi_K = 0$, the representation formula (9.8) and the relationship (9.7) imply the local representation formula:

$$\langle f, \varphi_K \rangle_K = \langle f, \varphi_K - I_M \varphi_K \rangle_K = \langle R_M, \varphi_K \rangle_K = \int_K (u' - U'_M)\varphi'_K.$$

Since $(u - U_M)_{|K} = (u - I_M u)_{|K} \in H_0^1(K)$, we deduce

$$\|f\|_{H^{-1}(K)} = \sup_{\varphi \in H_0^1(K), \|\varphi'\|_K = 1} \int_K (u - U_M)' \varphi'_K = \|u' - U'_M\|_K.$$

Consequently, the claimed identity follows from squaring, summing over all mesh elements $K \in M$ and $\|v\|^2_{(0,1)} = \sum_{K \in M} \|v\|^2_K$. □

The proofs of Lemmata 9.1 and 9.2 are based upon special features of problem (9.1) and its discretization (9.4). However, similar results can be shown for various well-posed problems. Then

- the equalities are often replaced by equivalences and
- the local dual norms usually depend also on the approximate solution U_M.

Thus, for both adaptive approximation and adaptive solution, the error can be split into local contributions, which may be used to guide the adaptive choices.

For this purpose, these local contributions should be computationally accessible. While the local best error $\inf_{p \in \mathbb{P}_1} \|u' - p'\|_K$ corresponds to a discrete optimization which can be approximated by means of numerical integration, the indicator $\|f\|_{H^{-1}(K)}$ corresponds to an infinite-dimensional optimization and so its accessibility is in general less clear and will depend in general on a priori knowledge of the functional f. We illustrate this by relating to the more common form of the so-called element residual.

If f has additional regularity, for example, at least $f \in L^2(0, 1)$, then the Poincaré-Friedrichs inequality

$$\forall \varphi \in H_0^1(K) \quad \|\varphi\|_K \leq \frac{h_K}{\pi} \|\varphi'\|_K$$

with $h_K :=$ length of K implies

$$\|f\|_{H^{-1}(K)} \leq \frac{h_K}{\pi} \|f\|_K. \tag{9.9}$$

The right-hand side can be approximated with the help of numerical integration, supposing, for example, that f has point values. One thus may consider the surrogate indicator

$$\frac{h_K}{\pi} \|f\|_K \tag{9.10}$$

instead of $\|f\|_{H^{-1}(K)}$. It is worth noting that, although (9.9) is sharp in that its constant cannot be improved, it may however entail overestimation. To see this, consider the element $K = (0, h_K)$ and recall that the eigenfunctions

$$\varphi_l(x) := \sqrt{\frac{2}{h_K}} \sin\left(\frac{l\pi}{h_K} x\right), \quad x \in K, \quad l \in \mathbb{N},$$

of the 1-dimensional Laplacian with homogeneous Dirichlet boundary values on K are a complete orthonormal system in $L^2(K)$ and orthogonal in $H_0^1(K)$. Consequently, writing $c_l = \int_K f \varphi_l$ for $l \in \mathbb{N}$, we have

$$\|f\|_K^2 = \sum_{l=1}^{\infty} c_l^2 \quad \text{and} \quad \|f\|_{H^{-1}(K)} = \sum_{l=1}^{\infty} \frac{h_K}{l\pi} c_l^2.$$

Hence, if $f = \varphi_1$, then we have equality in (9.9) and the more f is oscillatory or contains oscillatory modes, the more severe the overestimation is. These observations suggest that the local quantity to be approximated is $\|f\|_{H^{-1}(K)}$, while $\pi^{-1} h_K \|f\|_K$ is just one possible choice for a surrogate, under the assumption $f \in L^2(0, 1)$ and with the aforementioned drawbacks.

9.2.3 *Adaptive Mesh Refinement with Bisection*

The section introduces bisection and presents associated algorithms for adaptive approximation and adaptive solution.

This mesh refinement technique is characterized by the following basic operation. Given an element or interval $K = [a, b]$, we let

$$\texttt{bisect}(K) := \big\{[a, m], [m, b]\big\} \quad \text{with} \quad m = \frac{1}{2}(a + b),$$

which bisects K, i.e., subdivides it into two intervals of equal length. Given a mesh M and some element $K \in M$,

$$\texttt{refine}(M, K) := \big(M \setminus \{K\}\big) \cup \texttt{bisect}(K) \tag{9.11}$$

outputs a new mesh, where the element K is replaced by the two intervals bisecting it. We then may consider the following algorithm for mesh refinement:

$$\begin{aligned}
&M_1 := (0,1);\ n := 1;\\
&\textbf{while}\ (n < N)\\
&\quad \texttt{mark}\ \text{some}\ K_n \in M_n;\\
&\quad M_{n+1} := \texttt{refine}(M_n, K_n);\\
&\quad n := n+1;\\
&\textbf{end while}
\end{aligned} \tag{9.12}$$

where the rule for `mark` has to be specified and the cardinality of $\#M_n = n$ coincides with the iteration count.

Let us first discuss the class of generated meshes and then present two examples for the marking rule.

The arising elements are the dyadic intervals of $(0, 1)$. More precisely, for any $n < N$ and $K \in M_n$, there exist $\ell \in \mathbb{N}_0$ and $k \in \{1, \dots, 2^\ell\}$ such that $K = [(k-1)2^{-\ell}, k2^{-\ell}]$. Conversely, any dyadic interval can be generated by means of a suitable marking rule for sufficiently large N.

It will be convenient to exploit the tree structure of the dyadic intervals of $(0, 1)$. Before specifying it, let us recall a few notions around binary trees. A binary tree T is a directed graph in which every node has at most two children and, except for the root, a unique parent. The tree T is (in)finite, whenever its cardinality $\#T$ is (in)finite. If every node in T has either 0 or 2 children, it is called full. The nodes with at least one child are the internal nodes of T, while those with 0 children are leaves and collected in $\mathcal{L}(T)$. An ancestor of a node K is either the parent or (recursively) an ancestor of the parent. We let $\mathcal{A}(K)$ denote the set of all ancestor of K. Similarly, we define a descendant of K as either a child of K or (recursively) a descendant of a child of K and write $\mathcal{D}(K)$ for all descendants. A subtree of T is subset of T that is itself a tree.

The dyadic intervals of $(0, 1)$ form an infinite full binary tree T_∞, where the interval $(0, 1)$ is the root and, for any dyadic interval or node K, its two children are given by `bisect`(K). There is a one-to-one correspondence between the meshes that can be generated by Algorithm (9.12) and the subtrees in

$$\mathbb{T} := \{T \mid T \text{ is a finite, full subtree of } T_\infty \text{ with root } (0,1)\}$$

In fact, given such a mesh M, then the set $T := \cup_{K \in M} \mathcal{A}(K)$ of all ancestors of its elements forms a subtree $T \in \mathbb{T}$ and, vice versa, given $T \in \mathbb{T}$, its leaves $M = \mathcal{L}(T)$ are a mesh that can be generated by means of a suitable marking rule. Notice that T records the bisections generating the mesh M from $(0, 1)$, in a manner which respects their hierarchy but otherwise ignores the order in which they occurred. In summary, bisection meshes, i.e. meshes that can be generated by Algorithm (9.12), are given by

$$\mathbb{B}(0,1) := \{\mathcal{L}(T) \mid T \in \mathbb{T}\}. \tag{9.13}$$

We now give an example for the marking rule in Algorithm (9.12). Suppose the target function u is known. Then the local best errors in Lemma 9.1 can be considered accessible. In order to devise a marking criterion, observe that the Poincare-Wirtinger inequality

$$\forall v \in H^1(K) \text{ with } \int_K v = 0 \qquad \|v\|_K \le \frac{h_K}{\pi}\|v'\|_K$$

leads to the a priori bound

$$\varepsilon(K) := \inf_{p\in\mathbb{P}_1} \|u' - p'\|_K = \left\| u' - \frac{1}{|K|}\int_K u' \right\| \le \frac{h_K}{\pi}\|u''\|_K.$$

Similarly to (9.9), this bound may be accurate but may also overestimate, where the former will hold asymptotically under suitable assumptions. Interestingly, if it applies and it is accurate before and after bisection in that

$$\varepsilon(K) \approx h_K\|u''\|_K \quad \text{and} \quad \varepsilon(K_i) \approx h_{K_i}\|u''\|_{K_i}, \quad i = 1, 2,$$

for $\{K_1, K_2\} = \texttt{bisect}(K)$, we have

$$\varepsilon(K_1)^2 + \varepsilon(K_2)^2 \approx \frac{1}{4}\varepsilon(K)^2 \tag{9.14}$$

thanks to $h_{K_i} = \frac{1}{2}h_K$ for $i = 1, 2$. Hence, if (9.14) applies for all elements of the current mesh, the larger is the local error on an element, the more the bisection of that element reduces the global error. We are thus led to mark an element with maximum local error:

$$\texttt{mark} \text{ some } K_n \in M_n \text{ with } \varepsilon(K_n) = \max_{K\in M_n} \varepsilon(K). \tag{9.15}$$

This marking rule is called also greedy or maximum strategy.

Next, we turn to the case where the target function u is given only implicitly by the boundary value problem (9.1). Lemma 9.2 suggests to use the local error indicators

$$\eta(K) := \|f\|_{H^{-1}(K)}, \qquad K \in T_\infty,$$

which are accessible through the data f. It is one of the particularities of problem (9.1) that $\varepsilon(K) = \eta(K)$ for all $K \in T_\infty$ and so, in this special case, the only difference between the two local quantities is their computation.

In view of the similarities between Lemmata 9.1 and 9.2, one may use the following counterpart of the marking rule (9.15): given a parameter $\theta in(0, 1]$ and an iteration counter j,

$$\texttt{mark} \text{ the subset } \hat{M}_j := \{K \in M_j \mid \eta(K) \geq \theta \max_{K' \in M_j} \eta(K')\}. \tag{9.16}$$

The use of the parameter θ and a subset for elements to be bisected is motivated by the following fact. In general, the computation of the local error indicators requires to solve the discrete problem and therefore the adaptive solution of boundary value problems assumes the more involved structure

$$\begin{aligned}
&M_1 := (0, 1);\ \ j := 1;\\
&\textbf{while}\ (j < J)\\
&\quad U_j := \texttt{solve}(f, M_j);\\
&\quad \{\eta(K)\}_{K \in M_j} := \texttt{estimate}(f, M_j, U_j);\\
&\quad \hat{M}_j := \texttt{mark}\big(M_j, \{\eta(K)\}_{K \in M_j}\big);\\
&\quad M_{j+1} := \texttt{refine}(M_j, \hat{M}_j);\\
&\quad j := j + 1;\\
&\textbf{end while},
\end{aligned} \tag{9.17}$$

where the module $\texttt{refine}(M, \hat{M})$ is an iterative generalization of (9.11):

$$\begin{aligned}
&\textbf{for}\ K \in \hat{M}\\
&\quad M := \texttt{refine}(M, K);\\
&\textbf{end for}.
\end{aligned}$$

Since solve requires at least $\#M_j$ operations, we see that the total number of operations is at least of order $\sum_{j=1}^{J} \#M_j$. Marking always only one element leads to $\#M_j = j$ and therefore to at least $\sum_{j=1}^{J} j = \frac{1}{2}J(J + 1) = \frac{1}{2}\#M_J(1 + \#M_J)$ operations. Consequently, the optimal order $\#M_J$ for the cost is out of reach. If the parameter $\theta < 1$, then the strategy (9.16) allows for marking of several elements, keeping the spirit of (9.15). There are other variants of the maximum strategy, e.g., the so-called bulk chasing introduced by Dörfler [15].

Let us conclude this section by classifying Algorithm (9.12) with (9.15), an example for adaptive approximation with bisection, and Algorithm (9.17) with (9.16), an example for adaptive solution with bisection. Both methods pick approximants from the spaces

$$V_n = \bigcup_{M \in \mathbb{B}_n} V(M) \tag{9.18}$$

where $n \in \mathbb{N}$ and $\mathbb{B}_n := \{M \in \mathbb{B}(0,1) \mid \#M \leq n\}$ with $\mathbb{B}(0,1)$ from (9.13). Indeed, the final approximant of (9.12) is in V_N and the one of (9.17) is in $V_{\#M_J}$. This observation suggests to compare the error of the approximants with the global best errors

$$b_n := \inf_{v \in V_n} \|u - v\|_V. \tag{9.19}$$

It is important to notice that, for $n \geq 3$, the space V_n is not linear. To see this, we set, for any dyadic interval $K = [a,b] \in T_\infty$,

$$\varphi_K(x) := \begin{cases} \dfrac{x-a}{m-a}, & x \in [a,m], \\ \dfrac{b-x}{b-m}, & x \in [m,b], \\ 0, & \text{otherwise} \end{cases} \quad \text{with} \quad m = \frac{1}{2}(a+b) \tag{9.20}$$

and observe

$$\varphi_{[0,\frac{1}{2^n-1}]}, \varphi_{[\frac{2^n-1-1}{2^n-1},1]} \in V_n, \quad \text{but} \quad \varphi_{[0,\frac{1}{2^n-1}]} + \varphi_{[\frac{2^n-1-1}{2^n-1},1]} \notin V_n. \tag{9.21}$$

Hence, the aforementioned algorithms are examples of nonlinear approximation.

9.3 Abstract Adaptive Tree Approximation

Algorithm (9.12) with marking strategy (9.15) is an example of adaptive tree approximation. Algorithms of this kind already appeared in Birman/Solomjak [7]. A groundbreaking twist to design and theory of such algorithms was Binev/DeVore [4] and Binev [3]. These two works have strongly influenced our presentation and are represented by Theorem 9.2.

9.3.1 Setting, Goal, and Examples

We first introduce a setting for adaptive tree approximation. There are the following two ingredients:

- a 'master tree' T_∞ in which every node has exactly two children and, except the root K_*, one parent,
- a function $e : T_\infty \to \mathbb{R}_0^+$ assigning an 'error' to each node or element of the master tree.

We consider only a binary master tree instead of k-ary one, because all our examples are based upon bisection. The infinite tree T_∞ and e induce the 'meshing trees'

$$\mathbb{T} := \{T \mid T \text{ is a finite, full subtree of } T_\infty \text{ and } T \ni K_*\},$$

the 'global errors'

$$E(T) := \sum_{K \in \mathcal{L}(T)} e(K),$$

for $T \in \mathbb{T}$, and the 'best global errors'

$$b_n := \inf_{T \in \mathbb{T}, \#\mathcal{L}(T) \le n} E(T). \tag{9.22}$$

The best errors b_n and best meshing trees, i.e. trees $T^\star \in \mathbb{T}$ with $\#\mathcal{L}(T^\star) \le n$ and $E(T^\star) = b_n$, can be found by exploring all possibilities. Let us get an idea about the number of competing meshing trees. To this end, recall that

$$\#T = 2\#\mathcal{L}(T) - 1 \tag{9.23}$$

for any finite full binary tree and that there is a one-to-one correspondence between meshing trees with exactly n leaves and not necessarily full binary trees with $n-1$ nodes. In view of the recursion formula for Catalan numbers, the number of latter is the Catalan number C_{n-1}, where

$$C_n = \frac{(2n)!}{(n+1)!\,n!} \approx \frac{4^n}{\sqrt{\pi}\, n^{3/2}} \quad \text{as} \quad n \to \infty,$$

and the asymptotic equivalence is a consequence of Stirling's approximation. Consequently, even if we suppose that the research can be restricted to meshing trees with exactly n leaves, the number of possibilities grows exponentially with n.

We are interested in a cheaper alternative, with a growth of the number of operations close to the minimal order. Hereafter the evaluations of e are counted with unit cost. To obtain such an improvement, we shall rely on additional properties of e and relax the notion of best errors/meshing trees. More precisely, we shall use algorithms of the form

> $T_1 := \{K_*\};\ n := 1;$
>
> **while** $(n < N)$
>
>   `mark` some $K_n \in \mathcal{L}(T_n)$;
>
>   let T_{n+1} be the smallest tree in $\mathbb{T}$ containing T and the children of K_n;
>
> $n := n + 1;$
>
> **end while**

(9.24)

and aim at devising marking strategies ensuring the following two objectives under suitable assumptions on e: a total operation count close to $O(N)$ and

$$E(T_N) \leq C b_{\lfloor N/C \rfloor}, \tag{9.25}$$

where $\lfloor \cdot \rfloor$ indicates the floor function and $C \geq 1$ is independent of N. In this case T_N is a near best meshing tree.

In what follows, we shall use three 'running' examples. The first two immediately arise from the discussion in Sect. 9.1, while the third one is motivated by oscillation in multidimensional a posteriori analyses. In all three examples, the master tree T_∞ is given by the dyadic intervals of the unit interval (0, 1). Thus the meshing trees $\mathbb{T}$ correspond to $\mathbb{B}(0, 1)$ from (9.13) and Algorithm (9.24) is Algorithm (9.12).

Example 9.1 (H_0^1-Approximation) Given any function $u \in H_0^1(0, 1)$, set

$$e(K) := \inf_{p \in \mathbb{P}_1} \|u' - p'\|_K^2$$

for any dyadic interval $K \in T_\infty$. In view of the best error localization in Lemma 9.1, we then have

$$E(T) = \left(\inf_{v \in V(\mathcal{L}(T))} \|u' - v'\|_V \right)^2$$

for any tree $T \in \mathbb{T}$ corresponding to a bisection mesh and the best global errors in (9.22) coincide with those in (9.19).

Example 9.2 (Surrogate Indicator) Consider the model problem (9.1) with $f \in L^2(0, 1)$ and set

$$e(K) := h_K^2 \|f\|_K^2$$

for any dyadic interval $K \in T_\infty$. Then we have

$$E(T) = \sum_{K \in \mathcal{L}(T)} h_K^2 \|f\|_K^2$$

for any tree $T \in \mathbb{T}$ corresponding to a bisection mesh. The global best errors in (9.22) thus provide upper bounds for the errors of corresponding Galerkin approximations (9.4). Moreover, Algorithm (9.24) combined with one Galerkin solve on the final mesh can be viewed as an adaptive solution of (9.1) with the surrogate indicator (9.10).

Example 9.3 (Oscillation) Given a function $f \in L^2(0,1)$, denote by $\bar{f}_K := |K|^{-1} \int_K f$ its mean value on the interval K and set

$$e(K) := h_K^2 \|f - \bar{f}_K\|_K^2$$

for any dyadic interval $K \in T_\infty$. Then, for any tree $T \in \mathbb{T}$ corresponding to a bisection mesh,

$$E(T) = \sum_{K \in \mathcal{L}(T)} h_K^2 \|f - \bar{f}_K\|_K^2$$

is the so-called oscillation of f on $\mathcal{L}(T)$. Such a term typically spoils the equivalence of error and a posteriori error estimator and thus represents a defect of the a posteriori analysis or estimator. In this case, Algorithm (9.24) aims at constructing bisection meshes that effectively reduce this defect.

9.3.2 Local Error Reduction and Maximum Strategy

We have motivated the maximum strategy (9.15) by the approximate local reduction (9.14). It is the purpose of this section to show that this combination indeed leads to near best approximations.

In the setting of the previous section, the maximum strategy becomes

$$\texttt{mark} \text{ some } K_n \in \mathcal{L}(T_n) \text{ with } e(K_n) = \max_{K \in \mathcal{L}(T_n)} e(K). \tag{9.26}$$

If we organize the created nodes, e.g., in a heap, the total number of operations can be kept within $O(N \log N)$. The logarithmic factor can be avoided if we organize the created nodes in dyadic bins and mark elements K_n such that $e(K_n) \geq \frac{1}{2} \max_{K \in \mathcal{L}(T_n)} e(K)$. This slightly modified marking rule leads only to minor changes in what follows.

We say that e is monotone whenever we have the following: if $K \in T_\infty$ is any element and K_1, K_2 are its children, then

$$e(K_1) + e(K_2) \leq \alpha e(K) \tag{9.27}$$

with $\alpha = 1$. If we can choose even $\alpha \in [0, 1)$, we say that e reduces locally with factor α.

Applying Algorithm (9.24) with strategy (9.26) to Example 9.1 gives Algorithm (9.12) with strategy (9.15). The errors of Example 9.1 are monotone but may not reduce locally. The monotonicity readily follows from

$$e(K_1) + e(K_2) = \inf_{p \in \mathbb{P}_1} \|u' - p'\|_{K_1}^2 + \inf_{p \in \mathbb{P}_1} \|u' - p'\|_{K_2}^2 \leq \inf_{p \in \mathbb{P}_1} \|u' - p'\|_K^2 = e(K).$$

To see that local reduction in general does not hold, we first observe that, if $K = [a, b]$ is a dyadic interval and $m = \frac{1}{2}(a+b)$, then φ_K from (9.20) satisfies

$$\varphi_K' = \frac{2}{|K|^{\frac{1}{2}}} H_K \quad \text{where} \quad H_K(x) := \begin{cases} |K|^{-\frac{1}{2}}, & x \in [a, m], \\ -|K|^{\frac{1}{2}}, & x \in [a, m], \\ 0, & x \notin K, \end{cases} \tag{9.28}$$

is the Haar function of K. Notice also that H_K has mean 0 and is L^2-normalized on any interval containing K, i.e. for any ancestor of K. Consider now Example 9.1 for $u = \varphi_{K_0} \in H_0^1(0, 1)$, where K_0 is any dyadic interval with $|K_0| < 1/4$. Let K be an ancestor but not a parent of K_0. Then one of the children K_1, K_2 is also an ancestor of K_0 and so we have

$$e(K) = \inf_{p \in \mathbb{P}_1} \|u' - p'\|_K^2 = \frac{4}{|K_0|} \|H_{K_0}\|_K^2 = \frac{4}{|K_0|} = e(K_1) + e(K_2).$$

Hence, inequality (9.27) holds in general only with $\alpha = 1$ for Example 9.1.

However, the errors in Examples 9.2 and 9.3 satisfy (9.27) with $\alpha = \frac{1}{4}$ thanks to the meshsize reduction

$$\forall i = 1, 2 \quad h_{K_i} = \frac{1}{2} h_K. \tag{9.29}$$

In the case of Example 9.2, we have even equality. We shall not exploit here this special feature, which, e.g., does not hold for Example 9.3. In fact, if $f = H_K$, the bisection of the dyadic interval K into K_1 and K_2 yields a complete elimination of the error:

$$e(K) = h_K^2 \|f - \bar{f}_K\|_K^2 = h_K^2 \|H_K\|_K^2 = |K|^2 \quad \text{and} \quad e(K_1) = 0 = e(K_2).$$

Let us now turn to the task that local error reduction and maximum strategy ensure that the output tree T_N of (9.24) is near best. This requires to bound the global error $E(T_N)$ in terms of the best error b_m for some suitable $m \le n$. To this end, we take a best meshing tree $T_m^\star \in \mathbb{T}$ with $\#\mathcal{L}(T_m^\star) \le m$ and $E(T_m^\star) = b_m$ and handle the contributions to $E(T_N)$ depending on their relationship to $T_m^\star$. If a leaf $K \in \mathcal{L}(T_N)$ of the final tree happens to be also a leaf of the best meshing tree $T_m^\star$, no estimation is necessary. For leaves of T_N that are leaves of $T_m^\star$ or descendants thereof, we shall use the following consequence of (9.27), which holds for all $\alpha \in [0, 1]$ and follows from binary tree induction (see below): if T is any finite subtree of T_∞ rooted at K, then

$$\sum_{K' \in \mathcal{L}(T)} e(K') \le e(K). \tag{9.30}$$

The remaining, critical leaves of T_N are ancestors of leaves in the best meshing tree $T_m^\star$. For these leaves, we shall use two auxiliary statements: the first one hinges on local error reduction, while the second one relies on monotonicity and the maximum strategy.

Lemma 9.3 (Error Below a Node) *Let T be any finite subtree of T_∞, with root K. Then local reduction for e with $\alpha \in (0, 1)$ implies*

$$\sum_{K' \in T} e(K') \le \frac{e(K)}{1-\alpha}.$$

It is important to note that here the sum involves the whole tree and not only its leaves as in (9.30). If the master tree T_∞ is given by dyadic intervals, the intervals involved in the sum may overlap and the number of the overlapping layers is not bounded.

Proof We use a binary tree induction to prove

$$\sum_{K' \in T} e(K') \le e(K) \left(\sum_{i=0}^{\#T-1} \alpha^i \right), \tag{9.31}$$

which implies the claim thanks to $\sum_{i=0}^{\infty} \alpha^i = (1-\alpha)^{-1}$. If $\#T = 1$, then (9.31) is an equality and, if $\#T = 2$, it follows from the positivity of e and (9.27). To prove the induction step, assume that $\#T = n + 1$ and that (9.31) holds for all finite subtrees with cardinality $\le n$. Let K_1 and K_2 denote the children of the root K of T and write T_i for the subtree consisting of all descendants of K_i in T. Then $\#T_i \le n$ and applying (9.31) to T_i, $i = 1, 2$, and (9.27) yield

$$\begin{aligned}
\sum_{K' \in T} e(K') &\le e(K) + \sum_{K' \in T_1} e(K') + \sum_{K' \in T_2} e(K') \\
&\le e(K) + \left(\sum_{i=0}^{n-1} \alpha^i \right) \big(e(K_1) + e(K_2) \big) \\
&\le e(K) + \left(\sum_{i=1}^{n} \alpha^i \right) e(K) = \left(\sum_{i=0}^{n} \alpha^i \right) e(K).
\end{aligned}$$

□

Lemma 9.4 (Internal Errors) *Assume that e is monotone and that the maximum strategy* (9.26) *is used in Algorithm* (9.24)*. Then any internal node $K \in T_N \setminus \mathcal{L}(T_N)$ of the output tree satisfies $e(K) \ge \max_{K' \in \mathcal{L}(T_N)} e(K')$.*

Proof We first show that $t_n := \max_{K' \in \mathcal{L}(T_n)} e(K')$ is decreasing in $n = 1, \dots, N$. Fix $n \in \{1, \dots, N-1\}$ and let $K_{n,1}$ and $K_{n,2}$ denote the children of K_n. Then (9.27) implies $\max\{e(K_{n,1}), e(K_{n,2})\} \leq e(K_n)$, whence

$$t_{n+1} = \max_{K' \in \mathcal{L}(T_{n+1})} e(K') = \max\{e(K_{n,1}), e(K_{n,2}), \max_{K' \in \mathcal{L}(T_n)\setminus\{K_n\}} e(K')\}$$
$$\leq \max_{K' \in \mathcal{L}(T_n)} e(K') = t_n.$$

Let $K \in T_N \setminus \mathcal{L}(T_N)$ be an internal node of T_N. Then there exists $n \in \{1, \dots, N-1\}$ such that $K_n = K$. Consequently, the monotonicity of $n \mapsto t_n$ gives

$$e(K) = e(K_n) = t_n \geq t_N = \max_{K' \in \mathcal{L}(T_N)} e(K').$$

□

After these preparations, we are ready for the main result of this section.

Theorem 9.1 (Local Reduction and Maximum Strategy) *Assume that e reduces locally with factor* $\alpha \in (0, 1)$ *and use the maximum strategy* (9.26) *in Algorithm* (9.24). *Then the output tree* T_N *satisfies*

$$E(T_N) \leq \frac{1}{1-\alpha} \min_{m=1}^{N} \left(\frac{N}{N-m+1} b_m \right).$$

Proof Let $m \leq N$ and take a meshing tree $T_m^\star \in \mathbb{T}$ such that $\#\mathcal{L}(T_m) \leq m$ and $E(T_m^\star) = b_m$. We subdivide the leaves $\mathcal{L}(T_N)$ of the output tree into two groups:

$$L_1 := \mathcal{L}(T_N) \cap \big(T_m^\star \setminus \mathcal{L}(T_m^\star)\big) \quad \text{and} \quad L_2 := \mathcal{L}(T_N) \setminus L_1.$$

Let us consider first L_2. An element in L_2 is either a leaf of $T_m^\star$ or a descendant thereof. Thus, if we define $L^\star := \mathcal{L}(T_m^\star) \cap T_N$ and denote by $T_{K^\star}$ the largest subtree of T_N rooted at $K^\star \in L^\star$, we can write

$$L_2 = \bigcup_{K^\star \in L^\star} \mathcal{L}(T_{K^\star}).$$

This representation, inequality (9.30), and $E(T_m^\star) = b_m$ yield

$$\sum_{K \in L_2} e(K) = \sum_{K^\star \in L^\star} \sum_{K \in \mathcal{L}(T_{K^\star})} e(K) \leq \sum_{K \in L^\star} e(K^\star) \leq b_m. \tag{9.32}$$

If L_1 is empty, this implies $E(T_N) \leq b_m \leq N/(N-m+1)b_m$.

It remains to consider the critical case $L_1 \neq \emptyset$. Writing $t_N = \max_{K \in \mathcal{L}(T_N)} e(K)$ and recalling (9.23), we have

$$\sum_{K \in L_1} e(K) \leq \#L_1\, t_N \leq \#\big(T_m^\star \setminus \mathcal{L}(T_m^\star)\big) t_N \leq (m-1) t_N. \tag{9.33}$$

Since we are using the maximum strategy (9.26), we can apply Lemma 9.4 to obtain $e(K) \geq t_N$ for any internal node $K \in T_N \setminus \mathcal{L}(T_N)$. We shall use this to deduce a lower bound of $E(T_m^*)$ in terms of t_N. Writing $\tilde{T}_{K^\star} := T_{K^\star} \setminus \mathcal{L}(T_{K^\star})$, we obtain the following representation of internal nodes of T_N which are not internal to $T_m^\star$:

$$D := \big(T_N \setminus \mathcal{L}(T_N)\big) \setminus \big(T_m^\star \setminus \mathcal{L}(T_m^\star)\big) = \bigcup_{K^\star \in L^\star} \tilde{T}_{K^\star}.$$

Thus, Lemma 9.3 provides

$$\#D\, t_N \leq \sum_{K^\star \in L^\star} \sum_{K \in \tilde{T}_{K^\star}} e(K) \leq \frac{1}{1-\alpha} \sum_{K^\star \in L^\star} e(K^\star) \leq \frac{b_m}{1-\alpha}. \tag{9.34}$$

In order to determine $\#D$, we notice that if all $m-1$ internal nodes of $T_m^\star$ are internal to T_N, then L_1 is empty. Since we consider the case $L_1 \neq \emptyset$, there are at most $m-2$ nodes internal to $T_m^\star$ and T_N and we have $\#D \geq (N-1) - (m-2) = N - m + 1$. Inserting this inequality in (9.34) and recalling (9.33), we arrive at

$$\sum_{K \in L_1} e(K) \leq \frac{m-1}{N-m+1} \frac{b_m}{1-\alpha}. \tag{9.35}$$

Finally, combining the two bounds (9.32) and (9.35) for two types of leaves of T_N, we conclude the claimed inequality

$$E(T_N) = \sum_{K \in L_1} e(K) + \sum_{K \in L_2} e(K) \leq \frac{N}{N-m+1} \frac{b_m}{1-\alpha}.$$

□

In light of our discussion at the beginning of this section, we readily have the following applications of Theorem 9.1.

Corollary 9.1 (Surrogate Indicator and Oscillation) *Applying Algorithm* (9.24) *with the maximum strategy* (9.26) *to Examples 9.2 and 9.3 generates output trees with*

$$E(T_N) \leq \frac{4}{3} \min_{m=1}^{N} \left(\frac{N}{N-m+1} b_m \right).$$

In the next section, we shall see that the success of the maximum strategy in these examples hinges on meshsize reduction.

9.3.3 Bare Local Error Monotonicity

The local errors of H_0^1-approximation in Example 9.1 do not reduce locally. As a consequence, Theorem 9.1 does not apply to this case. It is the purpose of this section to provide a remedy.

Let us first see that this missing applicability of Theorem 9.1 is not just a technical issue, but has a deeper reason.

Example 9.4 (Maximum Strategy for H_0^1-Approximation) In the setting of Example 9.1, consider the function

$$u = 2^{L/2-1}\varphi_{[2^L-1,1]} + 2^{l/2-1}\sqrt{1-\epsilon} \sum_{K\subset[0,1/2],|K|=2^{-l}} \varphi_K,$$

where $l, L \in \mathbb{N}$, $\epsilon \in (0, 1)$, φ_K is the hat function from (9.20), and the sum involves only dyadic intervals. The function u thus consists of scaled hat functions of level l in $[0, 1/2]$ and one scaled hat function of level L in $[1/2, 1]$. The scalings are such that its derivative is

$$u' = H_{[2^L-1,1]} + \sqrt{1-\epsilon} \sum_{K\subset[0,1/2],|K|=2^{-l}} H_K,$$

where H_K is the Haar function of K; see (9.28). In view of the properties of the Haar functions, we have the following local errors

$$e(K) = \begin{cases} 1 + 2^{-\ell+1}(1-\epsilon), & \text{if } K = K_*, \\ 1, & \text{if } K \supset [2^L - 1, 1] \text{ and } K \neq K_*, \\ \dfrac{|K|}{2^l}(1-\epsilon), & \text{if } K \subset [0, 1/2], \\ 0, & \text{otherwise.} \end{cases}$$

Therefore, if we assume $0 < \epsilon < \frac{1}{2}$ and N sufficiently big, Algorithm (9.24) with the maximum strategy (9.26) refines, after the first bisection, in three stages as follows. Due to $2(1-\epsilon) > 1$, first, $2^{l-1} - 1$ bisections are operated in $[0, 1/2]$ such that all leaf intervals, except $[1/2, 1]$, have length 2^l. Then $L - 1$ times the right-most interval is bisected as $1 > 1 - \epsilon$. Finally, the algorithm returns to $[0, 1/2]$ and reaches global error 0 after another 2^{l-1} bisections. Thus, terminating the algorithm at the end of the second stage, the output tree T_N satisfies

$$E(T_N) = 2^{l-1}(1-\epsilon) \quad \text{and} \quad \#\mathcal{L}(T_N) = 2^{l-1} + L - 1. \tag{9.36}$$

An alternative way of refinement can be done in two stages after the first bisection. First, operate $2^l - 1$ bisections in $[0, 1/2]$ such that all leaf intervals, except $[1/2, 1]$, have length 2^{l+1}. Then bisect $L - 1$ times the right-most interval to reach global error zero. Here, the tree T corresponding to the end of the first stage satisfies

$$E(T) = 1 \quad \text{and} \quad \#\mathcal{L}(T) = 2^l. \tag{9.37}$$

The existence of (9.37) is not compatible with (9.36) being near best. In fact, for any constant C in (9.25), we can choose L such that

$$\#\mathcal{L}(T) = 2^l \leq \frac{(2^{l-1} + L - 1)}{C} = \frac{\#\mathcal{L}(T_N)}{C},$$

for all l. Consequently, (9.25) entails

$$2^{l-1}(1 - \epsilon) = E(T_N) \leq CE(T) = C,$$

which is a contradiction for sufficiently large l.

Notably, Example 9.4, slightly modified, and its conclusion apply also to adaptive solution by Algorithm (9.17) with the marking strategy (9.16).

Example 9.4 shows that the use of the maximum strategy (9.26) in Algorithm (9.24) does not ensure near best output trees if the local errors are barely monotone. In other words: in this case, the only chance for near best output trees is to modify the marking strategy. The following modification is due to Peter Binev; see [3]. We first associate new quantities to the nodes by setting

$$\nu(K_*) := e(K_*)$$

at the root K_* and, assuming that ν is already defined for the parent $K^\dagger$ of $K \in T_\infty$,

$$\nu(K) := \begin{cases} \dfrac{e(K)\nu(K^\dagger)}{e(K) + \nu(K^\dagger)}, & \text{if } e(K) + \nu(K^\dagger) > 0, \\ 0, & \text{otherwise.} \end{cases} \tag{9.38}$$

The new marking strategy is then simply a maximum strategy on these new quantities:

$$\texttt{mark} \text{ some } K_n \in \mathcal{L}(T_n) \text{ with } \nu(K_n) = \max_{K \in \mathcal{L}(T_n)} \nu(K). \tag{9.39}$$

The number of operations with this new strategy is essentially the same as with the original maximum strategy (9.26).

Let us now discuss first properties of ν and then prepare the proof of the counterpart of Theorem 9.1. We have

$$\forall K \in T_\infty \quad \nu(K) \leq e(K), \tag{9.40}$$

$$\forall K \in T_\infty \setminus \{K_*\} \quad \nu(K) \leq \nu(K^\dagger), \tag{9.41}$$

and

$$\forall K \in \mathcal{T} \quad e(K) = 0 \iff \nu(K) = 0 \tag{9.42}$$

whenever e is monotone. Moreover, if $e(K) + \nu(K^\dagger) > 0$, then (9.38) implies

$$\frac{1}{\nu(K)} = \frac{1}{e(K)} + \frac{1}{\nu(K^\dagger)}, \tag{9.43}$$

and, by induction, that

$$\frac{1}{\nu(K)} = \frac{1}{e(K)} + \sum_{K' \in \mathcal{A}(K)} \frac{1}{e(K')}. \tag{9.44}$$

The last formula gives a first hint on the effects of ν. In Example 9.4, for a dyadic interval K containing $[2^L - 1, 1]$, we have

$$\nu(K) = \frac{e(K_*)}{1 + |\log_2 |K|| \, e(K_*)} \quad \text{instead of} \quad e(K) = 1.$$

Similar changes apply to the other dyadic intervals. As a consequence, the three stages with the original maximum strategy (9.26) become mixed with the new strategy (9.39).

When we use the modified maximum strategy (9.39) in Algorithm (9.24), the output tree is constructed in terms of ν, while we would like to deduce that it is near best in terms of e. The necessary links are provided by the following two lemmas.

Lemma 9.5 (Upper Bound) *Assume that the local errors e are monotone and let T be a finite subtree of T_∞ with root K_*. If $\nu(K) \leq t$ for all leaves $K \in \mathcal{L}(T)$ with $t \geq 0$, then*

$$\sum_{K \in \mathcal{L}(T)} e(K) \leq \#T \, t.$$

Proof Without loss of generality, we can eliminate all leaves of T with $e(K) = 0$. Then the monotonicity of e ensures that e never vanishes on T. Exploiting also that T contains the root K_* of T_∞, we have (9.44) for any leaf $K \in \mathcal{L}(T)$. Multiplying it with $e(K)\nu(K)$ yields the relationship

$$e(K) = \nu(K) \left(1 + \sum_{K' \in \mathcal{A}(K)} \frac{e(K)}{e(K')} \right).$$

Using $\nu(K) \le t$ and summing over all leaves, we get

$$\sum_{K\in\mathcal{L}(T)} e(K) \le t \left(\#\mathcal{L}(T) + \sum_{K\in\mathcal{L}(T)} \sum_{K'\in\mathcal{A}(K)} \frac{e(K)}{e(K')} \right).$$

Thanks to the monotonicity of e, the inequality (9.30) applies and a reordering of the sum provides

$$\sum_{K\in\mathcal{L}(T)} \sum_{K'\in\mathcal{A}(K)} \frac{e(K)}{e(K')} = \sum_{K'\in T\setminus\mathcal{L}(T)} \frac{1}{e(K')} \sum_{K\in\mathcal{L}(T)\cap\mathcal{D}(K')} e(K) \le \#\big(T \setminus \mathcal{L}(T)\big),$$

which finishes the proof. □

Lemma 9.6 (Lower Bound) *Assume that the local errors e are monotone and let T be a finite subtree of T_∞ with root K, not necessarily K_*. If $\nu(K) \ge t$ for all nodes $K \in T$ with $t \ge 0$, then*

$$e(K) \ge \#T\, t.$$

Proof We can assume $t > 0$ without loss of generality and, by (9.40), e and ν never vanish on T. We proceed by a double binary tree induction, one to show the claimed inequality and another one to show an improved inequality if the parent $K^\dagger$ exists. If $\#T = 1$, then the claimed inequality follows from (9.40):

$$e(K) \ge \nu(K) \ge t.$$

If the parent $K^\dagger$ exists, this can be improved with the help of (9.43) to

$$e(K) = \nu(K)\frac{e(K)}{\nu(K)} = \nu(K)\left(1 + \frac{e(K)}{\nu(K^\dagger)}\right) \ge t\left(1 + \frac{e(K)}{\nu(K^\dagger)}\right).$$

To show the induction steps, we denote by K_1 and K_2 the children K of T, write T_i for the subtree consisting of all descendants of K_i in T, and assume that

$$e(K_i) \ge t\left(\#T_i + \frac{e(K_i)}{\nu(K)}\right). \tag{9.45}$$

Then

$$\begin{aligned} e(K) &= \big[e(K_1) + e(K_2)\big]\frac{e(K)}{e(K_1) + e(K_2)} \\ &\ge t\left((\#T_1 + \#T_2)\frac{e(K)}{e(K_1) + e(K_2)} + \frac{e(K)}{\nu(K)}\right). \end{aligned} \tag{9.46}$$

Proceeding with the monotonicity of e and (9.40) yields

$$e(K) \geq t(\#T_1 + \#T_2 + 1) = t\#T,$$

which corresponds to the claimed inequality. If the parent $K^\dagger$ exists, we can exploit also (9.43) for the last term in (9.46) and obtain

$$e(K) \geq t\left(\#T + \frac{e(K)}{\nu(K^\dagger)}\right),$$

which justifies assumption (9.45). Thus, the proof is complete. □

Example 9.4 and the following theorem show that the modified maximum strategy is superior to the original one.

Theorem 9.2 (Local Monotonicity and Modified Maximum Strategy) *Assume that e is monotone and use the modified maximum strategy* (9.39) *in Algorithm* (9.24). *Then the output tree T_N satisfies*

$$E(T_N) \leq \min_{m=1}^{N} \left(\frac{N}{N-m+1} b_m\right)$$

Proof The proof is very similar to the proof of Theorem 9.1; we shall focus on the differences.

Let $m \leq N$, take a meshing tree $T_m^\star \in \mathbb{T}$ such that $\#\mathcal{L}(T_m) \leq m$ and $E(T_m^\star) = b_m$, and subdivide the leaves $\mathcal{L}(T_N)$ of the output tree into two groups:

$$L_1 := \mathcal{L}(T_N) \cap \big(T_m^\star \setminus \mathcal{L}(T_m^\star)\big) \quad \text{and} \quad L_2 := \mathcal{L}(T_N) \setminus L_1.$$

Since the argument in the proof of Theorem 9.1 concerning L_2 involves neither the strategy nor the reduction, we obtain as there

$$\sum_{K \in L_2} e(K) \leq b_m \tag{9.47}$$

and are left only with the case $L_1 \neq \emptyset$. We set $t_N = \max_{K \in \mathcal{L}(T_N)} \nu(K)$ and let T be the minimal tree with root K_* and leaves L_1. Since $T \subset T_m^\star \setminus \mathcal{L}(T_m^\star)$, Lemma 9.5 gives

$$\sum_{K \in L_1} e(K) \leq \#T\, t_N \leq \#\big(T_m^\star \setminus \mathcal{L}(T_m^\star)\big)t_N = (m-1)t_N, \tag{9.48}$$

which establishes (9.33) for the modified maximum strategy. Moreover, thanks to (9.41), Lemma 9.4 carries over to the modified maximum strategy and we have $\nu(K) \geq t_N$ for any internal node $K \in T_N \setminus \mathcal{L}(T_N)$. As in the proof of Theorem 9.1, we write

$$D := \big(T_N \setminus \mathcal{L}(T_N)\big) \setminus \big(T_m^\star \setminus \mathcal{L}(T_m^\star)\big) = \bigcup_{K^\star \in L^\star} \tilde{T}_{K^\star},$$

where we can take $L^\star = \mathcal{L}(T_m^\star) \cap (T_N \setminus \mathcal{L}(T_N))$ and $\tilde{T}_{K^\star}$ is the maximal subtree of $T_N \setminus \mathcal{L}(T_N)$ rooted at $K^\star \in L^\star$. Lemma 9.6 then provides

$$\#D\, t_N \leq \sum_{K^\star \in L^\star} e(K^\star) \leq b_m, \tag{9.49}$$

which is an improvement of (9.34). Since $\#D \geq N - m + 1$, the two inequalities (9.48) and (9.49) yield

$$\sum_{K \in L_1} e(K) \leq \frac{m-1}{N-m+1} b_m.$$

Adding this inequality and (9.47) finishes the proof. □

Theorem 9.2 implies in particular that Algorithm (9.24) with the modified maximum strategy performs equally well for all examples of Sect. 9.3.1.

Corollary 9.2 (H_0^1-Approximation, Surrogate Indicators, and Oscillation) *Applying Algorithm* (9.24) *with the modified maximum strategy* (9.39) *to Examples 9.1–9.3 generates output trees with*

$$E(T_N) \leq \min_{m=1}^{N} \left(\frac{N}{N-m+1} b_m \right).$$

Notice that the inclusion of H_0^1-approximation is due to a necessary change of the marking strategy.

9.4 Convergence Rates with Bisection

In the preceding section we have seen that Algorithm (9.24) with the modified maximum strategy (9.39) fully exploits the approximation potential offered by 1-dimensional bisection, whenever the local errors are monotone. This however says nothing about the convergence speed for a given function, which would allow for some comparison with other approximation methods.

To provide some information of this type, we derive convergence rates for bisection. Results in this direction are already contained in Birman/Solomjak [7], but here we shall focus on the approach of DeVore [10], see also DeVore [11, Section 3.3].

9.4.1 Convergence Speed and Quasi-Seminorms

We start by presenting the mathematical structure underlying convergence speed. In light of Corollary 9.2, we are interested in quantifying the convergence of the respective sequence $\big(b_n(u)\big)_n$ to 0, where, e.g.,

$$b_n(u) = \inf_{v\in V_n} \|u - v\|_V$$

with V_n from (9.18). We say that $b_n(u)$ converges to 0 with rate $r > 0$ whenever there exists a constant C such that

$$\forall n \in \mathbb{N} \quad b_n(u) \le Cn^{-r}. \tag{9.50}$$

The best constant in this inequality is given by

$$|u|_r := \sup_{n\in\mathbb{N}} n^r b_n(u).$$

If all the spaces V_n were linear, every $b_n(u)$ and so $|u|_r$ would be seminorms in $u \in V$ and the functions with rate r would form a linear subspace of V given by the condition $|u|_r < \infty$. However, at the end of Sect. 9.2.3, we have seen that V_n is not linear for all $n \ge 3$. Interestingly, the nonlinearity is confined by

$$\forall n \ge 3 \quad V_n + V_n \subset V_{2(n-1)}. \tag{9.51}$$

In fact, in view of (9.23), each function of V_n can be associated to a full binary tree with $2n - 1$ nodes comprising $(0, 1)$, $(0, 1/2)$ and $(1/2, 1)$. Therefore, the sum of two such functions is associated to a tree with $4n - 5$ nodes and so $2(n - 1)$ leaves. If $u_1, u_2 \in V$, then the inclusion (9.51) implies

$$\inf_{v\in V_{2(n-1)}} \|u_1 + u_2 - v\|_V \le \|u_1 + u_2 - (v_1 + v_2)\|_V \le \|u_1 - v_1\|_V + \|u_2 - v_2\|_V$$

for all $v_1, v_2 \in V_n$ and so

$$\inf_{v\in V_{2(n-1)}} \|u_1 + u_2 - v\|_V \le \inf_{v\in V_n} \|u_1 - v\|_V + \inf_{v\in V_n} \|u_2 - v\|_V.$$

As a consequence, we obtain a triangle-like inequality

$$|u_1 + u_2|_r \le 2^r\big(|u_1|_r + |u_1|_r\big).$$

Since $|\cdot|_r$ is also positively homogeneous, we say that $|\cdot|_r$ is a quasi-seminorm. Thus, the functions with rate $r > 0$ form a linear subspace of V, but, in general, the constant in (9.50) may have to be given in terms of a quasi-seminorm.

9.4.2 Smoothness Spaces with Maximal Functions

Convergence rates have to be established with the help of some kind of regularity. In view of the preceding section, this amounts to finding a smoothness (quasi-)norm that bounds, or is even equivalent to, $|\cdot|_r$. Here we shall use smoothness norms allowing for p-integrability with $p < 1$ and involving maximal functions.

For $0 < p < \infty$, set

$$L^p(0,1) := \left\{ g : (0,1) \to \mathbb{R} \mid g \text{ measurable}, \int_0^1 |g|^p < \infty \right\}$$

with

$$\|g\|_{L^p} := \left(\int_0^1 |g|^p \right)^{\frac{1}{p}},$$

which is a norm for $p \geq 1$ and a quasi-norm for $p < 1$, with constant $\max\{1, 2^{\frac{1}{p}-1}\}$ in the generalized triangle inequality.

Given $g \in L^1(0,1)$, its maximal function of Hardy-Littlewood is defined by

$$Mg(x) := \sup_{x \in I \subset (0,1)} \frac{1}{|I|} \int_I |g|, \quad x \in (0,1). \tag{9.52}$$

where the sup is taken over all subintervals $I \subset (0,1)$ containing x. In view of Lebesgue's differentiation theorem, this seminorm satisfies

$$|g(x)| \leq Mg(x)$$

for almost every $x \in (0,1)$ and

$$\|Mg\|_{L^1} < \infty \iff \int_0^1 |g| \max\{0, \log |g|\} < \infty; \tag{9.53}$$

see Theorem 6.7 in Bennett/Sharpley [2]. Thus, the requirement $\|Mg\|_{L^1} < \infty$ is a little bit more than $g \in L^1(0,1)$, but less than $g \in L^p(0,1)$ for any $p > 1$. This is also expressed by the inequalities

$$\|g\|_{L^1} \leq \|Mg\|_{L^1} \leq c \frac{p}{p-1} \|g\|_{L^p}, \tag{9.54}$$

where $c > 0$ is constant and the second inequality follows, e.g., from Theorem 6.7 and (6.14) in [2, Ch. 4].

In order to have similar strengthening of $g \in L^p(0,1)$ for any $p \in (0,\infty)$ at our disposal, we define

$$M_q g(x) := \left[M(|g|^q)(x)\right]^{\frac{1}{q}} = \sup_{x\in I\subset(0,1)} \left(\frac{1}{|I|}\int_I |g|^q\right)^{\frac{1}{q}}, \quad x \in (0,1), \tag{9.55}$$

which is a (quasi-)seminorm.

There are more general constructions of maximal-type functions involving a smoothness parameter s and integrability parameter p; see, e.g., DeVore/Sharpley [13]. For our purposes, the following simple version of the flat (maximal) function is sufficient. Given $0 < p < \infty$, $s \in (0,1]$, and $g \in L^p(0,1)$, define

$$g^\flat_{s,p}(x) := \sup_{x\in I\subset(0,1)} \frac{1}{|I|^s} \inf_{c\in\mathbb{R}} \left(\frac{1}{|I|}\int_I |g-c|^p\right)^{\frac{1}{p}}, \quad x \in (0,1).$$

The following relationship between $g^\flat_{s,p}$ for different integrabilities will be useful. For any $q \in (0,p)$, the Hölder inequality and [13, Theorem 4.3] imply, for every $x \in (0,1)$,

$$g^\flat_{s,q}(x) \le g^\flat_{s,p}(x) \le c M_\rho(g^\flat_{s,q})(x) \quad \text{with} \quad \rho = \left(s + \frac{1}{p}\right)^{-1}, \tag{9.56}$$

where c depends on s, q, and p.

Moreover, we set

$$\mathcal{C}^s_p(0,1) := \{g \in L^p(0,1) \mid g^\flat_{s,p} \in L^p(0,1)\}$$

with

$$|g|_{\mathcal{C}^s_p} := \|g^\flat_{s,p}\|_{L^p}, \quad \|g\|_{\mathcal{C}^s_p} := \|g\|_{L^p} + \|g^\flat_{s,p}\|_{L^p},$$

which are (semi)norms for $p \ge 1$ and quasi-(semi)norms for $p < 1$. We then have

$$\forall p \in (1,\infty) \quad \mathcal{C}^1_p(0,1) = W^{1,p}(0,1),$$

where $W^{1,p}(0,1)$ is the space of functions whose first weak derivative is p-integrable; see, e.g., [13, Theorem 6.2]. In other words: the spaces $\mathcal{C}^s_p$ provide a fractional generalization of p-integrable weak derivative covering the whole range of integrability, which corresponds to certain Triebel-Lizorkin spaces; see, e.g., Triebel [23]. This generalization is closely related with, but different from, the fractional smoothness provided by Besov spaces $B^{s,q}_p(0,1)$: although

$$\forall s \in (0,1),\, p \in (0,\infty) \quad B^{s,p}_p(0,1) \subset \mathcal{C}^s_p(0,1) \subset B^{s,\infty}_p(0,1), \tag{9.57}$$

see [13, (12.2) and (12.3)], there is no $q \in [p, \infty]$ such that $B_p^{s,q}(0,1) = \mathcal{C}_p^s(0,1)$ due to [13, Corollary 7.4]. Moreover, it is worth mentioning that, if $s \in (0,1)$ and $0 < q < p < \infty$ such that $s - \frac{1}{q} \geq -\frac{1}{p}$, then [13, Theorem 12.5] shows

$$\mathcal{C}_q^s(0,1) \subset L^p(0,1). \tag{9.58}$$

9.4.3 Convergence Rates

The meshes generated by Algorithm (9.12) can gain in grading with growing cardinality: indeed, given $n \in \mathbb{N}$, we have

$$\max_{M \in \mathbb{B}(0,1):\#M \leq n} \frac{\max_{K \in M} h_K}{\min_{K \in M} h_K} \leq 2^{n-2} \quad \text{for} \quad n \geq 2.$$

This property leads to an advantage in the handling of singularities or, more generally, to relatively weak smoothness assumptions for a given decay rage of a best error. To quantify this advantage in the following results, we measure smoothness with the flat function of the preceding section.

Theorem 9.3 (**H_0^1-Approximation**) *For any $r > 0$, the best H_0^1-error in V_n from* (9.18) *satisfies*

$$\inf_{v \in V_n} \|u' - v'\|_V \leq 2^{\frac{1}{p}} \|(u')^\flat_{r,2}\|_{L_p} n^{-r},$$

where $p = (\frac{1}{2} + r)^{-1}$.

Proof We need to construct suitable bisection meshes. To this end, we invoke Algorithm (9.12) with the strategy

$$\texttt{mark} \text{ some } K_n \in M_n \text{ with } \inf_{p \in \mathbb{P}_1} \|u' - p'\|_{K_n} > t$$

where the threshold $t > 0$ will be specified later. Since $u' \in L^2(0,1)$, we have that

$$\inf_{p \in \mathbb{P}_1} \|u' - p'\|_K \leq \|u'\|_K \to 0 \quad \text{as} \quad |K| \to 0 \tag{9.59}$$

Thus, choosing N large enough, we arrive at a mesh M_t for which no element is marked. Its global error satisfies

$$\inf_{v \in V(M_t)} \|u' - v'\|_V \leq \sqrt{\#M_t}\, t. \tag{9.60}$$

Let us assume that the threshold t is so small that $\#M_t > 1$. Then, each element of the mesh M_t has a parent $K^\dagger$ and, since $K^\dagger$ was bisected, its local error has to verify $\inf_{p \in \mathbb{P}_1} \|u' - p'\|_{K^\dagger} > t$. Observe however that the parents overlap and that this overlapping may become unbounded as $\#M_t$ grows. We handle it by means of the flat function. Using that the mean value provides the best constant in L^2, we deduce

$$t < \inf_{p \in \mathbb{P}_1} \|u' - p'\|_{K^\dagger} = \left(\int_{K^\dagger} \left| u' - \overline{(u')}_{K^\dagger} \right|^2 \right)^{\frac{1}{2}} \leq |K^\dagger|^{\frac{1}{2}+r} \inf_K (u')^\flat_{r,2},$$

where we restrict the inf to the overlapping-free mesh element $K \in M_t$. We note $\frac{1}{2} + r = \frac{1}{p}$, raise to the power p, and exploit $|K^\dagger| = 2|K|$ to get

$$t^p \leq 2|K| \inf_K |(u')^\flat_{r,2}|^p \leq 2 \int_K |(u')^\flat_{r,2}|^p$$

Summing over all $K \in M_t$, we arrive at

$$\#M_t t^p \leq \sum_{K \in M_t} \inf_{p \in \mathbb{P}_1} \|u' - p'\|^p_{K^\dagger} \leq 2 \|(u')^\flat_{r,2}\|^p_{L_p}. \tag{9.61}$$

Given $n \in \mathbb{N}$, the choice

$$t = \frac{2^{\frac{1}{p}} \|(u')^\flat_{r,2}\|_{L_p}}{n^{\frac{1}{p}}}$$

yields $\#M_t \leq n$ and, with the help of (9.60),

$$\inf_{v \in V(M_t)} \|u' - v'\|_V \leq 2^{\frac{1}{p}} \|(u')^\flat_{r,2}\|_{L_p} n^{\frac{1}{2} - \frac{1}{p}},$$

which proves the claimed inequality. □

The mixture of integrabilities in $\|(u')^\flat_{r,2}\|_{L_p}$ arises from the fact that the error norm is L^2-based, while p provides the correct summability in space. In any case, these mixture can be avoided at the price of a possible small overestimation. Indeed, (9.56) yields

$$|u'|_{\mathcal{C}^r_p} = \|(u')^\flat_{r,p}\|_{L_p} \leq \|(u')^\flat_{r,2}\|_{L_p} \leq \|M_p (u')^\flat_{r,p}\|_{L_p}. \tag{9.62}$$

Lemma 9.1 holds also without constraining the boundary values and so does Theorem 9.3. This gives us the opportunity to compare bisection with other types of mesh refinement by exploiting that the approximation of any function in $H^1(0, 1)$ by continuous piecewise functions corresponds to the approximation of any function in $L^2(0, 1)$ by piecewise constants. We consider the rates with respect to the number of intervals.

Theorem 9.3 without boundary values ensures rate $r \in (0, 1]$ if $M_p(u')^\flat_{r,p} \in L^p(0, 1)$ with $\frac{1}{p} = \frac{1}{2} + r$. In light of (9.62), this is slightly stronger than $u' \in \mathcal{C}^r_p(0, 1)$.

Classical non-adaptive meshes with intervals of equal length achieve rate 1 if and only if $u' \in H^1(0, 1) = C^1_2(0, 1)$ and rate $r \in (0, 1)$ if and only if $u' \in B^{r,\infty}_2(0, 1)$; see, e.g., [12, Ch. 12, Theorem 2.4]. The space $B^{r,\infty}_2(0, 1)$ is slightly bigger than $H^r(0, 1)$, which implies $o(n^{-r})$-convergence.

Meshes with arbitrary intervals or free breakpoints achieve rate $r \in (0, 1)$ if $u' \in B^{r,p}_p(0, 1)$, again with $\frac{1}{p} = \frac{1}{2} + r$; see [12, Ch. 12, Theorem 8.2]. Since that theorem offers also an accompanying Bernstein inequality, this requirement is quite sharp.

We thus see that, in order to ensure rate r, all three refinement techniques need a 'derivative of order r'. However, they differ in the required integrability of that derivative. The embeddings in (9.57) and the characterization (9.53) suggest that the integrability requirements of bisection and free breakpoints are very close. In addition, considering also (9.58), these two adaptive techniques appear as almost borderline cases. The difference in the integrability requirement between classical global refinement and the two adaptive refinements grows with r and, for $r = 1$, contrasts 2-integrability with $\frac{2}{3}$-integrability.

Let us illustrate these observations with an example.

Example 9.5 (Power Functions) Given any $\rho > \frac{1}{2}$ with $\rho \neq 1$, consider the function

$$u_\rho(x) := \frac{1}{\rho} x^\rho, \quad \text{with derivative} \quad u'_\rho(x) = x^{\rho-1}, \quad x \in (0, 1).$$

The restriction on ρ gives exactly all powers ρ for which $u_\rho \in H^1(0, 1) \setminus \mathbb{P}_1$. Using the Poincaré inequality in every interval except the left-most for $\rho \in (\frac{1}{2}, \frac{3}{2}]$, we see that the best error with classical global refinement decays like

$$\begin{cases} n^{-1} & \text{if } \rho > \frac{3}{2}, \\ |\log n|^{\frac{1}{2}} n^{-1} & \text{if } \rho = \frac{3}{2}, \\ n^{-(\rho - \frac{1}{2})} & \text{if } \rho > \frac{1}{2}. \end{cases}$$

Since $u'_\rho \in H^1(0, 1)$ if and only if $\rho > \frac{3}{2}$ and $u'_\rho \in B^{s,\infty}_2(0, 1)$ if and only if $\rho \geq s - \frac{1}{2}$, these rates are consistent with [12, Ch. 12, Theorem 2.4] and sharp. Therefore, depending on ρ, the best error decay rate with classical global refinement can be arbitrarily small.

We turn to bisection and analyze first the flat function $(u'_\rho)^\flat_{1,2/3}$ with for $\rho \in (\frac{1}{2}, \frac{3}{2}]$. In this case u'_ρ has its strongest variation close to 0 and therefore the critical intervals in the definition of the flat functions contain 0:

$$(u'_\rho)^\flat_{1,\frac{2}{3}}(x) \le \sup_{b\in[x,1]} \frac{1}{b}\left(\frac{1}{b}\int_0^b |y^{\rho-1} - x^{\rho-1}|^{\frac{2}{3}} dy\right)^{\frac{3}{2}},$$

where we take $x^{\rho-1}$ as approximating constant. We split the integral into the two parts $(0, x)$ and (x, b) and consider the critical part $(0, x)$ first. For $y \in (0, x)$, we have

$$|y^{\rho-1} - x^{\rho-1}| \le (\rho-1)y^{\rho-2}(x-y) \le (\rho-1)y^{\rho-2}x$$

and therefore, upon noting $(\rho-2)\frac{2}{3} > -1$,

$$\frac{1}{b}\int_0^x |y^{\rho-1} - x^{\rho-1}|^{\frac{2}{3}} dy \le (\rho-1)^{\frac{2}{3}} x^{-\frac{1}{3}} \int_0^x y^{(\rho-2)\frac{2}{3}} dy = \frac{(\rho-1)^{\frac{2}{3}}}{(\rho-2)\frac{2}{3}+1} x^{(\rho-1)\frac{2}{3}}$$

Similarly, using

$$|y^{\rho-1} - x^{\rho-1}| \le (\rho-1)x^{\rho-2}(y-x) \le (\rho-1)x^{\rho-2}b$$

for $y \in (x, b)$, we obtain

$$\frac{1}{b}\int_x^b |y^{\rho-1} - x^{\rho-1}|^{\frac{2}{3}} dy \le (\rho-1)^{\frac{2}{3}} x^{(\rho-1)\frac{2}{3}}.$$

Taking both parts together, we obtain

$$(u'_\rho)^\flat_{1,\frac{2}{3}}(x) \le c_\rho \sup_{b\in[x,1]} \frac{1}{b} x^{\rho-1} \le x^{\rho-2}.$$

This entails

$$M_{\frac{2}{3}}(u'_\rho)^\flat_{1,\frac{2}{3}}(x) \le c_\rho \left(\frac{1}{x}\int_0^x y^{(\rho-2)\frac{2}{3}} dy\right)^{\frac{3}{2}} \le c_\rho x^{\rho-2}$$

whence

$$\|M_{\frac{2}{3}}(u'_\rho)^\flat_{1,\frac{2}{3}}\|_{L^{\frac{2}{3}}} < \infty$$

for all $\rho > \frac{1}{2}$. Consequently, Theorem 9.3 ensures best error decay rate 1 for bisection, irrespective of ρ. Since the approximation spaces with free breakpoints are even larger, the same holds for them.

Let us now explore the decay rates for upper bounds for best errors that may be viewed as upper bounds of the best error analyzed in Theorem 9.3. Let us start with the surrogate indicator $h_K \|f\|_K$ of Example 9.2. Here we can establish the error decay rate 1 whenever the indicator is meaningful.

Lemma 9.7 (Surrogate Indicator) *For $f \in L^2(0,1)$, we have*

$$\inf_{M \in \mathbb{B}_n} \left(\sum_{K \in M} h_K^2 \|f\|_K^2 \right)^{\frac{1}{2}} \le 2\|f\|_{L^2} n^{-1}.$$

Proof We mark elements in Algorithm (9.12) such that the resulting meshes are almost uniform, i.e., such that, for all $n \in \mathbb{N}$, we have $\max_{K \in M_n} h_K / \min_{K \in M_n} h_K \le 2$. The meshes are thus independent of f or 'non-adaptive'. For any $n \in \mathbb{N}$, we have $\max_{K \in M_n} h_K \le \frac{2}{n}$ and so

$$\left(\sum_{K \in M} h_K^2 \|f\|_K^2 \right)^{\frac{1}{2}} \le 2\|f\|_{L^2} n^{-1}.$$

□

If $f = u''$ where u is the solution of (9.1), the assumption $f \in L^2(0,1)$ means $u' \in H^1(0,1)$. Thus, Lemma 9.7 is consistent with the fact that there is no advantage of adaptive refinement in terms of asymptotic speed if $u' \in H^1(0,1)$. More specifically, in the context of Example 9.5, the requirement $u'' = f \in L^2(0,1)$ entails $\rho > \frac{3}{2}$ and so excludes the cases where adaptive refinement is favorable in terms of asymptotic convergence speed.

A partial remedy within functions of this disappointing result can be obtained by using the alternative surrogate indicator $h_K^{1/2} \|f\|_{L^1(K)}$. This new indicator scales like the original $h_K \|f\|_{L^2(K)}$, but requires only $f \in L^1(0,1)$ and is sharper. In fact, in view of

$$\forall \varphi \in H_0^1(K) \quad \sup_K |\varphi| \le \|\varphi'\|_{L^1(K)} \le h_K^{1/2} \|\varphi'\|_{L^2(K)},$$

we have

$$\|f\|_{H^{-1}(K)} \le h_K^{1/2} \|f\|_{L^1(K)} \le h_K \|f\|_{L^2(K)}.$$

Notice that, after applying numerical integration, both indicators are equivalent.

Proposition 9.1 (Modified Surrogate Indicator) *For $f \in L^1(0,1)$, we have*

$$\inf_{M \in \mathbb{B}_n} \left(\sum_{K \in M} h_K \|f\|_{L^1(K)}^2 \right)^{\frac{1}{2}} \le c\|f\|_{L^1} n^{-1}.$$

Proof The proof is along the lines of the proof of Theorem 3 in the primer [20]. This argument is similar to the proof of the above Theorem 9.3, but handles the overlapping by means of a geometric series arising from the scaling factor $h_K^{1/2}$. □

Consequently, we have again the error decay rate 1 whenever the indicator is meaningful. In the case that $f = u''$, the weaker assumption $f \in L^1(0,1)$ now includes also cases where adaptive refinement leads to an improved asymptotic convergence speed. In particular, for Example 9.5, Proposition 9.1 ensures error decay rate 1 for $\rho \in (1, \frac{3}{2}]$. Apart from its increased sharpness, these observations substantiate that the surrogate indicator $h_K^{1/2}\|f\|_{L^1(K)}$ is superior to the classical one $h_K\|f\|_{L^2(K)}$.

We conclude this section by deriving decay rate for the oscillation in Example 9.3. In view of Lemma 9.7, the decay rate is at least 1 whenever the oscillation is meaningful.

Proposition 9.2 (Oscillation) *For $f \in L^2(0,1)$ and any $s > 0$, we have*

$$\inf_{M\in\mathbb{B}_n}\left(\sum_{K\in M} h_K^2\|f-\bar{f}_K\|_K^2\right)^{\frac{1}{2}} \le 2^{\frac{1}{p}}\|f_{s,2}^\flat\|_{L_p} n^{-1-s},$$

with $p = (\frac{3}{2}+s)^{-1}$.

Proof The proof is along the lines of the above proof of Theorem 9.3. This time we apply Algorithm (9.12) with the strategy

$$\texttt{mark some } K_n \in M_n \text{ with } h_{K_n}\|f-\bar{f}_{K_n}\|_{K_n} > t$$

where the threshold $t > 0$ will be specified below. Since $h_K\|f-\bar{f}_K\|_K \to 0$ as $|K| \to 0$, we can construct a mesh M_t for which no element is marked and so

$$\left(\sum_{K\in M_t} h_K^2\|f-\bar{f}_K\|_K^2\right)^{\frac{1}{2}} \le \sqrt{\#M_t}\,t. \tag{9.63}$$

Let us assume that the threshold t is so small that $\#M_t > 1$. Then, if $K \in M_t$, its parent $K^\dagger$ satisfies

$$t \le h_{K^\dagger}\|f-\bar{f}_K\|_{K^\dagger} \le |K^\dagger|^{3/2+s}\inf_K f_{s,2}^\flat,$$

where we have applied the definition $f_{s,2}^\flat$. Using $\frac{3}{2}+s = \frac{1}{p}$ and $|K^\dagger| = 2|K|$, we get

$$t^p \le 2|K|\inf_K |f_{s,2}^\flat|^p \le 2\int_K |f_{s,2}^\flat|^p$$

and, by summing over all $K \in M_t$,

$$\#M_t t^p \leq 2\|f_{s,2}^{\flat}\|_{L^p}^p.$$

Then, letting $t = 2^{\frac{1}{p}}\|f_{s,2}^{\flat}\|_{L^p} n^{-\frac{1}{p}}$ finishes the proof as for Theorem 9.3. □

We thus see that adaptive bisection leads to higher decay rates of the oscillation under regularity assumptions that are weaker than $f \in H^s(0,1)$ with $s \in (0,1]$, very much in the spirit of the discussion following Theorem 9.3.

9.5 Comments on the Multivariate Case

This section briefly outlines the multivariate case with the help of the available literature, following the lines of this contribution.

The article [24] establishes a prototypical generalization of the best error localization in Lemma 9.1. It concerns the best error in H_0^1 with continuous piecewise polynomials of fixed maximal degree over simplicial meshes in arbitrary dimension. A best error localization for L^2 and the reaction-diffusion norm can be found in Tantardini/Veeser/Verfürth [22]. More results of this type involving other norms or other finite element spaces are in preparation. In all these results, mesh conformity is assumed and the equivalence constants involves the shape regularity coefficient of the mesh. These two assumptions (or variants like limited non-conformity) often appear also when discretizing a PDE or in a posteriori analyses, that is error localizations in the vein of Lemma 9.2; see the monographs Ainsworth/Oden [1], Verfürth [25] or the introduction in Sections 2 and 3 of the primer [20].

Shape regularity puts constraints on iterated subdivision of elements, while mesh conformity may propagate refinement, questioning its locality. Recursive bisection of triangles as in Mitchell [19] and of simplices as in Kossaczký [17] or Maubach [18] generate shape regular conforming meshes. Furthermore, in view of results as Lemma 2.5 in Binev/Dahmen/DeVore [6], the refinement is essentially local. An account of the 2-dimensional case is given in Sections 1.3 and 6 of the primer [20], while the general d-dimensional case is reviewed in Section 4 of the survey [21].

Multidimensional domains usually cannot be meshed with a single element. As a consequence, the initial mesh consists typically of various elements, which then correspond to various roots of binary trees. We thus have a master forest instead of a master tree, a difference which however does not have any important impact on the theory exposed in Sect. 9.3. Requiring conformity of meshes however does have an impact, but can be handled by means of the aforementioned results. In this context, the results concerning bisection in Diening/Kreuzer/Stevenson [14] are also of interest.

Convergence rates for continuous piecewise affine functions and 2-dimensional bisection are derived in Binev et al. [5]. There, smoothness assumptions are given within Besov spaces. A generalization to continuous functions that are piecewise polynomial (of fixed maximal degree) is given in Gaspoz/Morin [16].

References

1. Ainsworth, M., Oden, J.T.: A Posteriori Error Estimation in Finite Element Analysis. Pure and Applied Mathematics. Wiley-Interscience, New York (2000)
2. Bennett, C., Sharpley, R.: Interpolation of Operators. Pure and Applied Mathematics, vol. 29. Academic, Boston (1988)
3. Binev, P.: Tree approximation for *hp*-adaptivity. Interdisciplinary Mathematics Institut, University of South Carolina (2015). Preprint 2015:07
4. Binev, P., DeVore, R.: Fast computation in adaptive tree approximation. Numer. Math. **97**, 193–217 (2004)
5. Binev, P., Dahmen, W., DeVore, R., Petrushev, P.: Approximation classes for adaptive methods. Serdica Math. J. **28**, 391–416 (2002)
6. Binev, P., Dahmen, W., DeVore, R.: Adaptive finite element methods with convergence rates. Numer. Math. **97**, 219–268 (2004)
7. Birman, M.Š, Solomjak, M.Z.: Piecewise polynomial approximations of functions of classes ${W_p}^{\alpha}$. Mat. Sb. (N.S.) **73**(115), 331–355 (1967)
8. Brenner, S.C., Scott, L.R.: The Mathematical Theory of Finite Element Methods. Texts in Applied Mathematics, vol. 15. Springer, New York (2008)
9. Carstensen, C., Feischl, M., Page, M., Praetorius, D.: Axioms of adaptivity. Comput. Math. Appl. **67**, 1195–1253 (2014)
10. DeVore, R.A.: A note on adaptive approximation. In: Proceedings of China-U.S. Joint Conference on Approximation Theory, vol. 3, pp. 74–78 (1987)
11. DeVore, R.A.: Nonlinear approximation. Acta Numer. **7**, 51–150 (1998)
12. DeVore R.A., Lorentz, G.G.: Constructive Approximation. Grundlehren der Mathematischen Wissenschaften [Fundamental Principles of Mathematical Sciences], vol. 33. Springer, Berlin (1993)
13. DeVore, R.A., Sharpley, R.C.: Maximal functions measuring smoothness. Mem. Am. Math. Soc. **47**, (1984)
14. Diening, L., Kreuzer, C., Stevenson, R.: Instance optimality of the adaptive maximum strategy. Found. Comput. Math. **16**, 33–68 (2016)
15. Dörfler, W.: A convergent adaptive algorithm for Poisson's equation. SIAM J. Numer. Anal. **33**, 1106–1124 (1996)
16. Gaspoz, F.D., Morin, P.: Approximation classes for adaptive higher order finite element approximation. Math. Comput. **83**, 2127–2160 (2014)
17. Kossaczký, I.: A recursive approach to local mesh refinement in two and three dimensions. J. Comput. Appl. Math. **55**, 275–288 (1994)
18. Maubach, J.M.: Local bisection refinement for n-simplicial grids generated by reflection. SIAM J. Sci. Comput. **16**, 210–227 (1995)
19. Mitchell, W.F.: A comparison of adaptive refinement techniques for elliptic problems. ACM Trans. Math. Softw. **15**, 326–347 (1990)
20. Nochetto, R.H., Veeser, A.: Primer of adaptive finite element methods. In: Multiscale and Adaptivity: Modeling, Numerics and Applications. Lecture Notes in Mathematics, vol. 2040, pp. 125–225. Springer, Berlin (2012)
21. Nochetto, R.H., Siebert, K.G. Veeser, A.: Theory of adaptive finite element methods: an introduction. In: Multiscale, Nonlinear and Adaptive Approximation, pp. 409–542. Springer, Berlin (2009)

22. Tantardini, F., Veeser, A., Verfürth, R.: Robust localization of the best error with finite elements in the reaction-diffusion norm. Constr. Approx. **42**, 313–347 (2015)
23. Triebel, H.: Theory of Function Spaces. II. Monographs in Mathematics, vol. 84. Birkhäuser, Basel, (1992)
24. Veeser, A.: Approximating gradients with continuous piecewise polynomial functions. Found. Comput. Math. **16**, 723–750 (2016)
25. Verfürth, R.: A Posteriori Error Estimation Techniques for Finite Element Methods. Numerical Mathematics and Scientific Computation. Oxford University Press, Oxford (2013)

Chapter 10
Defective Boundary Conditions for PDEs with Applications in Haemodynamics

Luca Formaggia and Christian Vergara

Abstract This works gives an overview of the mathematical treatment of state-of-the-art techniques for partial differential problems where boundary data are provided only in terms of averaged quantities. A condition normally indicated as "defective boundary condition". We present and analyze several procedures by which this type of problems can be handled.

10.1 Introduction

In many applications of practical relevance, it could happen that only average data is available on a portion of the boundary. For instance the space average of the solution or of the stress. Here with stress we mean the solution dependent quantity contained in the boundary term emerging from integration by parts when the weak formulation is derived. Depending on the problem at hand, it could represent several physical quantities, e.g. a heat flux, the elastic traction, the normal Cauchy stress, just to provide some examples.

This situation often occurs on the so-called *artificial boundaries*, i.e. portions of the boundary introduced by an artificial cut of the physical domain, as it happens, for instance, in a pipe. On such boundaries, often there are no strong physical arguments that can be used to devise suitable boundary conditions.

In practical situations one may provide boundary information on artificial boundaries by (1) the acquisition of some measurements or (2) the coupling with reduced models (typically based on the solution of another differential problem) able to give a suitable description of what happens in the cut region. However, in many contexts both techniques provide just averaged quantities. An example is *hemodynamics*, where non-invasive measurements (like Echo-Color Doppler) of blood velocity or pressure, as well as the coupling with reduced models, are

L. Formaggia (✉) · C. Vergara
MOX, Dipartimento di Matematica, Politecnico di Milano, Milano, Italy
e-mail: luca.formaggia@polimi.it; christian.vergara@polimi.it

D. A. Di Pietro et al. (eds.), *Numerical Methods for PDEs*,
SEMA SIMAI Springer Series 15, https://doi.org/10.1007/978-3-319-94676-4_10

often used to provide boundary information to full three-dimensional simulations [3, 4, 12, 32–34]. For the case of general hydraulic networks, see also [25], while another context where the coupling with a lumped parameter model leads to a defective condition is that of heat transfer in a pipe [16].

From the mathematical viewpoint, defective problems are not well posed since the data on the artificial boundaries are insufficient to guarantee uniqueness of the solution. Many approaches have been developed so far to fill this gap: some of them take inspiration from engineering principles and practices, others have a more mathematical foundation. In any case, suitable hypotheses are introduced in order to make the defective problems solvable.

In this review, we describe the main techniques to prescribe defective boundary conditions. To better highlight the mathematical principles behind them, we first treat the case of the Poisson equation. Then, we address the case where such strategies were originally developed, i.e. fluid-dynamics, focussing to the Stokes problem. Finally we provide some examples taken from real haemodynamic studies.

10.2 Defective Poisson Problem

In this section, we address the simple case of a scalar Poisson problem. This will allow us to introduce all the key-points at the basis of numerical methods for the prescription of defective data.

To begin with, we consider the following defective problem on a bounded domain $\Omega \subset \mathbb{R}^d$ with $d = 2$ or 3, with Lipschitz boundary:

$$-\nabla \cdot (\mu \nabla u) = f \qquad \text{in } \Omega, \tag{10.1a}$$

$$u = 0 \qquad \text{on } \Gamma, \tag{10.1b}$$

$$\int_\Sigma u \, d\Sigma = Q, \tag{10.1c}$$

with $\Sigma = \partial\Omega \setminus \Gamma$, $f \in L^2(\Omega)$, $Q \in \mathbb{R}$, and $\mu : \Omega \to \mathbb{R}$ bounded away from zero, i.e. $\mu \in L^\infty(\Omega)$ such that $0 < \mu_0 \leq \mu(\mathbf{x})$ for almost all $\mathbf{x} \in \Omega$ and for a suitable scalar μ_0.

Notice that in (10.1c) we are prescribing only the average value of u over Σ, thus a defective condition. Alternatively, we could consider the defective problem obtained by (10.1a)–(10.1b) together with

$$\int_\Sigma \mu \frac{\partial u}{\partial \mathbf{n}} \, d\Sigma = P, \tag{10.2}$$

with $P \in \mathbb{R}$ given, $\mathbf{n}$ the outward unit vector to Ω, and $\frac{\partial u}{\partial \mathbf{n}} = \nabla u \cdot \mathbf{n}$ the derivative normal to the boundary.

Condition (10.2) prescribes the average of the stress, thus it is again a defective condition. In what follows, we will refer to problems given by (10.1) and (10.1a), (10.1b), (10.2) as *mean solution* and *mean stress* problems, respectively. Of course, in both cases the solution is not unique. For this reason, suitable hypothesis should be introduced in order to find a reasonable solution of such problems. This will be discussed in the next sections. To make the exposition simpler, we will address only the case of one defective condition. The results may be readily extended to the case where defective conditions are applied to several non-overlapping parts Σ_i of $\partial\Omega$.

10.2.1 Empirical Methods

A simplest choice to make problem (10.1) solvable is to select a-priori a profile of u on Σ that satisfies (10.1c). Thus, problem (10.1) is transformed into a standard Dirichlet problem,

$$-\nabla\cdot(\mu\nabla u) = f \qquad \text{in } \Omega, \tag{10.3a}$$

$$u = 0 \qquad \text{on } \Gamma, \tag{10.3b}$$

$$u = g \qquad \text{on } \Sigma, \tag{10.3c}$$

where $g \in H_{00}^{1/2}(\Sigma)$ satisfies

$$\int_{\Sigma} g\, d\Sigma = Q.$$

Now, the solution of problem (10.3) is clearly unique. However, such a solution is heavily influenced by the choice of the datum g. Let g be an educated guess of the "real" solution $u = g_{ex}$ on Σ, of which we actually know the average Q. Thus, the error e committed by solving (10.3) satisfies

$$\|e\|_{H^1(\Omega)} \leq C\|g - g_{ex}\|_{H^{1/2}(\Sigma)},$$

which of course annihilates only for $g = g_{ex}$. The fact that $\int_{\Sigma}(g - g_{ex})d\Sigma = 0$ does not help so much, since $\|g - g_{ex}\|_{H^{1/2}(\Sigma)}$ could still be arbitrarily large. Thus, in absence of any further information about the solution at Σ, this method could lead to not negligible errors.

Analogously, for problem given by (10.1a), (10.1b), (10.2), one could think to prescribe the following Neumann condition together with (10.1a), (10.1b):

$$\mu\frac{\partial u}{\partial \mathbf{n}} = h \qquad \text{on } \Sigma,$$

with h satisfying

$$\int_{\Sigma} h\, d\Sigma = P.$$

Similar conclusions found for the mean solution problem hold as well in this case since the choice of h is arbitrary.

In the next subsections, we will consider four alternative strategies which are mathematically more justified.

10.2.2 Lagrange Multiplier Approach

We note that problem (10.1) could be equivalently written as the following constrained minimization problem: find $u \in V = \{v \in H^1(\Omega) : v|_{\Gamma} = 0\}$ such that functional

$$J(v) = \frac{1}{2} \int_{\Omega} \mu\, (\nabla v)^2\, d\mathbf{x} - \int_{\Omega} f v\, d\mathbf{x} \tag{10.4}$$

is minimized in V under the constraint (10.1c).

This problem can be rewritten as an unconstrained problem by introducing the corresponding Lagrangian functional: find $u \in V$ and $\lambda \in \mathbb{R}$ such that the following Lagrangian functional

$$L(v, \xi) = J(v) + \xi \left(\int_{\Sigma} v\, d\Sigma - Q \right)$$

has a stationary point (in fact a saddle point) in $V \times \mathbb{R}$. The associated variational problem is: find $u \in V$ and $\lambda \in \mathbb{R}$ such that for all $(v, \xi) \in V \times \mathbb{R}$

$$(\mu \nabla u, \nabla v) + b(v, \lambda) = (f, v), \tag{10.5a}$$

$$b(u, \xi) = \xi\, Q, \tag{10.5b}$$

where $b(v, \xi) = \xi \int_{\Sigma} v d\Sigma$ and $(v, w) = \int_{\Omega} v w\, d\Omega$ denotes the $L^2(\Omega)$ inner product.

This formulation is the *Lagrange multiplier formulation* of the mean solution problem (10.1), and is in fact the extension to the defective case of the Lagrange multiplier technique to enforce Dirichlet boundary conditions proposed and analyzed, for instance, in [2].

We have the following result.

Proposition 10.1 *Assume that* $f \in L^2(\Omega)$. *Then, the problem given by* (10.5) *admits a unique solution* $(u, \lambda) \in V \times \mathbb{R}$.

Proof We can use the theory illustrated, for instance, in [5]. In the case $|\Gamma| \neq 0$, thanks to bounds on μ, the bilinear form $(\mu \nabla v, \nabla w)$ is coercive and continuous with respect to the H^1-seminorm $|v|_{H^1(\Omega)} = \|\nabla v\|_{L^2(\Omega)}$, which is in this case equivalent to the H^1 norm thanks to Poincaré inequality. The b term is a bilinear and continuous form on $V \times \mathbb{R}$, indeed

$$|b(v, \xi)| \leq |\xi| \int_{\Sigma} |v| d\Sigma \leq C_{\Sigma} \sqrt{|\Sigma|} |\xi| \|v\|_V, \quad \forall (v, \xi) \in V \times \mathbb{R},$$

where C_Σ is the constant in the trace inequality $\|v\|_{L^2(\Sigma)} \leq C_\Sigma \|v\|_V$.

To prove that it satisfies the inf-sup condition it is sufficient to note that it is possible to construct a function $\phi \in H_{00}^{1/2}(\Sigma)$ so that $\int_\Sigma \phi d\Sigma = 1$. For a given $\xi \in \mathbb{R}$ we set $\phi_\xi = \xi\phi$ and find u_ξ solution of

$$\begin{aligned} -\nabla \cdot (\mu \nabla u_\xi) &= 0 && \text{in } \Omega, \\ u_\xi &= 0 && \text{on } \Gamma, \\ u_\xi &= \phi_\xi && \text{on } \Sigma. \end{aligned} \tag{10.6}$$

We have that $b(u_\xi, \xi) = \xi^2$ and, by standard regularity results, $\|u_\xi\|_V \leq C|\xi|$ for a constant C independent of ξ. Therefore, by combining the two previous relations and taking $\beta = 1/C > 0$, we can state that for all $\xi \in \mathbb{R}$, there exists $u_\xi \in V$ satisfying

$$b(u_\xi, \xi) \geq \beta |\xi| \|u_\xi\|_V.$$

The case $\Gamma = \emptyset$, i.e. $\Sigma = \partial\Omega$, can also be treated in a standard way, by proving that the bilinear form $a(u, v) = (\mu \nabla u, \nabla v)$ is coercive on the space

$$\hat{V} = \{v \in V = H^1(\Omega) : b(v, \xi) = 0, \forall \xi \in \mathbb{R}\}.$$

Indeed, for all $v \in \hat{V}$ we may write $a(v, v) \geq \mu_0 \|\nabla v\|^2_{L^2(\Omega)} = |||v|||^2$, where $|||v||| = \left(\|\nabla v\|^2_{L^2(\Omega)} + |\int_{\partial\Omega} v \, d\Omega|^2 \right)^{1/2}$ is a norm equivalent to $\|v\|_{H^1(\Omega)}$. Indeed, $|||v||| \leq C\|v\|_{H^1(\Omega)}$, thanks to trace inequality, so we are left to prove that there exists a constant $C > 0$ so that $|||v||| \geq C\|v\|_{H^1(\Omega)}$. To show it we proceed by contradiction. Negating the statement is equivalent to say that there exists a sequence $v_n \in H^1(\Omega)$ such that $\|v_n\|_{H^1(\Omega)} = 1$ while $|||v_n||| \to 0$. Since v_n is bounded in $H^1(\Omega)$ there exists a subsequence v_{n_k} weakly converging to a $v \in H^1(\Omega)$ and such that $v_{n_k} \to v$ in $L^2(\Omega)$. For the sake of simplicity, in the sequel we will use the subscript n for the subsequence. Weak convergence implies that $\|\nabla v\|^2_{L^2(\Omega)} = \lim_{n\to\infty} (\nabla v_n, \nabla v)_{L^2(\Omega)} \leq \lim_{n\to\infty} \|\nabla v_n\|_{L^2(\Omega)} \|\nabla v\|_{L^2(\Omega)}$. By which

$$\|\nabla v\|_{L^2(\Omega)} \leq \lim_{n\to\infty} \|\nabla v_n\|_{L^2(\Omega)}. \tag{10.7}$$

The hypothesis $|||v_n||| \rightarrow 0$ implies that $\|\nabla v_n\|_{L^2(\Omega)} \rightarrow 0$ thus, by (10.7), $\|\nabla v\|_{L^2(\Omega)} = 0$, i.e. $\|v\|_{H^1(\Omega)} = \|v\|_{L^2(\Omega)}$. The hypothesis on the norm of the elements of the sequence, the strong convergence of the subsequence in $L^2(\Omega)$ and the previous result imply $\|v\|_{L^2(\Omega)} = \lim_{n\to\infty} \|v_n\|_{L^2(\Omega)} = 1$. Now, $\|\nabla v\|_{L^2(\Omega)} = 0$, then $v = c$ where c is a constant, which is different from zero since $\|v\|_{L^2(\Omega)} = 1$. But then, since $|||v_n||| \rightarrow 0$ also implies $\lim_{n\to\infty} \left(\int_{\partial\Omega} v_n\right)^2 = \left(\int_{\partial\Omega} v\right)^2 = 0$, we have a contradiction because $\left(\int_{\partial\Omega} v\right)^2 = |\partial\Omega|^2 c^2 > 0$. □

From (10.5), it is easy to show that the Lagrange multiplier λ plays the role of a constant stress on Σ, i.e. the solutions u and λ satisfy

$$\lambda = -\mu \frac{\partial u}{\partial \mathbf{n}} \qquad \text{on } \Sigma.$$

Thus, this approach implicitly implies that the stress is constant on Σ. In other words, among all the possible solutions of problem (10.1), this technique selects the (unique) one with constant stress on Σ. We thus expect a great accuracy in those scenarios when the stress is almost constant over Σ. If we do not have further information, this technique is anyway optimal in the sense that it is the one that minimizes the energy functional (10.4) associated to the problem.

If we consider now a finite dimensional subspace $V_h = \text{span}(\varphi_1, \ldots, \varphi_{N_h})$ approximating V, h being the mesh size, for instance a finite element space corresponding to a triangulation $\mathcal{T}_h$ of Ω [9], the Galerkin approximation of (10.5) leads to the following algebraic problem

$$\begin{bmatrix} A & \mathbf{b} \\ \mathbf{b}^T & 0 \end{bmatrix} \begin{bmatrix} \mathbf{U} \\ \lambda_h \end{bmatrix} = \begin{bmatrix} \mathbf{f} \\ Q \end{bmatrix}, \tag{10.8}$$

where $A_{ij} = \int_\Omega \mu \nabla\varphi_j \cdot \nabla\varphi_i d\Omega$, $b_i = \int_\Sigma \varphi_i d\Sigma$, $f_i = \int_\Omega f\varphi_i d\Omega$, for $i = 1, \ldots, N_h$, and $j = 1, \ldots, N_h$, while U_i, $i = 1, \ldots, N_h$, are the unknown coefficients (degrees of freedom) of the linear expansion of the Galerkin solution by means of the basis functions φ_i, and λ_h the discrete Lagrange multiplier.

For the numerical solution of (10.8), we can consider a monolithic approach where the linear system is solved e.g. by a direct or an iterative method. However, this strategy is not *modular* in the sense that we cannot exploit pre-existing codes we may have at disposal for the numerical solution of the Poisson problem. Alternatively, if $|\Gamma| \neq 0$ then A is non-singular and one could consider, like it is done in [12] for a defective Stokes problem, the *Schur complement* equation related to (10.8), which reads

$$\mathbf{b}^T A^{-1} \mathbf{b} \lambda = Q - \mathbf{b}^T A^{-1} \mathbf{f}. \tag{10.9}$$

Notice that it is, in this very simple case, just a scalar equation, whose solution requires to solve two linear systems in A. In particular, we have the following algorithm:

1. Solve the linear system $A\mathbf{U}_1 = \mathbf{f}$;
2. Compute $\lambda_1 = Q - \mathbf{b}^T\mathbf{U}_1$;
3. Solve the linear system $A\mathbf{U}_2 = \mathbf{b}$;
4. Compute $\lambda_2 = \mathbf{b}^T\mathbf{U}_2$;
5. Compute λ from (10.9): $\lambda = \lambda_1/\lambda_2$;
6. Compute $\mathbf{U}$ from the first of (10.8): $\mathbf{U} = \mathbf{U}_1 - \lambda\mathbf{U}_2$.

The previous strategy may seem more expensive than the monolithic one (we need to solve 2 linear problems instead of 1), yet, not only the matrix in (10.8) is of larger size, but it is also indefinite, while A is, in this case, symmetric positive definite, so more suited for efficient solvers. Moreover, with the proposed algorithm we can exploit pre-existing solvers for the Poisson problem. In particular, the first linear system (point 1.) corresponds to (10.1a)–(10.1b) with a homogeneous Neumann condition on Σ, whereas the second one (point 3.) corresponds to (10.1a)–(10.1b) with $f = 0$ and

$$\mu\frac{\partial u}{\partial \mathbf{n}} = 1 \quad \text{on } \Sigma.$$

Remark 10.1 The previous algorithm can be extended to the case of more than one flow rate conditions (let say m), and requires the solution of $m + 1$ "classical" problems [12].

In the case $|\Gamma| = 0$ matrix A is singular and the standard Shur-complement procedure does not apply. However, since in this case $V = H^1(\Omega)$, we can take $v = 1$ in (10.5) to get

$$\lambda = |\partial\Omega|^{-1}\int_{\Omega} f\, d\Sigma. \tag{10.10}$$

We can then decompose the solution as $u = \mathring{u} + \overline{u}$, where $\overline{u}$ is a constant and $\mathring{u}$ is the unique solution in $H^1(\Omega)\setminus\mathbb{R} = \{\mathring{w} \in H^1(\Omega) : \int_\Omega \mathring{w}\, d\Omega = 0\}$ of

$$(\mu\nabla\mathring{u}, \nabla v) = (f, v) - (\lambda, v) \quad \forall v \in H^1(\Omega)\setminus\mathbb{R}. \tag{10.11}$$

Then, $\overline{u} = |\partial\Omega|^{-1}\left(Q - \int_{\partial\Omega}\mathring{u}\, d\Sigma\right)$.

Note that since in standard finite element approximation for this class of problems $1 \in V_h$, also λ_h can be computed by (10.10), while (10.11) can be approximated by standard means.

We now give an alternative proof for the well posedness of (10.5) which gives some useful insights for the next Section. First of all we rewrite the definition of $\hat{V}$ as $\hat{V} : \{w \in V : \int_\Sigma w = 0\}$. We have shown in the proof of Proposition 10.1

that for $v \in \hat{V}$ the seminorm $|v|_{H^1(\Omega)} = \|\nabla v\|_{L^2(\Omega)}$ is equivalent to $\|\nabla v\|_{H^1(\Omega)}$. Indeed, for a $v \in \hat{V}$ we have $|v|_{H^1(\Omega)} = |||v|||$. We can then consider the following problem: find $u \in V$ such that $\int_\Sigma u = Q$ and

$$(\mu \nabla u, \nabla v) = (f, v) \quad \forall v \in \hat{V}. \tag{10.12}$$

This problem can be found to be equivalent to the following differential problem: find $(u, \lambda) \in \hat{V} \times \mathbb{R}$ such that

$$\begin{aligned} & -\nabla \cdot (\mu \nabla u) = f && \text{in } \Omega, \\ & u = 0 && \text{on } \Gamma, \\ & \int_\Sigma u \, d\Sigma = Q, \\ & -\mu \frac{\partial u}{\partial n} = \lambda && \text{on } \Sigma. \end{aligned} \tag{10.13}$$

That is, problem (10.12) forces (in a weak sense) the conormal derivative $\mu \frac{\partial u}{\partial n}$ to be constant on Σ. If we have a solution of (10.12) we can recover λ as

$$\lambda = (f, v) - (\mu \nabla u, \nabla v) \tag{10.14}$$

for any $v \in V$ with $\int_\Gamma v = 1$.

Proposition 10.2 *Problem* (10.12) *is well posed, and the couple* (u, λ), *where* λ *is obtained by* (10.14), *is the unique solution of* (10.5). *Moreover,* $\|u\|_{H^1(\Omega)} \leq C\left(\|f\|_{L^2(\Omega)} + |Q|\right)$.

Proof First of all $\hat{V}$ is an Hilbert subspace of $H^1(\Omega)$ and equipped with the same topology. Let $a(z, v) = (\mu \nabla z, \nabla v)$. We have already seen in the proof of Proposition 10.1 that the form a is bilinear, continuous and coercive $\hat{V} \times \hat{V}$. It is always possible to find a $w \in V$ such that $\int_\Sigma w = Q$ and $\|w\|_{H^1(\Omega)} \leq C|Q|$. We the consider the problem: find $\hat{u} \in \hat{V}$ so that $a(\hat{u}, v) = F(v)$ for all $v \in \hat{V}$, where $F(v) = (f, v) - a(w, v)$. This is a classical elliptic problem by which well posedness is proved by standard application of Lax-Milgram Lemma and we have

$$\|\hat{u}\|_{H^1(\Omega)} \leq C(\|f\|_{L^2(\Omega)} + \|w\|_{H^1(\Omega)}) \leq C(\|f\|_{L^2(\Omega)} + |Q|).$$

We then set $u = \hat{u} + w$ and it is immediate to verify that u is a solution of (10.12), it satisfies $\int_\Gamma u d\Gamma = Q$, and it does not depend on the choice of w. Moreover,

$$\|u\|_{H^1(\Omega)} \leq \|\hat{u}\|_{H^1(\Omega)} + \|w\|_{H^1(\Omega)} \leq C(\|f\|_{L^2(\Omega)} + |Q|).$$

It is unique since if we have two solutions u_1 and u_2 of (10.12) and we set $y = u_1 - u_2$ we have $y \in \hat{V}$ and $a(y, v) = 0$ for all $v \in \hat{V}$, and this implies $y = 0$, thus $u_1 = u_2$. Now, by construction u satisfies (10.5b), while, since any $v \in V$ may be written as $v = \hat{v} + c\tilde{v}$, with $\hat{v} \in \hat{V}$ and $\tilde{v} \in \tilde{V}$ and $c = \int_\Gamma v d\Gamma$, we have that

$$\begin{aligned} a(u, v) + b(v, \lambda) - (f, v) &= a(u, \hat{v}) - (f, \hat{v}) + c\big(b(\tilde{v}, \lambda) + a(u, \tilde{v}) - (f, \tilde{v})\big) \\ &= c\big(b(\tilde{v}, \lambda) + a(u, \tilde{v}) - (f, \tilde{v})\big), \end{aligned}$$

which is zero $\forall v \in V$ if and only if $b(\tilde{v}, \lambda)+a(u, \tilde{v})-(f, \tilde{v}) = 0$. Since $b(\tilde{v}, \lambda) = \lambda$ we obtain (10.14). So the couple (u, λ) given by the solution of (10.12) and λ given by (10.14) are solutions of problem (10.5).

With analogous arguments we may verify that (u, λ) solution of (10.5) satisfies (10.12) and (10.14). □

Remark 10.2 One could think to apply the Lagrange multiplier approach also to the mean stress problem (10.1a)–(10.1b)–(10.2) by devising the following augmented problem

$$\begin{aligned} &(\mu \nabla u, \nabla v) + \lambda \int_\Sigma \mu \frac{\partial v}{\partial \mathbf{n}} d\Sigma = (f, v) \quad \forall v \in V, \\ &\xi \int_\Sigma \mu \frac{\partial u}{\partial \mathbf{n}} d\Sigma = \xi P \quad \forall \xi \in \mathbb{R}. \end{aligned}$$

However, the term $\int_\Sigma \lambda \mu \frac{\partial v}{\partial \mathbf{n}}$ is not well defined for $v \in V \subset H^1(\Omega)$ and $\lambda \in \mathbb{R}$ (unless $\Sigma = \partial\Omega$), so this formulation is, in general, not feasible. Indeed the integral should reinterpreted as a duality pairing $< \lambda, \mu \frac{\partial v}{\partial \mathbf{n}} >$ between $H^1_{00}(\Gamma)$ and its dual. Yet a non-zero constant function on Γ does not belong to $H^{1/2}_{00}(\Sigma)$. In practice, solving the stated problem numerically by means, for instance, finite elements, will give a solution that has an unwanted oscillations near the boundary of Σ, whose amplitude increases as the mesh is refined. We have similar difficulties for the mean stress problem in the context of Stokes equations, see for instance [12].

10.2.3 Penalization Methods

We start by observing that a way to overcome the introduction of the further unknown given by the Lagrange multiplier is to prescribe the flow rate condition (10.1c) not as a constrain but as a penalization. Let us consider a finite element space $V_h \subset V$. We propose to minimize at the discrete level the following functional

$$J(v_h) = \frac{1}{2}(\mu \nabla v_h, \nabla v_h) - (f, v_h) + \frac{1}{2}\gamma \left(\int_\Sigma u_h \, d\Sigma - Q\right)^2, \tag{10.16}$$

over $V_h \subset V$ and where $\gamma > 0$ is a penalization parameter. This leads to the following *penalization formulation*: find $u_h \in V_h$ such that

$$(\mu \nabla u_h, \nabla v_h) + \gamma \int_\Sigma u_h \, d\Sigma \int_\Sigma v_h \, d\Sigma = (f, v_h) + \gamma Q \int_\Sigma v_h \, d\Sigma \qquad \forall v_h \in V_h. \tag{10.17}$$

However, (10.17) is not consistent with (10.1a). It is easy to show that the truncation error is $\tau(v_h) = \int_\Sigma \mu \frac{\partial u}{\partial \mathbf{n}} v_h \, d\Sigma$, which of course does not go to zero when $h \to 0$.

To overcome this limitation, we adapt to the mean solution problem (10.1) the Nitsche penalization method introduced and analyzed in [30] for a standard Dirichlet problem, following the ideas in [42]. We remember that the Nitsche method is a strongly consistent penalization method, featuring an optimal convergence error. It consists in adding to the penalization term a consistency and, possibly, a symmetry term. The former is, as the name says, required to recover a consistent scheme, the latter is not strictly necessary but it maintains the symmetry of the original problem. To make the expressions more compact we use the notations: $a(u, v) = (\mu \nabla u, \nabla v)$, $< u, v >_\Sigma = |\Sigma|^{-1} \int_\Sigma u \int_\Sigma v$ and $|u|_\Sigma = \sqrt{< u, u >_\Sigma} = |\Sigma|^{-1/2} \left| \int_\Sigma u \right|$. It is evident that we have a Cauchy-Schwarz type inequality $< u, v >_\Sigma \leq |u|_\Sigma |v|_\Sigma$.

The Nitsche approximation of defective problem (10.1) is then: find $u_h \in V_h$ such that

$$\begin{aligned} &a(u_h, v_h) + \gamma h^{-1} < u_h, v_h >_\Sigma - < \mu \frac{\partial u_h}{\partial \mathbf{n}}, v_h >_\Sigma - < \mu \frac{\partial v_h}{\partial \mathbf{n}}, u_h >_\Sigma \\ &= (f, v_h) + \gamma h^{-1} < Q, v_h >_\Sigma - < Q, \mu \frac{\partial v_h}{\partial \mathbf{n}} >_\Sigma \qquad \forall v_h \in V_h. \end{aligned} \tag{10.18}$$

We can write it in a more compact form by introducing

$$a_h(u_h, v_h) = a(u_h, v_h) + \gamma h^{-1} < u_h, v_h >_\Sigma - < \mu \frac{\partial u_h}{\partial \mathbf{n}}, v_h >_\Sigma - < \mu \frac{\partial v_h}{\partial \mathbf{n}}, u_h >_\Sigma$$

and

$$F_h(v_h) = (f, v_h) + \gamma h^{-1} < Q, v_h >_\Sigma - < Q, \mu \frac{\partial v_h}{\partial \mathbf{n}} >_\Sigma,$$

as: find $u_h \in V_h$ such that

$$a_h(u_h, v_h) = F_h(v_h) \quad \forall v_h \in V_h. \tag{10.19}$$

Proposition 10.3 *Problem* (10.19) *is strongly consistent with the solution provided by the Lagrange multiplier approach in* (10.5)*.*

Proof The solution of the Lagrange multiplier approach satisfies (using the new notation) $a(u, v) + < \lambda, v >_\Sigma - (f, v) = 0$ and $< u, v >_\Sigma = < Q, v >_\Sigma$ for all

$v \in V$, thus $a(u, v_h) + < \lambda, v_h >_\Sigma - (f, v_h) = 0$ and $< u, v_h >_\Sigma = < Q, v_h >_\Sigma$ for all $v_h \in V_h$. Moreover, we have that $< \lambda, v_h >_\Sigma = -< \frac{\partial u}{\partial \mathbf{n}}, v_h >_\Sigma$, for all $v_h \in V_h$.

Consequently, $a_h(u, v_h) - F_h(v_h) = 0$ for all $v_h \in V_h$, since

$$\begin{aligned} a_h(u, v_h) - F_h(v_h) &= a(u, v_h) + \gamma h^{-1} < u, v_h >_\Sigma \\ &\quad - < \mu \frac{\partial u}{\partial \mathbf{n}}, v_h >_\Sigma - < \mu \frac{\partial v_h}{\partial \mathbf{n}}, u >_\Sigma \\ &\quad - (f, v_h) - \gamma h^{-1} < Q, v_h >_\Sigma + < Q, \mu \frac{\partial v_h}{\partial \mathbf{n}} >_\Sigma \\ &= -< \lambda - \mu \frac{\partial v_h}{\partial \mathbf{n}}, v_h >_\Sigma + \gamma h^{-1} < Q - u, v_h >_\Sigma \\ &\quad - < \mu \frac{\partial v_h}{\partial \mathbf{n}}, Q - u >_\Sigma = 0. \end{aligned}$$ □

In particular, we have the following "orthogonality property": $a_h(u - u_h, v_h) = 0$ for all $v_h \in V_h$.

Now, consider the following mesh dependent norm

$$\|v_h\|_h^2 = \|\sqrt{\mu}\nabla v_h\|_{L^2(\Omega)}^2 + \gamma h^{-1} |v_h|_\Sigma{}^2 = a(v_h, v_h) + \gamma h^{-1} |v_h|_\Sigma. \tag{10.20}$$

We give, without proof, the following result

Proposition 10.4 *There exist two positive constants C_* and C^* such that, for any $v \in V$*

$$C_* \|v\|_{H^1(\Omega)} \leq \|v_h\|_h \leq C^* h^{-1/2} \|v\|_{H^1(\Omega)}.$$

Moreover, it is immediate to verify that $|v|_\Sigma \leq \|v\|_{L^2(\Omega)}$. We now assume that the following inverse inequalities holds, which is true, for instance, for standard Lagrangian finite elements, see, for instance [9].

Assumption 10.1 *There are two positive constants C_Ω and C_Σ such that for any $v_h \in V_h$*

$$h\|\nabla v_h\|_{L^2(\Omega)} \leq C_\Omega \|v_h\|_{L^2(\Omega)} \quad \text{and} \quad h^{1/2} \|v_h\|_{L^2(\Sigma)} \leq C_\Sigma \|v_h\|_{L^2(\Omega)}, \tag{10.21}$$

and, consequently, since μ is bounded away from zero,

$$h \left| \mu \frac{\partial v_h}{\partial \mathbf{n}} \right|_\Gamma^2 \leq \|\mu \frac{\partial v_h}{\partial \mathbf{n}}\|_{L^2(\Sigma)}^2 \leq C_\Sigma^2 \mu_\Sigma (\mu \nabla v_h, \nabla v_h)^{1/2}, \tag{10.22}$$

where $\mu_\Sigma = \|\mu\|_{L^\infty(\Sigma)}$.

Thus, we have the following result.

Proposition 10.5 *If* $\gamma > C_{\Sigma}^{2}\mu_{\Sigma}$ *problem* (10.18) *is well posed.*

Proof We can use Lax-Milgram Lemma. Bilinearity and continuity of a_h as well as linearity and continuity of functional F_h are easily found thanks to standard inequalities. We now prove coercivity of a_h with respect to the norm $\|v_h\|_h$ given by (10.20). For any $v_h \in V_h$, we have that

$$a_h(v_h, v_h) \geq (\mu\nabla v_h, \nabla v_h)^2_{L^2(\Omega)} + \gamma h^{-1}|v_h|_{\Sigma}{}^2 - 2 < \mu\frac{\partial v_h}{\partial \mathbf{n}}, v_h >_{\Sigma}.$$

Now, for any $\epsilon > 0$, using Young's inequality and (10.22),

$$2 < \mu\partial v_h/\partial\mathbf{n}, v_h >_{\Sigma} \leq \epsilon h|\mu\partial v_h/\partial\mathbf{n}|_{\Sigma}{}^2 + \frac{1}{\epsilon h}|v_h|_{\Sigma}{}^2 \leq \epsilon C_{\Sigma}^2\mu_{\Sigma}(\mu\nabla v_h, \nabla v_h)$$
$$+ \frac{1}{\epsilon h}|v_h|_{\Sigma}{}^2,$$

that is

$$a_h(v_h, v_h) \geq (1 - \epsilon C_{\Sigma}^2\mu_{\Sigma})(\mu\nabla v_h, \nabla v_h) + h^{-1}(\gamma - \frac{1}{\epsilon})|v_h|_{\Sigma}{}^2.$$

The desired result is obtained if $1 - \epsilon C_{\Sigma}^2\mu_{\Sigma} > 0$ and $\gamma - \frac{1}{\epsilon} > 0$. If $\gamma > C_{\Sigma}^2\mu_{\Sigma}$, the latter inequality is satisfied by taking $\epsilon < \frac{1}{\mu_{\Sigma}C_{\Sigma}^2}$ and, consequently, we may find a constant $\alpha > 0$ so that $a_h(v_h, v_h) \geq \alpha\|v_h\|_h^2$. □

Proposition 10.6 *If u is the solution of* (10.5) *and u_h the solution of* (10.19), *under the same conditions of Proposition 10.5, we have that*

$$\|u - u_h\|_h \leq \frac{M}{\alpha}\inf_{v_h\in V_h}\|u - v_h\|_h,$$

where M and α are the continuity and coercivity constants of a_h.

Proof The result is rather classical and exploits the Galerkin orthogonality proved in Proposition 10.3 and the results of Proposition 10.5. Indeed, for any $v_h \in V_h$ we have

$$\|u - u_h\|_h^2 \leq \alpha^{-1}a_h(u - u_h, u - u_h) = \alpha^{-1}a_h(u - u_h, u - v_h)$$
$$\leq \alpha^{-1}M\|u - u_h\|_h\|u - v_h\|_h. \qquad \square$$

This result allows us to exploit interpolation inequalities to obtain optimal convergence rate of finite element approximations.

We consider now the defective mean stress problem (10.1a), (10.1b) and (10.2). In this case, in analogy with (10.16), we can consider the following functional to be minimized over V_h:

$$J(v_h) = \frac{1}{2}(\mu\nabla v_h, \nabla v_h) - (f, v_h) + \frac{1}{2}\gamma\left(\int_\Sigma \frac{\partial v_h}{\partial \mathbf{n}}\, d\Sigma - P\right)^2.$$

This leads to the following problem: find $u_h \in V_h$ such that

$$(\mu\nabla u_h, \nabla v_h) + \gamma \int_\Sigma \frac{\partial u_h}{\partial \mathbf{n}}\, d\Sigma \int_\Sigma \frac{\partial v_h}{\partial \mathbf{n}}\, d\Sigma = (f, v_h) + \gamma\frac{h}{|\Gamma|}P\int_\Sigma \frac{\partial v_h}{\partial \mathbf{n}}\, d\Sigma$$

$$\forall v_h \in V_h.$$

Like formulation (10.17), the previous one is not consistent. Indeed, the truncation error is again $\tau(v_h) = \int_\Sigma \mu\frac{\partial u_h}{\partial \mathbf{n}} v_h\, d\Sigma$. If one assumes that the normal stress is constant over Σ, and thus equal to P, consistency is recovered by adding the term $P\int_\Sigma v_h\, d\Sigma$ to the right hand side of the previous formulation. For convergence one should again take γ large enough, see [22, 39].

10.2.4 Augmented Lagrangian Formulation

It is quite natural now to consider an *augmented Lagrangian* formulation for the solution of (10.1). As done for the penalty formulations, we write it directly at the discrete level. In particular, the augmented Lagrangian formulation is obtained by finding a stationary point in $V_h \times \mathbb{R}$ of the following functional:

$$\begin{aligned} L_\gamma(v_h, \xi) = {} & \frac{1}{2}\int_\Omega \mu\,|\nabla v_h|^2\, d\mathbf{x} - (f, v_h) + \xi\left(\int_\Sigma v_h\, d\Sigma - Q\right) \\ & + \frac{1}{2}\gamma\left(\int_\Sigma v_h\, d\Sigma - Q\right)^2, \end{aligned}$$

which is equivalent to the following discrete formulation: find $u_h \in V_h$ and $\lambda_h \in \mathbb{R}$ such that $\forall v_h \in V_h$ and $\forall \xi \in \mathbb{R}$

$$(\mu\nabla u_h, \nabla v_h) + \lambda_h\int_\Sigma v_h\, d\Sigma + \gamma\int_\Sigma u_h\, d\Sigma \int_\Sigma v_h\, d\Sigma = (f, v_h) + \gamma Q\int_\Sigma v_h\, d\Sigma,$$

$$\xi\int_\Sigma u\, d\Sigma = \xi Q.$$

Again, the Lagrange multiplier λ_h has the physical meaning of a (constant) stress on Σ.

For its numerical solution, we can consider the following *Uzawa method*:

Given $\gamma > 0$, a stopping tolerance $\tau > 0$, a parameter $\rho > 0$, and $\lambda_h^{(0)} \in \mathbb{R}$, for $k = 1, 2, \ldots$:

1. Find $u_h^{(k)}$ solution of

$$\left(\mu \nabla u_h^{(k)}, \nabla v_h\right) + \lambda_h^{(k)} \int_\Sigma v_h \, d\Sigma + \gamma \int_\Sigma u_h^{(k)} \, d\Sigma \int_\Sigma v_h \, d\Sigma$$
$$= (f, v_h) + \gamma Q \int_\Sigma v_h \, d\Sigma, \ \forall v_h \in V_h;$$

2. Update the Lagrange multiplier:

$$\lambda_h^{(k+1)} = \lambda_h^{(k)} + \rho \left(Q - \int_\Sigma u_h^{(k)} \, d\Sigma \right);$$

3. Stop if $|\lambda_h^{(k+1)} = \lambda_h^{(k)}| \le \tau$.

The convergence of the previous method is guaranteed for $0 < \rho_0 \le \rho \le 2\gamma$, for a suitable ρ_0, see [15].

The previous method allows, unlike the penalization ones, to prescribe condition (10.1c) strongly, for any $\gamma > 0$. The presence of the penalization term improves convergence of the Uzawa method with respect to when it is applied to the classical Lagrange multiplier approach. Of course, this method is not of particular interest in case of a single flow rate, since, as highlighted in Sect. 10.2.2, the solution in this case is achieved in two steps with the Schur complement approach. However, in the case of defective conditions applied to several portion of the boundary, the algorithm based on the Schur complement requires $m + 1$ solutions of a Poisson problem, see Remark 10.1. Thus, the Uzawa algorithm could be competitive if it allows a satisfactory convergence in less than $m + 1$ iterations. We do not report here the extension of the algorithm to the case of more than one flow rate conditions, since it is straightforward.

10.2.5 Methods Based on Control Theory

The last strategy we present could be considered as the dual of the Lagrange multiplier approach. Indeed, in this case the flow rate condition (10.1c) is used to build the functional to be minimized, while the differential problem (10.1a), (10.1b) defines the constraint. This gives rise to the following *optimal control* problem: given $\alpha \ge 0$, find $z \in \mathbb{R}$ such that

$$z = \text{argmin}_{s \in \mathbb{R}} J(v(s), s) = \frac{1}{2} \left(\int_\Sigma v(s) d\Sigma - Q \right)^2 + \frac{\alpha}{2} (s - z_0)^2, \qquad (10.24)$$

where $v = v(s) \in V$ satisfies

$$(\mu \nabla v(s), \nabla \psi) = (f, \psi) + \int_\Sigma s\psi \, d\Sigma \quad \forall \psi \in V. \tag{10.25}$$

This corresponds to the weak form of the following differential problem

$$-\nabla \cdot (\mu \nabla v) = f \qquad \text{in } \Omega, \tag{10.26a}$$

$$v = 0 \qquad \text{on } \Gamma, \tag{10.26b}$$

$$\mu \frac{\partial v}{\partial \mathbf{n}} = s \qquad \text{on } \Sigma. \tag{10.26c}$$

The solution u is then recovered by setting $u = v(z)$. The term involving the parameter α is a *Tikhonov regularization* term [8] and z_0 a reference value. Notice also that z assumes the same meaning of the Lagrange multiplier λ. This is a control problem with control on the Neumann boundary and boundary observations on the same portion of the boundary.

If $|\Gamma| > 0$, $\alpha > 0$, $f \in L^2(\Omega)$ and for $\partial\Omega$ with Liptshitz boundary, the map $s \to v(s) : \mathbb{R} \to V$ is linear and continuous and the functional $J : \mathcal{C} \to \mathbb{R}$ defined in (10.24) is convex, coercive and differentiable. This implies that the previous constrained minimization problem admits a unique solution.

Proceeding us usual in control theory for PDEs [19, 27, 36], problem (10.24)–(10.26) is equivalent to the following *first order optimality* conditions (also referred to as *Karush-Kuhn-Tucker* (KKT) conditions): find $z \in \mathbb{R}$, $u \in V$ and $\lambda_u \in V$ such that

$$\text{State pbl}: \quad (\mu \nabla u, \nabla v) + z \int_\Sigma v \, d\Sigma = (f, v), \tag{10.27a}$$

$$\text{Adj pbl}: \quad (\mu \nabla v, \nabla \lambda_u) + \int_\Sigma u \, d\Sigma \int_\Sigma v \, d\Sigma = Q \int_\Sigma v \, d\Sigma, \tag{10.27b}$$

$$\text{Opt. cond}: \quad \int_\Sigma \lambda_u \, d\Sigma + \alpha(z - z_0) = 0, \tag{10.27c}$$

for all $v \in V$ and where λ_u is the solution of the adjoint problem. The optimality condition corresponds to setting to zero the Frechèt derivative of $J'(s)$.

For the numerical solutions of the previous problem, a monolithic approach could be considered, which corresponds to solve a linear system of the form

$$\begin{bmatrix} A_{uu} & 0 & \mathbf{a}_{uz} \\ A_{u\lambda} & A_{\lambda\lambda} & 0 \\ 0 & A_{z\lambda} & \alpha \end{bmatrix} \begin{bmatrix} \mathbf{U} \\ \boldsymbol{\Lambda} \\ z_h \end{bmatrix} = \begin{bmatrix} \mathbf{F}_u \\ \mathbf{F}_\lambda \\ \alpha z_0 \end{bmatrix},$$

where the various term derive, for instance, from a Galerkin discretization of (10.27).

However, this implies loss of modularity and the need to solve a relatively large system. Alternatively, one could consider a descent algorithm, the simplest one being a steepest descent, which gives rise to:

Given $z_h^{(0)} \in \mathbb{R}$, a suitable relaxation parameter $\beta_k > 0$, and a tolerance $\tau > 0$, for $k = 1, \ldots$:

1. Solve
$$A_{uu}\mathbf{U}^{(k)} = \mathbf{F}_u - \mathbf{a}_{uz} z_h^{(k-1)};$$
2. Solve
$$A_{\lambda\lambda}\boldsymbol{\Lambda}^{(k)} = \mathbf{F}_\lambda - A_{\lambda u}\mathbf{U}^{(k)};$$
3. Update the control variable
$$z_h^{(k)} = z_h^{(k-1)} - \beta_k \left(A_{z\lambda}\boldsymbol{\Lambda}^{(k)} + \alpha z_h^{(k)} \right);$$
4. Stop if $|z^{(k)} - z^{(k-1)}| \leq \tau$.

The first step is equivalent to solve the Poisson problem with Neumann data $z^{(k-1)}$, the second problem is again a Poisson problem with Neumann data $\int_\Sigma u^{(k)}\, d\Sigma$. Therefore, they can be both tackled with standard solvers. Other, more efficient solvers for the control problem may be found, for instance in [31].

It may seem that this method is less efficient than the other ones, yet it has the advantage of being rather flexible for more general problems. For instance, it may be used to implement defective Robin conditions $\int_\Sigma (u + \beta\mu\partial u/\partial\mathbf{n})\, d\Sigma = Q$ (which however may be implemented also by the Nitsche's penalization approach, see [11]).

10.3 Defective Boundary Condition for Stokes/Navier Stokes

We now briefly describe some extensions of the proposed techniques to the Stokes equations, which form the basis for the application to Navier-Stokes. Indeed, defective boundary problems have been originally studied in the context of fluid-dynamics [7, 21], in particular in hemodynamics where often the measures or the coupling with reduced models provide only average data on the artificial sections [12].

For the sake of exposition we consider the following steady Stokes problem (all the strategies reported can be extended to the case of unsteady Navier-Stokes). Let the velocity $\mathbf{u}$ and pressure p be solution of:

$$-\mu\Delta\mathbf{u}+\nabla p=\mathbf{f} \qquad \text{in } \Omega, \tag{10.28a}$$

$$\nabla\cdot\mathbf{u}=0 \qquad \text{in } \Omega, \tag{10.28b}$$

$$\mathbf{u}=\mathbf{0} \qquad \text{on } \Gamma, \tag{10.28c}$$

$$\int_{\Sigma}\mathbf{u}\cdot\mathbf{n}\,d\Sigma=Q, \tag{10.28d}$$

where the notation introduced Sect. 10.2 has been used. The previous is a *flow rate* defective problem, where only the average of normal component of the velocity is known. This is a typical situation when clinical measures are known in hemodynamics or geometrically reduced models are coupled at the artificial sections [33].

Alternative to the flow rate (10.28d), the following mean stress condition could be prescribed on the artificial sections:

$$\int_{\Sigma}(-p\mathbf{n}+\mu\nabla\mathbf{u}\,\mathbf{n})\,d\Sigma=-|\Sigma|P, \tag{10.29}$$

We refer to the defective problem give by (10.28a)–(10.28b)–(10.28c)–(10.29) as *mean stress* problem. Often, the viscosity term is neglected in condition (10.29) since it is negligible on artificial section with respect to the pressure. In this case we have

$$\int_{\Sigma}p\,d\Sigma=|\Sigma|P, \tag{10.30}$$

and we refer to the corresponding defective condition as *mean pressure* condition.

In the following subsections, we review the most classical approaches proposed so far for the two defective problems introduced above.

10.3.1 Empirical Methods

The most used strategy in the engineering community to prescribe the flow rate condition (10.28d) is to select a priori a velocity profile $\mathbf{g}$ such that $\int_{\Sigma}\mathbf{g}\cdot\mathbf{n}\,d\Sigma=Q$ and then prescribe the Dirichlet condition

$$\mathbf{u}=\mathbf{g} \qquad \text{on } \Sigma.$$

Classical choices for circular sections are the parabolic one, which works well for example in the carotids [6], the flat one, which is quite often used in the aorta [28], and the one based on the Womersley solution [20]. However, in practical situations the artificial sections are not circular, and a suitable morphing is needed [20]. In any case, the choice of the velocity profile in general influences the numerical solution and introduces an error inside the computational domain. In particular, it is known from the computational practice that the flow fully develops after a characteristic distance from the section. This means that inside the region identified by this length, an error due to the wrong choice of the velocity profile is associated to the solution, whereas outside this region the solution could be considered acceptable. This is the reason why in the engineering practice, the computational domain is extended at the section at hand of a length which is comparable with the characteristic length needed to the flow to fully develop. This characteristic length is known to increase for increasing Reynolds number Re [35]. In particular, for steady flows in a cylindrical domain, its value can be approximated by $0.058\, D\, Re$ (D being the diameter of the inlet section) [41]. In [38], it has been proved that the error features an exponential decay with respect to the distance from the section where the arbitrary profile is prescribed, with a constant which increases with Re.

Regarding the mean stress problem (10.28a)–(10.28b)–(10.28c)–(10.29), a classical empirical approach consists in selecting a constant stress aligned with the normal direction [11], i.e.

$$-p\mathbf{n} + \mu \nabla \mathbf{u}\, \mathbf{n} = -P\mathbf{n}.$$

This assumption is in general acceptable for example in hemodynamics, where the pressure mainly changes along the axial direction. The previous Neumann condition has been proposed also to treat the mean pressure problem (10.28a)–(10.28b)–(10.28c)–(10.30) [21]. However, in this case the corresponding weak formulation is not consistent with the defective condition [11]. To recover a consistent approximation, the *curl-curl* formulation of the Stokes problem should be considered since the corresponding natural condition is the pressure [7, 37].

10.3.2 Lagrange Multiplier Approach

The Lagrange multiplier approach for the flow rate problem (10.28) has been introduced in [12]. Following the idea reported in Sect. 10.2.2, the following augmented formulation is obtained: find $\mathbf{u} \in [V]^d$, $p \in Q = L^2(\Omega)$ and $\lambda \in \mathbb{R}$ such that for all $(v, q, \xi) \in [V]^d \times Q \times \mathbb{R}$,

$$(\mu \nabla \mathbf{u}, \nabla \mathbf{v}) - (p, \nabla \cdot \mathbf{v}) + b(\mathbf{v} \cdot \mathbf{n}, \lambda) = (\mathbf{f}, \mathbf{v}) \tag{10.31a}$$

$$(q, \nabla \cdot \mathbf{u}) = 0 \tag{10.31b}$$

$$b(\mathbf{u} \cdot \mathbf{n}, \xi) = \xi Q, \tag{10.31c}$$

where the notation is the same of Sect. 10.2.2. An inf-sup condition holds true for the previous augmented Stokes problem [37]. Again, the Lagrange multiplier has the physical meaning of constant normal stress on Σ [12].

Notice that the flow rate condition (10.28d) does not provide any information (neither defective) on the tangential velocity. To close the augmented problem, boundary conditions on the tangential velocity or stress should be considered. This choice is given by the choice of the functional space $[V]^d$. In particular, if its functions are not constrained in the tangential direction on Σ, a homogeneous Neumann condition is implicitly assumed for the tangential stress; otherwise, if they vanish in the tangential direction, they imply Dirichlet condition on the tangential velocity.

For its numerical solution, we can consider either a monolithic approach or an algorithm similar to that presented in Sect. 10.2.2 based on the Schur complement equation [12]. Analogously to the Poisson case, this algorithm is modular and consists in the solution of two Stokes problems with Neumann conditions on Σ. The extension to the Navier-Stokes case has been obtained in [37]. In both the cases, for m flow rate conditions this algorithm relies on the solution of $m + 1$ Stokes/Navier-Stokes problems. For this reason, in [38] an inexact splitting algorithm has been proposed to save computational time, consisting in the solution of just 1 Stokes/Navier-Stokes problem, where however an error near to Σ is introduced. The authors noticed that since the error is introduced by solving a null flow rate problem arising from the splitting by means of a homogeneous Dirichlet condition, the error is the smallest one provided by any empirical approach, since the Reynolds number at the section at hand is zero.

The extension of the Lagrange multiplier approach to the case of compliant walls has been addressed in [14], whereas the case of quasi-Newtonian fluid has been analyzed in [10], where an error analysis for the numerical approximation is also given.

10.3.3 Penalization Methods

The Nitche's approach reported in Sect. 10.2.3 may be extended to the Stokes problem.

In [42], a consistent penalization method for the mean flux problem as been designed. It reads: find $\mathbf{u}_h \in [V_h]^d$ and $p_h \in Q_h$, such that for all $\mathbf{v}_h \in [V_h]^d$ and $q_h \in Q_h$,

$$
\begin{aligned}
&(\mu\nabla\mathbf{u}_h, \nabla\mathbf{v}_h) - (p_h, \nabla\cdot\mathbf{v}_h) + \gamma \int_\Sigma \mathbf{u}_h\cdot\mathbf{n}\, d\Sigma \int_\Sigma \mathbf{v}_h\cdot\mathbf{n}\, d\Sigma \\
&\quad -\tfrac{1}{|\Sigma|}\int_\Sigma \mu\nabla\mathbf{u}_h\cdot\mathbf{n}\, d\Sigma \int_\Sigma \mathbf{v}_h\cdot\mathbf{n}\, d\Sigma \\
&\quad -\tfrac{1}{|\Sigma|}\int_\Sigma \mathbf{u}_h\cdot\mathbf{n}\, d\Sigma \int_\Sigma \mu\nabla\mathbf{v}_h\cdot\mathbf{n}\, d\Sigma + \tfrac{1}{|\Sigma|}\int_\Sigma p_h\, d\Sigma \int_\Sigma \mathbf{v}_h\cdot\mathbf{n}\, d\Sigma \\
&\quad = (\mathbf{f}, \mathbf{v}_h) + \gamma Q \int_\Sigma \mathbf{v}_h\cdot\mathbf{n}\, d\Sigma - \tfrac{1}{|\Sigma|} Q \int_\Sigma \mu\nabla\mathbf{v}_h\cdot\mathbf{n}\, d\Sigma, \\
&-(q_h, \nabla\cdot\mathbf{u}_h) + \frac{1}{|\Sigma|}\int_\Sigma q_h\, d\Sigma \int_\Sigma \mathbf{u}_h\cdot\mathbf{n}\, d\Sigma = \frac{1}{|\Sigma|} Q \int_\Sigma q_h\, d\Sigma.
\end{aligned}
$$

In [42], it has been proved that, if it exists a constant c such that $\mu\nabla\mathbf{u}\,\mathbf{n} - p\mathbf{n} = c\mathbf{n}$ on Σ, then the previous formulation is consistent with (10.28) and that if $\gamma = \widehat{\gamma}/(h|\Sigma|)$ with $\widehat{\gamma}$ large enough, the solution is unique. The arguments are similar to those illustrated for the Poisson problem.

Referring to the notation introduced in Sect. 10.2, the algebraic problem related to the Galerkin approximation of the previous Nitsche formulation is

$$\begin{bmatrix} A^Q & (B^Q)^T \\ B^Q & 0 \end{bmatrix}\begin{bmatrix} \mathbf{U} \\ \mathbf{P} \end{bmatrix} = \begin{bmatrix} \mathbf{F}^Q \\ \mathbf{G}^Q \end{bmatrix},$$

where $A_{ij}^Q = (\mu\nabla\varphi_j, \varphi_i) + \gamma\int_\Sigma \varphi_j \cdot \mathbf{n}\,d\Sigma \int_\Sigma \varphi_i \cdot \mathbf{n}\,d\Sigma - \frac{1}{|\Sigma|}\int_\Sigma \mu\nabla\varphi_j \cdot \mathbf{n}\,d\Sigma \int_\Sigma \varphi_i \cdot \mathbf{n}\,d\Sigma - \frac{1}{|\Sigma|}\int_\Sigma \varphi_i \cdot \mathbf{n}\,d\Sigma \int_\Sigma \mu\nabla\varphi_j \cdot \mathbf{n}\,d\Sigma$, $B_{kj}^Q = -(\psi_k, \nabla\cdot\varphi_j) + \frac{1}{|\Sigma|}\int_\Sigma \psi_k\,d\Sigma \int_\Sigma \varphi_j \cdot \mathbf{n}\,d\Sigma$, $\mathbf{P}$ collects the pressure unknowns, $F_i^Q = (\mathbf{f}, \varphi_i) + \gamma Q\int_\Sigma \varphi_i \cdot \mathbf{n}\,d\Sigma - \frac{1}{|\Sigma|}Q\int_\Sigma \mu\nabla\varphi_i \cdot \mathbf{n}\,d\Sigma$, $G_k^Q = \frac{1}{|\Sigma|}Q\int_\Sigma \psi_k\,d\Sigma$ and where ψ_k are the basis function for the pressure approximation. The previous linear system preserves the saddle-point nature of the classical Stokes problem.

A Nitsche formulation has been proposed for the mean stress problem in [39]. The corresponding weak formulation reads: find $\mathbf{u}_h \in [V_h]^d$ and $p_h \in Q_h$, such that for all $\mathbf{v}_h \in [V_h]^d$ and $q_h \in Q_h$,

$$\begin{aligned}
&(\mu\nabla\mathbf{u}_h, \nabla\mathbf{v}_h) - (p_h, \nabla\cdot\mathbf{v}_h) - \gamma\int_\Sigma \mu\nabla\mathbf{u}_h \cdot \mathbf{n}\,d\Sigma \int_\Sigma \mu\nabla\mathbf{v}_h \cdot \mathbf{n}\,d\Sigma \\
&\quad + \tfrac{1}{|\Sigma|}\textstyle\int_\Sigma p_h\,d\Sigma \int_\Sigma \nabla\mathbf{v}_h\,\mathbf{n}\,d\Sigma = (\mathbf{f}, \mathbf{v}_h) - P\displaystyle\int_\Sigma \mathbf{v}_h \cdot \mathbf{n}\,d\Sigma \\
&\quad + \gamma P|\Sigma| \textstyle\int_\Sigma \mu\nabla\mathbf{v}_h \cdot \mathbf{n}\,d\Sigma, \\
&-(q_h, \nabla\cdot\mathbf{u}_h) - \gamma\displaystyle\int_\Sigma p_h\,d\Sigma \int_\Sigma q_h\,d\Sigma + \gamma\int_\Sigma \nabla\mathbf{u}_h \cdot \mathbf{n}\,d\Sigma \int_\Sigma q_h\,d\Sigma \\
&\quad = -\gamma P|\Sigma| \textstyle\int_\Sigma q_h\,d\Sigma.
\end{aligned}$$

In [39], it has been proved that, if it exists a constant c such that $\mu\nabla\mathbf{u}\,\mathbf{n} - p\mathbf{n} = c\mathbf{n}$ on Σ, then the previous formulation is consistent with (10.28a)–(10.28b)–(10.28c)–(10.29) and that if $\gamma = \widehat{\gamma}h/|\Sigma|$ with $\widehat{\gamma}$ large enough, we have again a unique solution. The corresponding algebraic problem related to the Galerkin approximation of the Nitsche formulation reads

$$\begin{bmatrix} A^P & (B^P)^T \\ B^P & C^P \end{bmatrix}\begin{bmatrix} \mathbf{U} \\ \mathbf{P} \end{bmatrix} = \begin{bmatrix} \mathbf{F}^P \\ \mathbf{G}^P \end{bmatrix},$$

where $A_{ij}^P = (\mu\nabla\varphi_j, \nabla\varphi_i) - \gamma\int_\Sigma \mu\nabla\varphi_j \cdot \mathbf{n}\,d\Sigma \int_\Sigma \mu\nabla\varphi_i \cdot \mathbf{n}\,d\Sigma$, $B_{kj}^P = -(\psi_k, \nabla\cdot\varphi_j) + \frac{1}{|\Sigma|}\int_\Sigma \psi_k\,d\Sigma \int_\Sigma \nabla\varphi_j\,\mathbf{n}\,d\Sigma$, $C_{kl}^P = -\gamma\int_\Sigma \psi_l\,d\Sigma \int_\Sigma \psi_k\,d\Sigma$, $F_j^P = (\mathbf{f}, \varphi_j) - P\int_\Sigma \varphi_j \cdot \mathbf{n}\,d\Sigma + \gamma P|\Sigma|\int_\Sigma \mu\nabla\varphi_j \cdot \mathbf{n}\,d\Sigma$ and $G_k^P = -\gamma P|\Sigma|\int_\Sigma \psi_k\,d\Sigma$.

An alternative formulation consistent with the mean pressure problem (10.28a)–(10.28b)–(10.28c)–(10.30) has been proposed in [39].

10.3.4 Augmented Lagrangian Formulation

Following the idea reported in Sect. 10.2.4, we can introduce also for the flow rate problem (10.28) an augmented Lagrangian formulation where both a Lagrange multiplier and a penalization term are introduced. This allows to prescribe strongly the flow rate condition (10.28d) and to improve the convergence in an Uzawa-like algorithm, see Sect. 10.2.4. In particular, this formulation reads: find $\mathbf{u}_h \in [V_h]^d$, $p_h \in Q_h$ and $\lambda_h \in \mathbb{R}$ such that $\forall \mathbf{v}_h \in [V_h]^d$, $q_h \in Q_h$ and $\xi \in \mathbb{R}$,

$$
\begin{aligned}
&(\mu \nabla u_h, \nabla v_h) - (p_h, \nabla \cdot \mathbf{v}_h) + \lambda_h \int_\Sigma \mathbf{v}_h \cdot \mathbf{n}\, d\Sigma + \gamma \int_\Sigma \mathbf{u}_h \cdot \mathbf{n}\, d\Sigma \int_\Sigma \mathbf{v}_h \cdot \mathbf{n}\, d\Sigma \\
&\qquad = (\mathbf{f}, \mathbf{v}_h) + \gamma Q \int_\Sigma \mathbf{v}_h \cdot \mathbf{n}\, d\Sigma, \\
&(q_h, \nabla \cdot \mathbf{u}_h) = 0, \\
&\xi \int_\Sigma u_h \, d\Sigma = \xi Q.
\end{aligned}
$$

10.3.5 Methods Based on Control

As observed in Sect. 10.2.5, these techniques are based on minimizing a functional related to the flow rate condition (10.28d) under the constrain given by the Stokes problem. In analogy of what observed in Sect. 10.2.5, the control variable z is here the constant normal component of the normal stress [13]

$$
-p\mathbf{n} + \nabla \mathbf{u}\, \mathbf{n} = z\mathbf{n} \qquad \text{on } \Sigma.
$$

Referring to the notation of Sect. 10.2, this leads to the following first order optimality conditions: find $z \in \mathbb{R}$, $\mathbf{u} \in [V]^d$, $p \in Q$, $\lambda_u \in [V]^d$ and $\lambda_p \in Q$ such that

$$
\begin{aligned}
&\text{State pbl :} \quad && (\mu \nabla u, \nabla v) - (p, \nabla \cdot \mathbf{v}) + z \int_\Sigma \mathbf{v} \cdot \mathbf{n}\, d\Sigma = (\mathbf{f}, \mathbf{v}), \\
& && (q, \nabla \cdot \mathbf{u}) = 0 \\
&\text{Adj pbl :} \quad && (\mu \nabla v, \nabla \lambda_u) - (\lambda_p, \nabla \cdot \mathbf{v}) + \int_\Sigma \mathbf{u} \cdot \mathbf{n}\, d\Sigma \int_\Sigma \mathbf{v} \cdot \mathbf{n}\, d\Sigma = Q \int_\Sigma \mathbf{v} \cdot \mathbf{n}\, d\Sigma, \\
& && (q, \nabla \cdot \lambda_u) = 0 \\
&\text{Opt. cond :} \quad && \int_\Sigma \lambda_u \cdot \mathbf{n}\, d\Sigma + \alpha(z - z_0) = 0,
\end{aligned}
$$

for all $\mathbf{v} \in [V]^d$ and $q \in Q$. Existence and unicity of the solution under the constraint that the normal stress is constant and aligned along the normal component are provided in [13].

In [17, 18, 24], the complete normal stress is chosen as control variable $\mathbf{z}$

$$-p\mathbf{n} + \nabla\mathbf{u}\,\mathbf{n} = \mathbf{z} \qquad \text{on } \Sigma.$$

This allows one to treat also cases where the normal stress is not supposed to be aligned with the axial direction, e.g. for not orthogonal artificial sections. Existence of a solution is provided [24]. Alternatively, the value of the velocity $\mathbf{u}$ on Σ could be used as control variables $\mathbf{z}$, see [24].

The optimal control approach has been proposed also for the mean pressure problem (10.28a)–(10.28b)–(10.28c)–(10.30) in [13]. In this case, the control variable is set equal to the flow rate or to the complete normal stress on Σ, see also [24] for the latter case.

The case of fluid problem in compliant vessels has been addressed in [14].

10.4 Some Applications to Hemodynamics

We present here two examples of applications in haemodynamics on real geometries reconstructed from radiological images acquired at Ospedale Ca' Granda-Policlinico di Milano, Italy. For both numerical experiments, we have considered the incompressible Navier-Stokes equations in rigid domains and a flow rate condition (10.28d) at the inlet. For the prescription of the flow rate condition, we have used the Lagrange multipliers approach presented in Sect. 10.3.2 and the algorithm based on the Schur complement equation introduced in [37] for Navier-Stokes equations. We also used $P2 - P1$ Finite Elements and the Backward Difference Formula of order 2 (BDF2) for the space and time discretization, respectively.

The results have been obtained with the parallel Finite Element library *LIFEV* developed at MOX—Politecnico di Milano, INRIA—Paris, CMCS—EPF Lausanne, and Emory University—Atlanta (www.lifev.org). The linear system arising at each time step has been solved with GMRes preconditioned with an Additive Shwartz preconditioner.

10.4.1 The Case of a Stenotic Carotid

In the first numerical experiment, we consider a stenotic carotid due to the presence of an atheromasic plaque at the bifurcation. We prescribed the flow rates depicted in Fig. 10.1, left, at the inlet (Common Carotid Artery, CCA) and at one of the two outlets, namely at the Internal Carotid Artery (ICA). As a comparison, we considered also the case where a parabolic profile fitting the flow rate is used instead of the Lagrange multipliers approach.

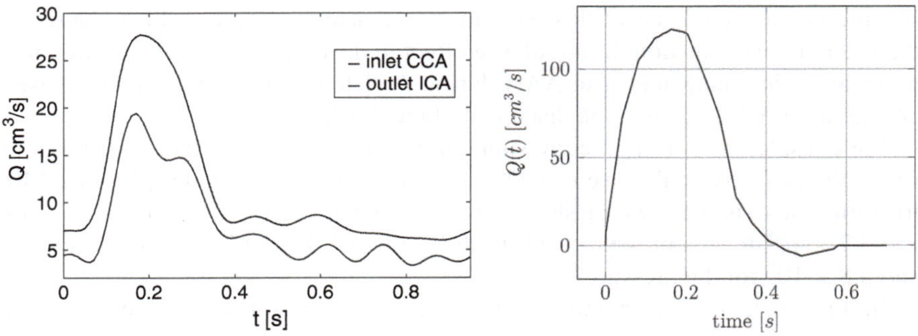

Fig. 10.1 Flow waveforms prescribed at the inlet. Left: carotid simulation. Right: AAA simulation

Fig. 10.2 Blood velocity field vectors in the case of a parabolic profile at the ICA (left) and streamlines obtained with the Lagrange multiplier approach at the ICA (right). Image by B. Guerciotti

In Fig. 10.2 we observe that the numerical result obtained when a parabolic profile is prescribed blows up. This is due to the swirling nature of velocity pattern in the ICA, induced by the stenosis, which is not able to fit the parabolic profile prescribed at the outlet. Instead, the Lagrange multiplier approach works well, adjusting the velocity profile at the ICA so that the matching with the inner velocity is stable.

10.4.2 The Case of an Aortic Abdominal Aneurysm

In the second numerical experiment, we consider a blood flow simulation in an aortic abdominal aneurysm (AAA). We prescribed the flow rates depicted in Fig. 10.1, right, at the inlet, and homogeneous Neumann conditions at the two outlets. The specific geometry related to this pathology, characterized by a sudden change of diameter, allows transition to turbulence effect to develop [1, 23, 26]. In particular, transitional flow may appear in AAA during the late systolic and diastolic phases, and are usually localized in the distal end of the AAA sac.

We consider here a *large eddy simulation* (LES) model for the description of such effects. In particular, we consider the eddy viscosity σ*-model* [29] (see [40] for further details) and two meshes: *mesh I* (about 1.1×10^6 dofs for the velocity and 3.7×10^5 for the pressure) and *mesh II* (about 2.0×10^6 dofs for the velocity and 6.8×10^5 for the pressure).

In Fig. 10.3, we report the results at four time instants for three different simulations, namely: (1) a no model simulation (i.e. without LES) in mesh I; (2) a

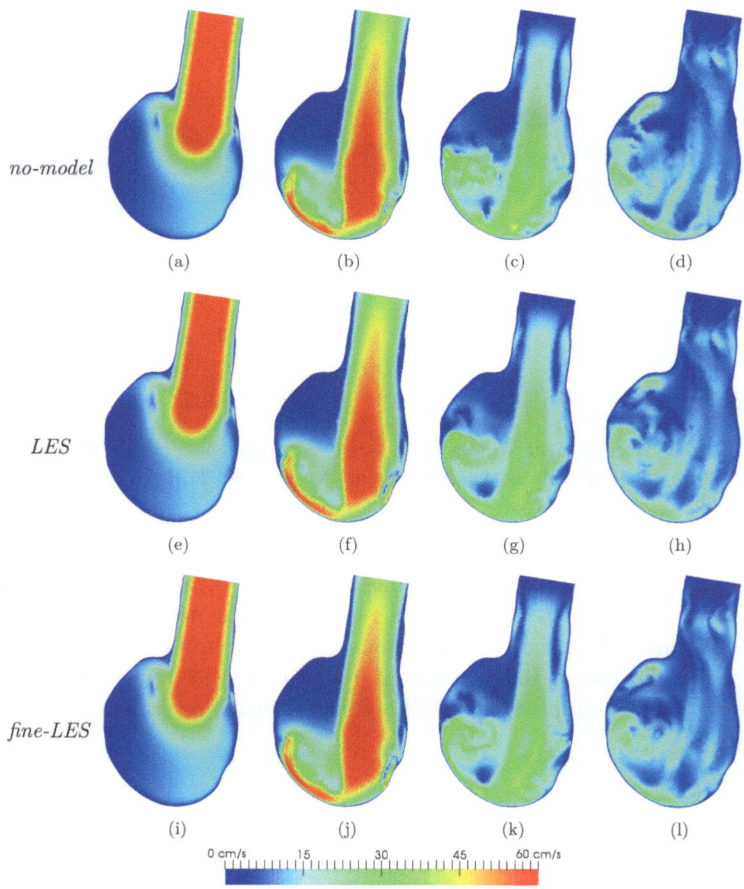

Fig. 10.3 Velocity pattern at four different instants, from left to right: systole $t = 0.16$ s, mid-deceleration $t = 0.29$ s, early diastole $t = 0.40$ s, late diastole $t = 0.49$ s. Top: no-model simulation; Middle: LES simulation; Bottom: LES simulation on a finer mesh. Image by D. Le Van

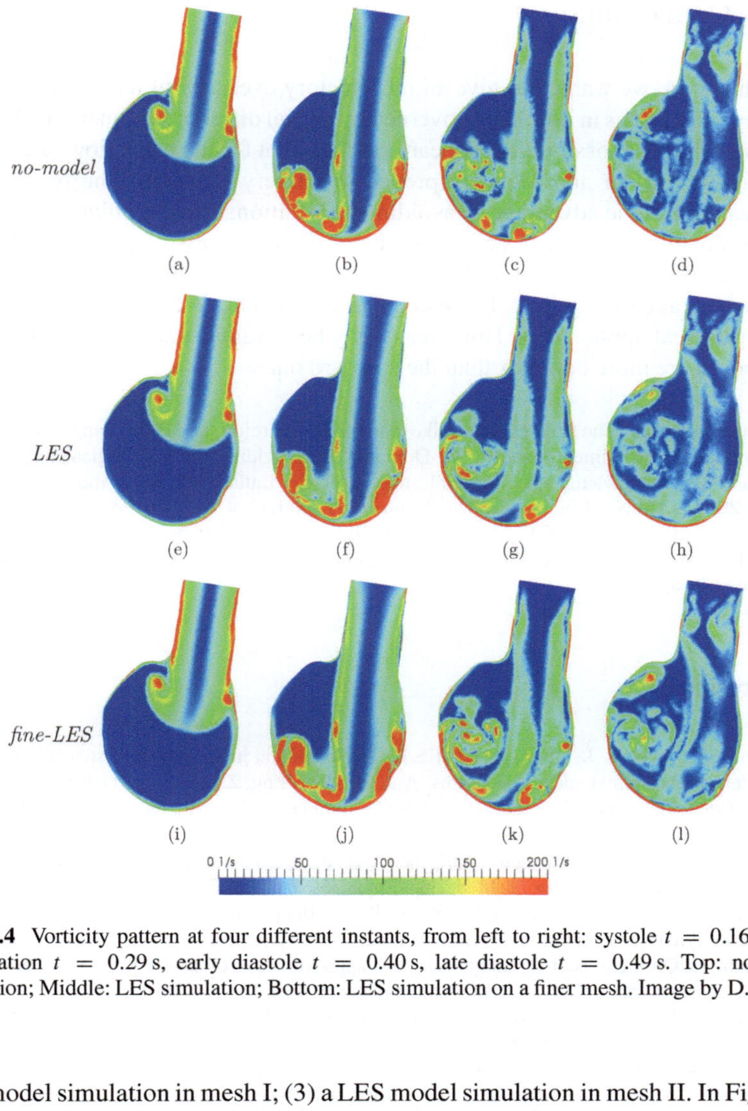

Fig. 10.4 Vorticity pattern at four different instants, from left to right: systole $t = 0.16$ s, mid-deceleration $t = 0.29$ s, early diastole $t = 0.40$ s, late diastole $t = 0.49$ s. Top: no-model simulation; Middle: LES simulation; Bottom: LES simulation on a finer mesh. Image by D. Le Van

LES model simulation in mesh I; (3) a LES model simulation in mesh II. In Fig. 10.4 we report for the same specific cases the vorticity patterns.

From these results, we observe some differences between the results obtained with LES and "no-model" simulations, whereas a good agreement between those obtained with LES a two different meshes, highlighting that probably LES simulation with mesh I is enough to have accurate results. Also in this case we have used the Lagrange multiplier approach to impose mean conditions at the inlet sections, proving that the method works well also in presence of turbulent models.

10.5 Conclusions

With this work we wanted to give an introductory overview of techniques to apply defective conditions in problems governed by partial differential equations. We have illustrated various possibilities. We can conclude that for the mean flow problem the Lagrange multiplier approach has proved to be very effective. The Nitsche type penalization has the advantage of avoiding an additional saddle point problem, and is more flexible since it can accommodate also mean stress condition (and in fact also defective conditions of Robin type), and is a valid alternative. The control approach is rather interesting, but also rather costly, and, so far, has not found much use in practical applications. However it may be advantageous if the constraints to be imposed are more complex than the standard ones.

Acknowledgements The authors would like to thank B. Guerciotti and D. Le Van for their support in the numerical experiments, and dr M. Domanin for providing the radiological images. They also gratefully acknowledge the financial support of the Italian MIUR by the grant PRIN12, number 201289A4LX, "Mathematical and numerical models of the cardiovascular system, and their clinical applications". CV has been partially supported also by the H2020-MSCA-ITN-2017, EU project 765374 "ROMSOC - Reduced Order Modelling, Simulation and Optimization of Coupled systems".

References

1. Asbury, C., Ruberti, J., Bluth, E., Peattie, R.: Experimental investigation of steady flow in rigid models of abdominal aortic aneurysms. Ann. Biomed. Eng. **23**(1), 29–39 (1995)
2. Babuŝka, I.: The finite element method with Lagrangian multipliers. Numer. Math. **20**(3), 179–192 (1973)
3. Blanco, P., Feijóo, R.: A dimensionally-heterogeneous closed-loop model for the cardiovascular system and its applications. Med. Eng. Phys. **35**(5), 652–667 (2013)
4. Blanco, P., Pivello, M., Urquiza, S., Feijòo, R.: On the potentialities of 3d-1d coupled models in hemodynamics simulations. J. Biomech. **42**, 919–930 (2009)
5. Brezzi, F.: On the existence, uniqueness and approximation of saddle point problems arising from Lagrange multipliers. RAIRO Anal. Numer. **8**, 129–151 (1974)
6. Campbell, I., Ries, J., Dhawan, S., Quyyumi, A., Taylor, W., Oshinski, J.: Effect of inlet velocity profiles on patient-specific computational fluid dynamics simulations of the carotid bifurcation. J. Biomech. Eng. **134**(5), 051001 (2012)
7. Conca, C., Pares, C., Pironneau, O., Thiriet, M.: Navier-Stokes equations with imposed pressure and velocity fluxes. Int. J. Numer. Methods Fluids **20**(4), 267–287 (1995)
8. Engl, H., Hanke, M., Neubauer, A.: Regularization of Inverse Problems. Springer, Netherlands (1996)
9. Ern, A., Guermond, J.: Theory and Practice of Finite Elements. Springer, Berlin (2004)
10. Ervin, V., Lee, H.: Numerical approximation of a quasi-Newtonian stokes flow problem with defective boundary conditions. SIAM J. Numer. Anal. **45**(5), 2120–2140 (2007)
11. Formaggia, L., Vergara, C.: Prescription of general defective boundary conditions in fluid-dynamics. Milan J. Math. **80**(2), 333–350 (2012)
12. Formaggia, L., Gerbeau, J., Nobile, F., Quarteroni, A.: Numerical treatment of defective boundary conditions for the Navier-Stokes equation. SIAM J. Numer. Anal. **40**(1), 376–401 (2002)

13. Formaggia, L., Veneziani, A., Vergara, C.: A new approach to numerical solution of defective boundary value problems in incompressible fluid dynamics. SIAM J. Numer. Anal. **46**(6), 2769–2794 (2008)
14. Formaggia, L., Veneziani, A., Vergara, C.: Flow rate boundary problems for an incompressible fluid in deformable domains: formulations and solution methods. Comput. Methods Appl. Mech. Eng. **199**(9–12), 677–688 (2009)
15. Fortin, M., Guénette, R., Pierre, R.: Numerical analysis of the modified EVSS method. Comput. Methods Appl. Mech. Eng. **143**, 79–95 (1997)
16. Fustinoni, C., Marengo, M., Zinna, S.: Integration of a lumped parameters code with a finite volume code: numerical analysis of an heat pipe. In: XXVII UIT Congress, p. UIT09-031 (2009)
17. Galvin, K., Lee, H.: Analysis and approximation of the cross model for quasi-Newtonian flows with defective boundary conditions. Appl. Math. Comput. **222**, 244254 (2013)
18. Galvin, K., Lee, H., Rebholz, L.: Approximation of viscoelastic flows with defective boundary conditions. J. Non-Newtonian Fluid Mech. **169–170**, 104113 (2012)
19. Gunzburger, M.: Perspectives in Flow Control and Optimization. Advances in Design and Control. Society for Industrial and Applied Mathematics, Philadelphia (2003)
20. He, X., Ku, D. Jr., Moore, J.: Simple calculation of the velocity profiles for pulsatile flow in a blood vessel using mathematica. Ann. Biomed. Eng. **21**, 45–49 (1993)
21. Heywood, J., Rannacher, R., Turek, S.: Artificial boundaries and flux and pressure conditions for the incompressible Navier-Stokes equations. Int. J. Numer. Methods Fluids **22**, 325–352 (1996)
22. Juntunen, M., Stenberg, R.: Nitsche's method for general boundary conditions. Math. Comput. **78**, 1353–1374 (2009)
23. Khanafer, K., Bull, J., Upchurch, G. Jr., Berguer, R.: Turbulence significantly increases pressure and fluid shear stress in an aortic aneurysm model under resting and exercise flow conditions. Ann. Vasc. Surg. **21**(1), 67–74 (2007)
24. Lee, H.: Optimal control for quasi-Newtonian flows with defective boundary conditions. Comput. Methods Appl. Mech. Eng. **200**, 2498–2506 (2011)
25. Leiva, J., Blanco, P., Buscaglia, G.: Partitioned analysis for dimensionally-heterogeneous hydraulic networks. Multiscale Model. Simul. **9**, 872–903 (2011)
26. Les, A., Shadden, S., Figueroa, C., Park, J., Tedesco, M., Herfkens, R., Dalman, R., Taylor, C.: Quantification of hemodynamics in abdominal aortic aneurysms during rest and exercise using magnetic resonance imaging and computational fluid dynamics. Ann. Biomed. Eng. **38**(4), 1288–1313 (2010)
27. Lions, J.: Optimal Control of Systems Governed by Partial Differential Equations. Springer, Berlin (1971)
28. Moireau, P., Xiao, N., Astorino, M., Figueroa, C.A., Chapelle, D., Taylor, C.A., Gerbeau, J.: External tissue support and fluid–structure simulation in blood flows. Biomech. Model. Mechanobiol. **11**(1–2), 1–18 (2012)
29. Nicoud, F., Toda, H.B., Cabrit, O., Bose, S., Lee, J.: Using singular values to build a subgrid-scale model for large eddy simulations. Phys. Fluids **23**(8), 085106 (2011)
30. Nitsche, J.: Uber ein variationsprinzip zur lozung von dirichlet-problemen bei verwendung von teilraumen, die keinen randbedingungen unterworfen sind. Abh. Math. Semin. Univ. Hambg. **36**, 9–15 (1970/1971)
31. Nocedal, J., Wright, S.: Sequential Quadratic Programming. Springer, Berlin (2006)
32. Quarteroni, A., Tuveri, M., Veneziani, A.: Computational vascular fluid dynamics: problems, models and methods. Comput. Vis. Sci. **2**, 163–197 (2000)
33. Quarteroni, A., Veneziani, A., Vergara, C.: Geometric multiscale modeling of the cardiovascular system, between theory and practice. Comput. Methods Appl. Mech. Eng. **302**, 193–252 (2016)
34. Quarteroni, A., Manzoni, A., Vergara, C.: The cardiovascular system: mathematical modeling, numerical algorithms, clinical applications. Acta Numer. **26**(1), 365–590 (2017)

35. Redaelli, A., Boschetti, F., Inzoli, F.: The assignment of velocity profiles in finite elements simulations of pulsatile flow in arteries. Comput. Biol. Med. **27**(3), 233–247 (1997)
36. Tröel, F.: Optimal Control of Partial Differential Equations. Theory, Methods and Applications. American Mathematical Society, Providence (2010)
37. Veneziani, A., Vergara, C.: Flow rate defective boundary conditions in haemodynamics simulations. Int. J. Numer. Meth. Fluids **47**, 803–816 (2005)
38. Veneziani, A., Vergara, C.: An approximate method for solving incompressible Navier-Stokes problems with flow rate conditions. Comput. Methods Appl. Mech. Eng. **196**(9–12), 1685–1700 (2007)
39. Vergara, C.: Nitsche's method for defective boundary value problems in incompressible fluid-dynamics. J. Sci. Comput. **46**(1), 100–123 (2011)
40. Vergara, C., Le Van, D., Quadrio, M., Formaggia, L., Domanin, M.: Large eddy simulations of blood dynamics in abdominal aortic aneurysms. Med. Eng. Phys. **47**, 38–46 (2017)
41. Whitaker, S.: Introduction to Fluid Mechanics. R.E. Krieger, Malabar (1984)
42. Zunino, P.: Numerical approximation of incompressible flows with net flux defective boundary conditions by means of penalty technique. Comput. Methods Appl. Mech. Eng. **198**(37–40), 3026–3038 (2009)

MIX
Papier aus verantwortungsvollen Quellen
Paper from responsible sources
FSC® C105338

If you have any concerns about our products,
you can contact us on
ProductSafety@springernature.com

In case Publisher is established outside the EU,
the EU authorized representative is:
Springer Nature Customer Service Center GmbH
Europaplatz 3, 69115 Heidelberg, Germany

Printed by Libri Plureos GmbH
in Hamburg, Germany